Chevrolet Nova Automotive Repair Manual

by John H Haynes
Member of the Guild of Motoring Writers
and P Ward

Models covered:
Chevrolet Nova Sedan, Coupe and Hatchback
with 262, 305, 307, 350 or 396 cu in V8 engine
*Custom, LN, SS and basic versions of above with
manual or automatic transmission*
Does not cover 6-cylinder models

(24059 - 12V7)

Haynes Publishing Group
Sparkford Nr Yeovil
Somerset BA22 7JJ England

Haynes North America, Inc
859 Lawrence Drive
Newbury Park
California 91320 USA
www.haynes.com

About this manual

Its purpose

The purpose of this manual is to help you get the best value from your vehicle. It can do so in several ways. It can help you decide what work must be done, even if you choose to have it done by a dealer service department or a repair shop; it provides information and procedures for routine maintenance and servicing; and it offers diagnostic and repair procedures to follow when trouble occurs.

We hope you use the manual to tackle the work yourself. For many simpler jobs, doing it yourself may be quicker than arranging an appointment to get the vehicle into a shop and making the trips to leave it and pick it up. More importantly, a lot of money can be saved by avoiding the expense the shop must pass on to you to cover its labor and overhead costs. An added benefit is the sense of satisfaction and accomplishment that you feel after doing the job yourself.

Using the manual

The manual is divided into Chapters. Each Chapter is divided into numbered Sections, which are headed in bold type between horizontal lines. Each Section consists of consecutively numbered paragraphs.

At the beginning of each numbered Section you will be referred to any illustrations which apply to the procedures in that Section. The reference numbers used in illustration captions pinpoint the pertinent Section and the Step within that Section. That is, illustration 3.2 means the illustration refers to Section 3 and Step (or paragraph) 2 within that Section.

Procedures, once described in the text, are not normally repeated. When it's necessary to refer to another Chapter, the reference will be given as Chapter and Section number. Cross references given without use of the word "Chapter" apply to Sections and/or paragraphs in the same Chapter. For example, "see Section 8" means in the same Chapter.

References to the left or right side of the vehicle assume you are sitting in the driver's seat, facing forward.

Even though we have prepared this manual with extreme care, neither the publisher nor the author can accept responsibility for any errors in, or omissions from, the information given.

NOTE

A **Note** provides information necessary to properly complete a procedure or information which will make the procedure easier to understand.

CAUTION

A **Caution** provides a special procedure or special steps which must be taken while completing the procedure where the Caution is found. Not heeding a Caution can result in damage to the assembly being worked on.

WARNING

A **Warning** provides a special procedure or special steps which must be taken while completing the procedure where the Warning is found. Not heeding a Warning can result in personal injury.

Acknowledgements

Special thanks are due to Sgt. Jose A. Ortiz, who kindly loaned us his Chevrolet Nova for several weeks. This is the car that was dismantled and reassembled in our workshop, and featured in the original photographs throughout this manual. We are also grateful to all the other American servicemen based in the UK who offered their cars for our project.

We are also grateful for the help and cooperation of Tomco Industries, 1435 Woodson Road, St. Louis, Missouri 63132, for their assistance with technical information and carburetor illustrations. Valley Forge Technical Information Services also originated certain illustrations. Wiring diagrams originated exclusively for Haynes North America, Inc. by George Edward Brodd.

A book in the Haynes Automotive Repair Manual Series

ISBN-10: 0-85696-693-2

ISBN-13: 978-0-85696-693-4

Contents

1975 Chevrolet Nova Custom Sedan

Introduction to the Chevrolet Nova

This manual covers the models in the Chevrolet Nova range fitted with V8 engines, manufactured during the period 1969 thru 1979. There are two basic body styles, the 4-door Sedan and the 2-door Coupe. There is also a Hatchback version of the Coupe, introduced in 1973, providing a compromise between car and station wagon: this model is referred to separately where necessary.

Historically, the Nova originated as an optional model in the 1962 thru 1968 Chevy II range. The original in-line 4 and 6-cylinder engines were supplemented by a 283 cu in.

V8 engine in 1964, and it is this version which subsequently became more popular. In 1968, the Chevy II range was completely redesigned, and the basic body styling of the Sedan and Coupe have since remained for the Nova models. Although this manual does not pretend to deal with any of the Chevy II models, a great deal of the technical content is also applicable to the 1968 Chevy II which is a very near relative of the 1969 Nova.

Originally, engines of 307, 350 and 396 cu in. were available, but the largest version was phased out during 1970. The 350 cu in.

engine has remained throughout the production years but the 307 cu in. engine was not used after 1973. An optional 262 cu in. (4.3 liter) was introduced for the 1975 models, but was replaced in 1976 with the 305 cu in engine. The 305 like the 350, remained in use until the Nova line was discontinued after the 1979 model year.

Different transmission combinations have been used, and according to the particular engine, model and year of manufacture, a 3- or 4-speed manual, or a fully automatic transmission may be fitted.

Buying parts

Replacement parts are available from many sources, which generally fall into one of two categories - authorized dealer parts departments and independent retail auto parts stores. Our advice concerning these parts is as follows:

Retail auto parts stores: Good auto parts stores will stock frequently needed components which wear out relatively fast, such as clutch components, exhaust systems, brake parts, tune-up parts, etc. These stores often supply new or reconditioned parts on an exchange basis, which can save a considerable amount of money. Discount auto parts stores are often very good places to buy materials and parts needed for general vehicle maintenance such as oil, grease, filters, spark plugs, belts, touch-up paint, bulbs, etc. They also usually sell tools and general accessories, have convenient hours, charge lower prices and can often be found not far from home.

Authorized dealer parts department: This is the best source for parts which are unique to the vehicle and not generally available elsewhere (such as major engine parts, transmission parts, trim pieces, etc.).

Warranty information: If the vehicle is still covered under warranty, be sure that any replacement parts purchased - regardless of the source - do not invalidate the warranty!

To be sure of obtaining the correct parts, have engine and chassis numbers available and, if possible, take the old parts along for positive identification.

Vehicle identification numbers

Modifications are a continuing and unpublicized process in vehicle manufacture. Spare parts manuals and lists are compiled on a numerical basis, the individual vehicle numbers being essential to identify correctly, the component required. The following identification numbers are applicable to the Nova range.

Vehicle identification number plate

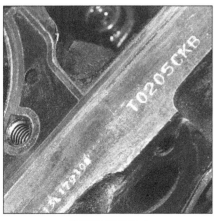

Engine number

Pre-1973 models:

Body type	Series	Style
X	11400	27 (Coupe); 69 (Sedan)

1973 and later models:

Body type	Series	Style
X (Nova)	IXX	17 (Hatchback); 27 (Coupe); 69 (Sedan)
XX (Custom)	IXY	17 (Hatchback); 27 (Coupe); 69 (Sedan)

Location of various unit identification numbers

Vehicle identification number plate	Top of instrument panel left, front **(see illustration)**
Body number, trim and paint plate	Upper left side of dash

Engine/transmission

Engine	On pad at front right side of cylinder block **(see illustration)**
3-speed Muncie	On boss above filler plug
4-speed Muncie	On right side of case, at lower rear of cover flange
3 - or 4- speed Saginaw	On lower right side of case adjacent to rear of cover
Powerglide	On left upper flange of converter opening of transmission housing, or boss on lower right side of converter housing
Hydra-matic 250/350	Right vertical surface of oil pan, or left upper flange of converter opening of transmission housing
Turbo Hydra-matic 200/400	On blue tag on right side of transmission, or left upper flange of converter opening of transmission housing
Rear axle number	On right or left axle tube, adjacent to carrier
Alternator	On top drive end of frame
Starter	Stamped on outer case, towards rear
Battery	On cell cover segment, top of battery

Body number plate

Vehicle Emission Control Information (VECI) label

Body number plate

A typical body number plate is shown in the **accompanying illustration**. It is normally mounted on the upper horizontal or vertical surface of the shroud.

Vehicle Emission Control Information (VECI) label

This decal is mounted on the radiator top panel or on the underside of the hood **(see illustration)**.

Tire pressure decal

This will be found on the driver's door edge for most models **(see illustration)**.

Tire pressure decal

Maintenance techniques, tools and working facilities

Maintenance techniques

There are a number of techniques involved in maintenance and repair that will be referred to throughout this manual. Application of these techniques will enable the home mechanic to be more efficient, better organized and capable of performing the various tasks properly, which will ensure that the repair job is thorough and complete.

Fasteners

Fasteners are nuts, bolts, studs and screws used to hold two or more parts together. There are a few things to keep in mind when working with fasteners. Almost all of them use a locking device of some type, either a lockwasher, locknut, locking tab or thread adhesive. All threaded fasteners should be clean and straight, with undamaged threads and undamaged corners on the hex head where the wrench fits. Develop the habit of replacing all damaged nuts and bolts with new ones. Special locknuts with nylon or fiber inserts can only be used once. If they

are removed, they lose their locking ability and must be replaced with new ones.

Rusted nuts and bolts should be treated with a penetrating fluid to ease removal and prevent breakage. Some mechanics use turpentine in a spout-type oil can, which works quite well. After applying the rust penetrant, let it work for a few minutes before trying to loosen the nut or bolt. Badly rusted fasteners may have to be chiseled or sawed off or removed with a special nut breaker, available at tool stores.

If a bolt or stud breaks off in an assembly, it can be drilled and removed with a special tool commonly available for this purpose. Most automotive machine shops can perform this task, as well as other repair procedures, such as the repair of threaded holes that have been stripped out.

Flat washers and lockwashers, when removed from an assembly, should always be replaced exactly as removed. Replace any damaged washers with new ones. Never use a lockwasher on any soft metal surface (such as aluminum), thin sheet metal or plastic.

Fastener sizes

For a number of reasons, automobile manufacturers are making wider and wider use of metric fasteners. Therefore, it is important to be able to tell the difference between standard (sometimes called U.S. or SAE) and metric hardware, since they cannot be interchanged.

All bolts, whether standard or metric, are sized according to diameter, thread pitch and length. For example, a standard 1/2 - 13 x 1 bolt is 1/2 inch in diameter, has 13 threads per inch and is 1 inch long. An M12 - 1.75 x 25 metric bolt is 12 mm in diameter, has a thread pitch of 1.75 mm (the distance between threads) and is 25 mm long. The two bolts are nearly identical, and easily confused, but they are not interchangeable.

In addition to the differences in diameter, thread pitch and length, metric and standard bolts can also be distinguished by examining the bolt heads. To begin with, the distance across the flats on a standard bolt head is measured in inches, while the same dimension on a metric bolt is sized in millimeters (the same is true for nuts). As a result, a standard wrench should not be used on a metric bolt and a metric wrench should not be used on a standard bolt. Also, most standard bolts have slashes radiating out from the center of the head to denote the grade or strength of the bolt, which is an indication of the amount of torque that can be applied to it. The greater the number of slashes, the greater the strength of the bolt. Grades 0 through 5 are commonly used on automobiles. Metric bolts have a property class (grade) number, rather than a slash, molded into their heads to indicate bolt strength. In this case, the higher the number, the stronger the bolt. Property class numbers 8.8, 9.8 and 10.9 are commonly used on automobiles.

Strength markings can also be used to distinguish standard hex nuts from metric hex nuts. Many standard nuts have dots stamped into one side, while metric nuts are marked with a number. The greater the number of dots, or the higher the number, the greater the strength of the nut.

Metric studs are also marked on their ends according to property class (grade). Larger studs are numbered (the same as metric bolts), while smaller studs carry a geometric code to denote grade.

It should be noted that many fasteners, especially Grades 0 through 2, have no distinguishing marks on them. When such is the case, the only way to determine whether it is standard or metric is to measure the thread pitch or compare it to a known fastener of the same size.

Standard fasteners are often referred to as SAE, as opposed to metric. However, it should be noted that SAE technically refers to a non-metric fine thread fastener only. Coarse thread non-metric fasteners are referred to as USS sizes.

Since fasteners of the same size (both

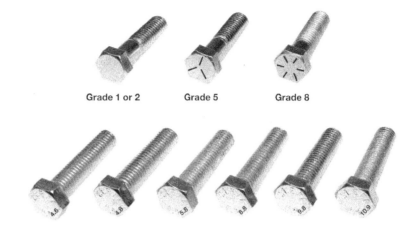

Grade 1 or 2   Grade 5   Grade 8

Bolt strength marking (standard/SAE/USS; bottom - metric)

standard and metric) may have different strength ratings, be sure to reinstall any bolts, studs or nuts removed from your vehicle in their original locations. Also, when replacing a fastener with a new one, make sure that the new one has a strength rating equal to or greater than the original.

Tightening sequences and procedures

Most threaded fasteners should be tightened to a specific torque value (torque is the twisting force applied to a threaded component such as a nut or bolt). Overtightening the fastener can weaken it and cause it to break, while undertightening can cause it to eventually come loose. Bolts, screws and studs, depending on the material they are made of and their thread diameters, have specific torque values, many of which are noted in the Specifications at the beginning of each Chapter. Be sure to follow the torque recommendations closely. For fasteners not assigned a specific torque, a general torque value chart is presented here as a guide. These torque values are for dry (unlubricated) fasteners threaded into steel or cast iron (not aluminum). As was previously mentioned, the size and grade of a fastener determine the amount of torque that can safely be applied to it. The figures listed here are approximate for Grade 2 and Grade 3 fasteners. Higher grades can tolerate higher torque values.

Fasteners laid out in a pattern, such as cylinder head bolts, oil pan bolts, differential cover bolts, etc., must be loosened or tightened in sequence to avoid warping the component. This sequence will normally be shown in the appropriate Chapter. If a specific pattern is not given, the following proce-

Metric thread sizes	Ft-lbs	Nm
M-6	6 to 9	9 to 12
M-8	14 to 21	19 to 28
M-10	28 to 40	38 to 54
M-12	50 to 71	68 to 96
M-14	80 to 140	109 to 154

Pipe thread sizes		
1/8	5 to 8	7 to 10
1/4	12 to 18	17 to 24
3/8	22 to 33	30 to 44
1/2	25 to 35	34 to 47

U.S. thread sizes		
1/4 - 20	6 to 9	9 to 12
5/16 - 18	12 to 18	17 to 24
5/16 - 24	14 to 20	19 to 27
3/8 - 16	22 to 32	30 to 43
3/8 - 24	27 to 38	37 to 51
7/16 - 14	40 to 55	55 to 74
7/16 - 20	40 to 60	55 to 81
1/2 - 13	55 to 80	75 to 108

dures can be used to prevent warping.

Initially, the bolts or nuts should be assembled finger-tight only. Next, they should be tightened one full turn each, in a criss-cross or diagonal pattern. After each one has been tightened one full turn, return to the first one and tighten them all one-half turn, following the same pattern. Finally, tighten each of them one-quarter turn at a time until each fastener has been tightened to the proper torque. To loosen and remove the fasteners, the procedure would be reversed.

Component disassembly

Component disassembly should be done with care and purpose to help ensure that the parts go back together properly. Always keep track of the sequence in which parts are removed. Make note of special characteristics or marks on parts that can be installed more than one way, such as a grooved thrust washer on a shaft. It is a good idea to lay the disassembled parts out on a clean surface in the order that they were removed. It may also be helpful to make sketches or take instant photos of components before removal.

When removing fasteners from a component, keep track of their locations. Sometimes threading a bolt back in a part, or putting the washers and nut back on a stud, can prevent mix-ups later. If nuts and bolts cannot be returned to their original locations, they should be kept in a compartmented box or a series of small boxes. A cupcake or muffin tin is ideal for this purpose, since each cavity can hold the bolts and nuts from a particular area (i.e. oil pan bolts, valve cover bolts, engine mount bolts, etc.). A pan of this type is especially helpful when working on assemblies with very small parts, such as the carburetor, alternator, valve train or interior dash and trim pieces. The cavities can be marked with paint or tape to identify the contents.

Whenever wiring looms, harnesses or connectors are separated, it is a good idea to identify the two halves with numbered pieces of masking tape so they can be easily reconnected.

Gasket sealing surfaces

Throughout any vehicle, gaskets are used to seal the mating surfaces between two parts and keep lubricants, fluids, vacuum or pressure contained in an assembly.

Many times these gaskets are coated with a liquid or paste-type gasket sealing compound before assembly. Age, heat and pressure can sometimes cause the two parts to stick together so tightly that they are very difficult to separate. Often, the assembly can be loosened by striking it with a soft-face hammer near the mating surfaces. A regular hammer can be used if a block of wood is placed between the hammer and the part. Do not hammer on cast parts or parts that could be easily damaged. With any particularly stubborn part, always recheck to make sure that every fastener has been removed.

Avoid using a screwdriver or bar to pry

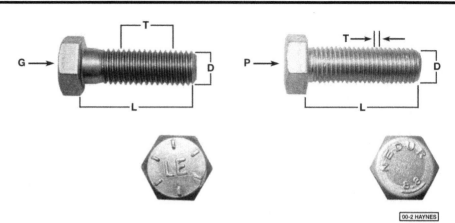

Standard (SAE and USS) bolt dimensions/grade marks

G   *Grade marks (bolt strength)*
L   *Length (in inches)*
T   *Thread pitch (number of threads per inch)*
D   *Nominal diameter (in inches)*

Metric bolt dimensions/grade marks

P   *Property class (bolt strength)*
L   *Length (in millimeters)*
T   *Thread pitch (distance between threads in millimeters)*
D   *Diameter*

apart an assembly, as they can easily mar the gasket sealing surfaces of the parts, which must remain smooth. If prying is absolutely necessary, use an old broom handle, but keep in mind that extra clean up will be necessary if the wood splinters.

After the parts are separated, the old gasket must be carefully scraped off and the gasket surfaces cleaned. Stubborn gasket material can be soaked with rust penetrant or treated with a special chemical to soften it so it can be easily scraped off. A scraper can be fashioned from a piece of copper tubing by flattening and sharpening one end. Copper is recommended because it is usually softer than the surfaces to be scraped, which reduces the chance of gouging the part. Some gaskets can be removed with a wire brush, but regardless of the method used, the mating surfaces must be left clean and smooth. If for some reason the gasket surface is gouged, then a gasket sealer thick enough to fill scratches will have to be used during reassembly of the components. For most applications, a non-drying (or semi-drying) gasket sealer should be used.

Hose removal tips

Warning: *If the vehicle is equipped with air conditioning, do not disconnect any of the A/C hoses without first having the system depressurized by a dealer service department or a service station.*

Hose removal precautions closely parallel gasket removal precautions. Avoid scratching or gouging the surface that the hose mates against or the connection may leak. This is especially true for radiator hoses. Because of various chemical reactions, the rubber in hoses can bond itself to the metal spigot that the hose fits over. To remove a hose, first loosen the hose clamps that secure it to the spigot. Then, with slip-joint pliers, grab the hose at the clamp and rotate it around the spigot. Work it back and forth

until it is completely free, then pull it off. Silicone or other lubricants will ease removal if they can be applied between the hose and the outside of the spigot. Apply the same lubricant to the inside of the hose and the outside of the spigot to simplify installation.

As a last resort (and if the hose is to be replaced with a new one anyway), the rubber can be slit with a knife and the hose peeled from the spigot. If this must be done, be careful that the metal connection is not damaged.

If a hose clamp is broken or damaged, do not reuse it. Wire-type clamps usually weaken with age, so it is a good idea to replace them with screw-type clamps whenever a hose is removed.

Tools

A selection of good tools is a basic requirement for anyone who plans to maintain and repair his or her own vehicle. For the owner who has few tools, the initial investment might seem high, but when compared to the spiraling costs of professional auto maintenance and repair, it is a wise one.

To help the owner decide which tools are needed to perform the tasks detailed in this manual, the following tool lists are offered: *Maintenance and minor repair, Repair/overhaul* and *Special.*

The newcomer to practical mechanics should start off with the *maintenance and minor repair* tool kit, which is adequate for the simpler jobs performed on a vehicle. Then, as confidence and experience grow, the owner can tackle more difficult tasks, buying additional tools as they are needed. Eventually the basic kit will be expanded into the *repair and overhaul* tool set. Over a period of time, the experienced do-it-yourselfer will assemble a tool set complete enough for most repair and overhaul procedures and will add tools from the special category when it is felt that the expense is justified by the frequency of use.

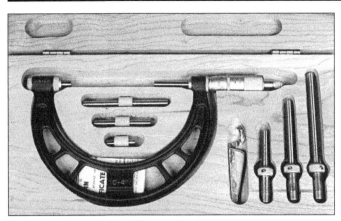
Micrometer set

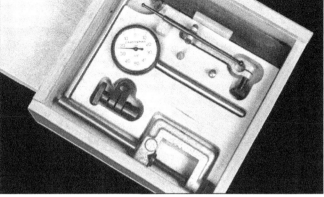
Dial indicator set

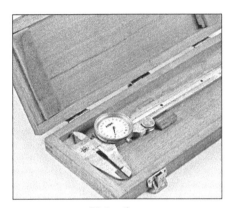
Dial caliper

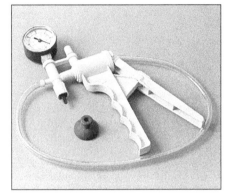
Hand-operated vacuum pump

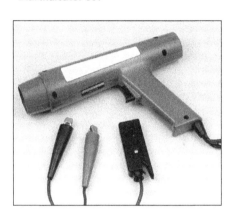
Timing light

Maintenance and minor repair tool kit

The tools in this list should be considered the minimum required for performance of routine maintenance, servicing and minor repair work. We recommend the purchase of combination wrenches (box-end and open-end combined in one wrench). While more expensive than open end wrenches, they offer the advantages of both types of wrench.

Combination wrench set (1/4-inch to 1 inch or 6 mm to 19 mm)
Adjustable wrench, 8 inch
Spark plug wrench with rubber insert
Spark plug gap adjusting tool
Feeler gauge set
Brake bleeder wrench
Standard screwdriver (5/16-inch x 6 inch)
Phillips screwdriver (No. 2 x 6 inch)
Combination pliers - 6 inch
Hacksaw and assortment of blades
Tire pressure gauge
Grease gun
Oil can
Fine emery cloth
Wire brush
Battery post and cable cleaning tool
Oil filter wrench
Funnel (medium size)
Safety goggles
Jackstands (2)
Drain pan

Note: *If basic tune-ups are going to be part of routine maintenance, it will be necessary to purchase a good quality stroboscopic timing light and combination tachometer/dwell meter. Although they are included in the list of special tools, it is mentioned here because they are absolutely necessary for tuning most vehicles properly.*

Repair and overhaul tool set

These tools are essential for anyone who plans to perform major repairs and are in addition to those in the maintenance and minor repair tool kit. Included is a comprehensive set of sockets which, though expensive, are invaluable because of their versatility, especially when various extensions and drives are available. We recommend the 1/2-inch drive over the 3/8-inch drive. Although the larger drive is bulky and more expensive, it has the capacity of accepting a very wide range of large sockets. Ideally, however, the mechanic should have a 3/8-inch drive set and a 1/2-inch drive set.

Socket set(s)
Reversible ratchet
Extension - 10 inch
Universal joint
Torque wrench (same size drive as sockets)
Ball peen hammer - 8 ounce
Soft-face hammer (plastic/rubber)

Standard screwdriver (1/4-inch x 6 inch)
Standard screwdriver (stubby - 5/16-inch)
Phillips screwdriver (No. 3 x 8 inch)
Phillips screwdriver (stubby - No. 2)
Pliers - vise grip
Pliers - lineman's
Pliers - needle nose
Pliers - snap-ring (internal and external)
Cold chisel - 1/2-inch
Scribe
Scraper (made from flattened copper tubing)
Centerpunch
Pin punches (1/16, 1/8, 3/16-inch)
Steel rule/straightedge - 12 inch
Allen wrench set (1/8 to 3/8-inch or 4 mm to 10 mm)
A selection of files
Wire brush (large)
Jackstands (second set)
Jack (scissor or hydraulic type)

Note: *Another tool which is often useful is an electric drill with a chuck capacity of 3/8-inch and a set of good quality drill bits.*

Special tools

The tools in this list include those which are not used regularly, are expensive to buy, or which need to be used in accordance with their manufacturer's instructions. Unless these tools will be used frequently, it is not very economical to purchase many of them.

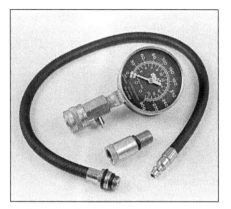

Compression gauge with spark plug hole adapter

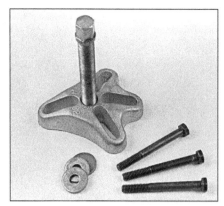

Damper/steering wheel puller

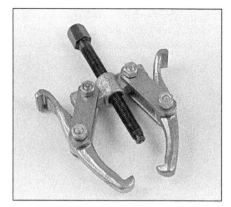

General purpose puller

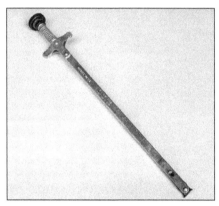

Hydraulic lifter removal tool

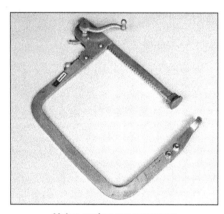

Valve spring compressor

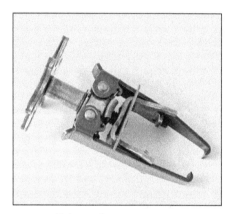

Valve spring compressor

Ridge reamer

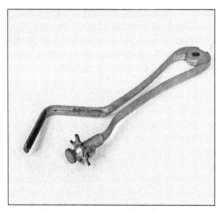

Piston ring groove cleaning tool

Ring removal/installation tool

A consideration would be to split the cost and use between yourself and a friend or friends. In addition, most of these tools can be obtained from a tool rental shop on a temporary basis.

This list primarily contains only those tools and instruments widely available to the public, and not those special tools produced by the vehicle manufacturer for distribution to dealer service departments. Occasionally, references to the manufacturer's special tools are included in the text of this manual. Generally, an alternative method of doing the job without the special tool is offered. How-ever, sometimes there is no alternative to their use. Where this is the case, and the tool cannot be purchased or borrowed, the work should be turned over to the dealer service department or an automotive repair shop.

Valve spring compressor
Piston ring groove cleaning tool
Piston ring compressor
Piston ring installation tool
Cylinder compression gauge
Cylinder ridge reamer
Cylinder surfacing hone
Cylinder bore gauge
Micrometers and/or dial calipers

Hydraulic lifter removal tool
Balljoint separator
Universal-type puller
Impact screwdriver
Dial indicator set
Stroboscopic timing light (inductive pick-up)
Hand operated vacuum/pressure pump
Tachometer/dwell meter
Universal electrical multimeter
Cable hoist
Brake spring removal and installation tools
Floor jack

Ring compressor

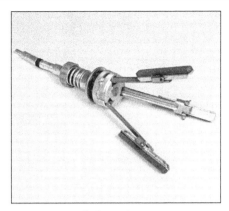

Cylinder hone

Brake hold-down spring tool

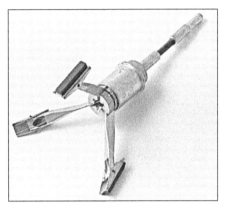

Brake cylinder hone

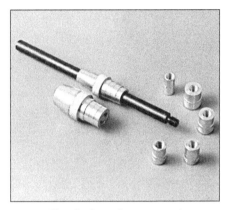

Clutch plate alignment tool

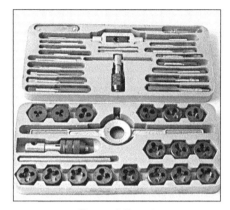

Tap and die set

Buying tools

For the do-it-yourselfer who is just starting to get involved in vehicle maintenance and repair, there are a number of options available when purchasing tools. If maintenance and minor repair is the extent of the work to be done, the purchase of individual tools is satisfactory. If, on the other hand, extensive work is planned, it would be a good idea to purchase a modest tool set from one of the large retail chain stores. A set can usually be bought at a substantial savings over the individual tool prices, and they often come with a tool box. As additional tools are needed, add-on sets, individual tools and a larger tool box can be purchased to expand the tool selection. Building a tool set gradually allows the cost of the tools to be spread over a longer period of time and gives the mechanic the freedom to choose only those tools that will actually be used.

Tool stores will often be the only source of some of the special tools that are needed, but regardless of where tools are bought, try to avoid cheap ones, especially when buying screwdrivers and sockets, because they won't last very long. The expense involved in replacing cheap tools will eventually be greater than the initial cost of quality tools.

Care and maintenance of tools

Good tools are expensive, so it makes sense to treat them with respect. Keep them clean and in usable condition and store them properly when not in use. Always wipe off any dirt, grease or metal chips before putting them away. Never leave tools lying around in the work area. Upon completion of a job, always check closely under the hood for tools that may have been left there so they won't get lost during a test drive.

Some tools, such as screwdrivers, pliers, wrenches and sockets, can be hung on a panel mounted on the garage or workshop wall, while others should be kept in a tool box or tray. Measuring instruments, gauges, meters, etc. must be carefully stored where they cannot be damaged by weather or impact from other tools.

When tools are used with care and stored properly, they will last a very long time. Even with the best of care, though, tools will wear out if used frequently. When a tool is damaged or worn out, replace it. Subsequent jobs will be safer and more enjoyable if you do.

How to repair damaged threads

Sometimes, the internal threads of a nut or bolt hole can become stripped, usually from overtightening. Stripping threads is an all-too-common occurrence, especially when working with aluminum parts, because aluminum is so soft that it easily strips out.

Usually, external or internal threads are only partially stripped. After they've been cleaned up with a tap or die, they'll still work. Sometimes, however, threads are badly damaged. When this happens, you've got three choices:

1) *Drill and tap the hole to the next suitable oversize and install a larger diameter bolt, screw or stud.*

2) *Drill and tap the hole to accept a threaded plug, then drill and tap the plug to the original screw size. You can also buy a plug already threaded to the original size. Then you simply drill a hole to the specified size, then run the threaded plug into the hole with a bolt and jam nut. Once the plug is fully seated, remove the jam nut and bolt.*

3) *The third method uses a patented thread repair kit like Heli-Coil or Slimsert. These easy-to-use kits are designed to repair damaged threads in straight-through holes and blind holes. Both are available as kits which can handle a variety of sizes and thread patterns. Drill the hole, then tap it with the special included tap. Install the Heli-Coil and the hole is back to its original diameter and thread pitch.*

Regardless of which method you use, be sure to proceed calmly and carefully. A little impatience or carelessness during one of these relatively simple procedures can ruin your whole day's work and cost you a bundle if you wreck an expensive part.

Working facilities

Not to be overlooked when discussing tools is the workshop. If anything more than routine maintenance is to be carried out, some sort of suitable work area is essential.

It is understood, and appreciated, that many home mechanics do not have a good workshop or garage available, and end up removing an engine or doing major repairs outside. It is recommended, however, that the overhaul or repair be completed under the cover of a roof.

A clean, flat workbench or table of com-fortable working height is an absolute neces-sity. The workbench should be equipped with a vise that has a jaw opening of at least four inches.

As mentioned previously, some clean, dry storage space is also required for tools, as well as the lubricants, fluids, cleaning sol-vents, etc. which soon become necessary.

Sometimes waste oil and fluids, drained from the engine or cooling system during normal maintenance or repairs, present a dis-posal problem. To avoid pouring them on the ground or into a sewage system, pour the used fluids into large containers, seal them with caps and take them to an authorized disposal site or recycling center. Plastic jugs, such as old antifreeze containers, are ideal for this purpose.

Always keep a supply of old newspa-pers and clean rags available. Old towels are excellent for mopping up spills. Many mechanics use rolls of paper towels for most work because they are readily available and disposable. To help keep the area under the vehicle clean, a large cardboard box can be cut open and flattened to protect the garage or shop floor.

Whenever working over a painted sur-face, such as when leaning over a fender to service something under the hood, always cover it with an old blanket or bedspread to protect the finish. Vinyl covered pads, made especially for this purpose, are available at auto parts stores.

Jacking and towing

Jacking

The jack supplied with the vehicle should only be used for raising the vehicle when changing a tire or placing jackstands under the frame. **Warning:** *Never work under the vehicle or start the engine while this jack is being used as the only means of support.*

The vehicle should be on level ground with the wheel blocked and the transmission in Park (automatic) or Reverse (manual). If the wheel is being replaced, loosen the wheel lug nuts on-half turn and leave them in place until the wheel is raised off the ground.

Place the hooked portion of the jack head in the slot in the bumper **(see illustra-tions)**. Operate the jack with a slow, smooth motion until the wheel is raised off the ground.

Lower the vehicle, remove the jack and tighten the lug nuts (if loosened or removed) in a criss-cross sequence.

Towing

In the event of a mechanical failure, the vehicle may be towed by another vehicle but special precautions must be taken with vehi-cles equipped with automatic transmission.

If the towing speed is likely to exceed 35 mph or the distance covered more than 50 miles, the driveshaft must be removed.

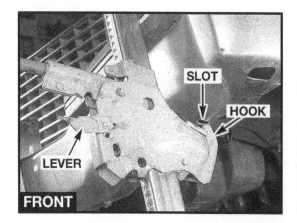
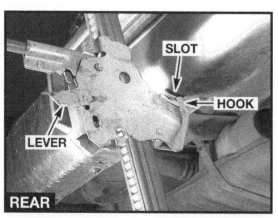

Make sure the hook on the jack head is securely engaged with the slot in the bumper, and the foot of the jack is resting on solid, level ground

Routine maintenance

Maintenance is essential for ensuring safety and desirable for the purpose of getting the best in terms of performance and economy from the car. Over the years the need for periodic lubrication - oiling, greasing and so on - has been drastically reduced if not totally eliminated. This has unfortunately tended to lead some owners to think that because no such action is required the items either no longer exist or will last for ever. This is a serious delusion. It follows therefore that the largest initial element of maintenance is visual examination. This may lead to repairs or renewals.

The service intervals and the procedures given are basically those recommended by the vehicle manufacturer.

Where any vehicle is used under very arduous conditions, in very dusty areas, in continual stop/start conditions or where the engine does not have a chance to thoroughly warm up, engine oil and filter should be changed at intervals of 3000 miles.

Note: *Where an item is referred to in the routine maintenance table, and that item is not applicable to a particular model, it is to be ignored.*

Every 250 miles or weekly - whichever comes first

Steering

Check the tire pressures
Examine tires for wear or damage
Is steering smooth and accurate?

Brakes

Check reservoir fluid level **(see illustration)**
Is there any fall off in braking efficiency?
Try an emergency stop. Is adjustment necessary?
Check parking brake operation

Lights, wipers and horns

Do all bulbs work at the front and rear?
Are the headlamp beams aligned properly?
Do the wipers and horns work?
Check windshield washer fluid level

Engine

Check the engine oil level and top-up if required **(see illustration)**
Check the coolant level and top-up if required **(see illustration)**
Check the battery electrolyte level and top-up to the level of the plates with distilled water as needed

Maintain the brake fluid level 1/4-inch below the top edge of the reservoir

---

Every 6000 miles or 4 months - whichever comes first

Lubricate suspension grease fittings
Change the engine oil and filter **(see illustration)**
Check tension of drivebelts
Check disc brake pad wear
Check transmission oil level (manual or automatic) at operating temperature
Check differential oil level, at operating temperature
Check power steering reservoir fluid level
Inspect all steering and suspension components for wear
Rotate the tires (front-to-rear) and at the same time remove stones and examine for cuts, damage and tread wear. If there is uneven tread wear, or steering vibration is

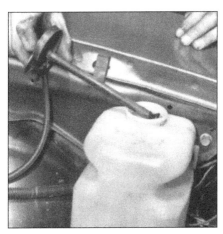

Check the engine coolant level and top-up if necessary

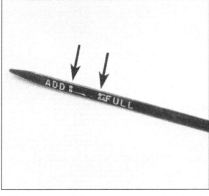

The engine oil level must be maintained between the marks at all times - it takes one quart of oil to raise the level from the ADD mark to the FULL mark

noticed, arrange for a wheel balance check
Clean contact breaker points and adjust dwell angle
Clean and re-gap spark plugs (not essential for 1975 and later models)
Inspect all brake lines and hoses for damage or deterioration
Lubricate the parking brake cables
Check the exhaust system for leaks and broken mountings
Check the air conditioning system and defroster operation
Lubricate all body moving parts (door hinges, locks etc)
Check operation of steering column lock
Check that neutral start switch (manual or automatic) only operates in correct mode
Check the condition of the seat belts and their mounting points
Check that all warning lights, instruments and buzzers operate

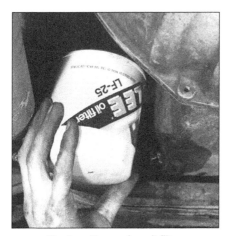

Replace the engine oil filter

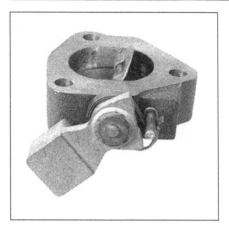

Typical manifold heat valve

Every 12000 miles or 12 months - whichever comes first

Check operation of the thermostatically controlled air cleaner

Check that the carburetor choke is operating correctly

Check that the manifold heat valve is operating **(see illustration)**

Check that the early fuel evaporation valve (EFE valve) is operating

Torque-tighten the carburetor mounting nuts/bolts

Check and adjust the ignition timing **(see illustration)**

Check and adjust the engine idle speed

Drain and refill 'Positraction' rear axles (when used for trailer hauling)

Inspect cooling system hoses for damage and deterioration, and clean pressure cap

Clean the exterior of the radiator core and air-conditioning condenser

Inspect drum brake linings

Inspect the throttle linkage for damage, kinks etc.

Arrange for the underside of the vehicle to be pressure washed (preferably after the winter, to remove dirt, salt or other contaminants)

Check condition of energy-absorbing bumper systems

Replace carburetor inlet fuel filter **(see illustration)**

Replace the PCV filter **(see illustration)**

Clean the EGR valve

Check operation of exhaust, crankcase and fuel emission and evaporation control systems

Replace spark plugs (except 1975 and later models)

Replace contact breaker points

Check ignition leads and distributor caps for cracks and other deterioration. Wipe the parts clean with a dry, lint-free cloth

Rotate the distributor cam lubricator through 180°

Replace the air cleaner element if operating in very dusty conditions

Replace the automatic transmission

The ignition timing mark on the torsional damper

fluid and replace the filter when operating under arduous conditions

Every 24000 miles or 24 months - whichever comes first

Check the condition of the fuel tank, lines and cap

Replace filter in evaporation control system (ECS)

Check hoses in air injector reactor (AIR) system

Replace the air cleaner element

Check and adjust mechanical valve lifters

Clean and repack the front wheel bearings

Replace the automatic transmission fluid and filter

Check the manual steering gear for lubricant leakage

Lubricate the clutch cross-shaft

Replace the engine antifreeze mixture and check the hoses

Replace distributor cam lubricator

Replace contact breaker points

Replace the PCV valve

Check the engine compression **(see illustration)**

Replace the spark plugs

Every 50000 miles or 3 years - whichever comes first

Replace all the rubber seals and hoses in the braking system

Replace the brake system fluid

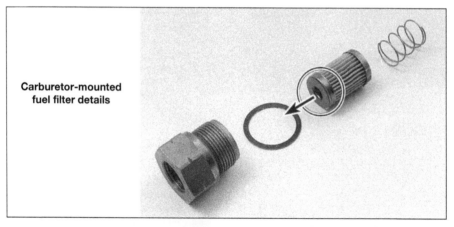

Carburetor-mounted fuel filter details

The Positive Crankcase Ventilation (PCV) filter on most models is located in the air cleaner housing

A gauge with a threaded fitting for the spark plug hole is preferred over the type that requires hand pressure to maintain the seal during the compression check

Booster battery (jump) starting

Observe these precautions when using a booster battery to start a vehicle:

a) *Before connecting the booster battery, make sure the ignition switch is in the Off position.*
b) *Turn off the lights, heater and other electrical loads.*
c) *Your eyes should be shielded. Safety goggles are a good idea.*
d) *Make sure the booster battery is the same voltage as the dead one in the vehicle.*
e) *The two vehicles MUST NOT TOUCH each other!*
f) *Make sure the transaxle is in Neutral (manual) or Park (automatic).*
g) *If the booster battery is not a maintenance-free type, remove the vent caps and lay a cloth over the vent holes.*

Connect the red jumper cable to the positive (+) terminals of each battery **(see illustration)**.

Connect one end of the black jumper cable to the negative (-) terminal of the booster battery. The other end of this cable should be connected to a good ground on the vehicle to be started, such as a bolt or bracket on the body.

Start the engine using the booster battery, then, with the engine running at idle speed, disconnect the jumper cables in the reverse order of connection.

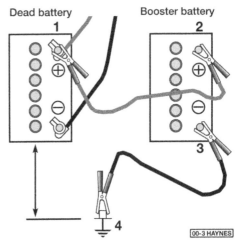

Make the booster battery cable connections in the numerical order shown (note that the negative cable of the booster battery is NOT attached to the negative terminal of the dead battery)

Recommended lubricants and fluids

Note: *Listed here are manufacturer recommendations at the time this manual was written. Manufacturers occasionally upgrade their fluid and lubricant specifications, so check with your local auto parts store for current recommendations.*

Component	Type of lubricant
Engine oil	
(use API grade SJ or better multigrade and fuel efficient oil)	
- 30° to 20°F	SAE 5W - 20
	SAE 5W - 30
0° to 60°F	SAE 5W - 30
	SAE 10W - 30
	SAE 10W - 40
20° to 100°F	SAE 20W - 20
	SAE 10W - 30
	SAE 10W - 40
	SAE 20W - 40
	SAE 20W - 50
Engine coolant	50/50 mixture of water and ethylene glycol-based antifreeze
Transmission	
Manual	SAE 80 GL5 gear oil
	SAE 90 GL5 gear oil
Automatic	Dexron-type automatic transmission fluid
Differential	
Standard	SAE 80 GL5 gear oil
	SAE 90 GL5 gear oil
Positraction (limited slip)	Add four fluid ounces of GM limited slip additive to the above specified lubricant
Steering	
Manual	EP chassis grease
Power	GM power steering fluid or equivalent
Chassis, parking brake cables	NLGI no. 2 chassis grease
Driveshaft universal joints	NLGI no. 2 chassis grease
Brake system	DOT 3 brake fluid

Safety first!

Regardless of how enthusiastic you may be about getting on with the job at hand, take the time to ensure that your safety is not jeopardized. A moment's lack of attention can result in an accident, as can failure to observe certain simple safety precautions. The possibility of an accident will always exist, and the following points should not be considered a comprehensive list of all dangers. Rather, they are intended to make you aware of the risks and to encourage a safety conscious approach to all work you carry out on your vehicle.

Essential DOs and DON'Ts

DON'T rely on a jack when working under the vehicle. Always use approved jackstands to support the weight of the vehicle and place them under the recommended lift or support points.

DON'T attempt to loosen extremely tight fasteners (i.e. wheel lug nuts) while the vehicle is on a jack - it may fall.

DON'T start the engine without first making sure that the transmission is in Neutral (or Park where applicable) and the parking brake is set.

DON'T remove the radiator cap from a hot cooling system - let it cool or cover it with a cloth and release the pressure gradually.

DON'T attempt to drain the engine oil until you are sure it has cooled to the point that it will not burn you.

DON'T touch any part of the engine or exhaust system until it has cooled sufficiently to avoid burns.

DON'T siphon toxic liquids such as gasoline, antifreeze and brake fluid by mouth, or allow them to remain on your skin.

DON'T inhale brake lining dust - it is potentially hazardous (see *Asbestos* below).

DON'T allow spilled oil or grease to remain on the floor - wipe it up before someone slips on it.

DON'T use loose fitting wrenches or other tools which may slip and cause injury.

DON'T push on wrenches when loosening or tightening nuts or bolts. Always try to pull the wrench toward you. If the situation calls for pushing the wrench away, push with an open hand to avoid scraped knuckles if the wrench should slip.

DON'T attempt to lift a heavy component alone - get someone to help you.

DON'T rush or take unsafe shortcuts to finish a job.

DON'T allow children or animals in or around the vehicle while you are working on it.

DO wear eye protection when using power tools such as a drill, sander, bench grinder, etc. and when working under a vehicle.

DO keep loose clothing and long hair well out of the way of moving parts.

DO make sure that any hoist used has a safe working load rating adequate for the job.

DO get someone to check on you periodically when working alone on a vehicle.

DO carry out work in a logical sequence and make sure that everything is correctly assembled and tightened.

DO keep chemicals and fluids tightly capped and out of the reach of children and pets.

DO remember that your vehicle's safety affects that of yourself and others. If in doubt on any point, get professional advice.

Asbestos

Certain friction, insulating, sealing, and other products - such as brake linings, brake bands, clutch linings, torque converters, gaskets, etc. - may contain asbestos. Extreme care must be taken to avoid inhalation of dust from such products, since it is hazardous to health. If in doubt, assume that they do contain asbestos.

Fire

Remember at all times that gasoline is highly flammable. Never smoke or have any kind of open flame around when working on a vehicle. But the risk does not end there. A spark caused by an electrical short circuit, by two metal surfaces contacting each other, or even by static electricity built up in your body under certain conditions, can ignite gasoline vapors, which in a confined space are highly explosive. Do not, under any circumstances, use gasoline for cleaning parts. Use an approved safety solvent.

Always disconnect the battery ground (-) cable at the battery before working on any part of the fuel system or electrical system. Never risk spilling fuel on a hot engine or exhaust component. It is strongly recommended that a fire extinguisher suitable for use on fuel and electrical fires be kept handy in the garage or workshop at all times. Never try to extinguish a fuel or electrical fire with water.

Fumes

Certain fumes are highly toxic and can quickly cause unconsciousness and even death if inhaled to any extent. Gasoline vapor falls into this category, as do the vapors from some cleaning solvents. Any draining or pouring of such volatile fluids should be done in a well ventilated area.

When using cleaning fluids and solvents, read the instructions on the container carefully. Never use materials from unmarked containers.

Never run the engine in an enclosed space, such as a garage. Exhaust fumes contain carbon monoxide, which is extremely poisonous. If you need to run the engine, always do so in the open air, or at least have the rear of the vehicle outside the work area.

If you are fortunate enough to have the use of an inspection pit, never drain or pour gasoline and never run the engine while the vehicle is over the pit. The fumes, being heavier than air, will concentrate in the pit with possibly lethal results.

The battery

Never create a spark or allow a bare light bulb near a battery. They normally give off a certain amount of hydrogen gas, which is highly explosive.

Always disconnect the battery ground (-) cable at the battery before working on the fuel or electrical systems.

If possible, loosen the filler caps or cover when charging the battery from an external source (this does not apply to sealed or maintenance-free batteries). Do not charge at an excessive rate or the battery may burst.

Take care when adding water to a non maintenance-free battery and when carrying a battery. The electrolyte, even when diluted, is very corrosive and should not be allowed to contact clothing or skin.

Always wear eye protection when cleaning the battery to prevent the caustic deposits from entering your eyes.

Household current

When using an electric power tool, inspection light, etc., which operates on household current, always make sure that the tool is correctly connected to its plug and that, where necessary, it is properly grounded. Do not use such items in damp conditions and, again, do not create a spark or apply excessive heat in the vicinity of fuel or fuel vapor.

Secondary ignition system voltage

A severe electric shock can result from touching certain parts of the ignition system (such as the spark plug wires) when the engine is running or being cranked, particularly if components are damp or the insulation is defective. In the case of an electronic ignition system, the secondary system voltage is much higher and could prove fatal.

Chapter 1 Engine

Contents

Specifications

1975 and earlier models

Engine - general

Type	8 cylinder, V-configuration, overhead valve, water-cooled
Firing order	1-8-4-3-6-5-7-2

**V8 ENGINE with
points-type ignition
Firing order
1-8-4-3-6-5-7-2**

**V8 ENGINE with
electronic (HEI) ignition**

**Cylinder location and
distributor rotation**

*The blackened terminal shown on the
distributor cap indicates the
Number One spark plug wire position*

1969/70 models

	307	350	350	396	396
Displacement (cu in.)	307	350	350	396	396
Horsepower (@ rpm)	200 @ 4600	255 @ 4800	300 @ 4800	350 @ 5200	375 @ 5600
Torque (lb ft @ rpm)	300 @ 2400	365 @ 3200	380 @ 3200	415 @ 3400	415 @ 3600
Bore (in.):					
1969	3.875	4	4	4.0938	4.0938
1970	3.875	4	4	4.125	4.125
Stroke	3.25	3.48	3.48	3.76	3.76
Compression ratio	9.0 : 1	9.0 : 1	10.25: 1	10.25: 1	11.0 : 1
Compression pressure (psi)	150	160	160	160	160
Max. variation between cylinders (psi)	20	20	20	20	20

1971 models

	307	350	350
Displacement (cu in.)	307	350	350
Horsepower (@ rpm)	200 @ 4600	245 @ 4800	270 @ 4800
Torque (lb ft @ rpm)	300 @ 2400	350 @ 2800	360 @ 3200
Bore (in.)	3.875	4	4
Stroke (in.)	3.25	3.48	3.48
Compression ratio	8.5 : 1	8.5 : 1	8.5 : 1
Compression pressure (psi)	150	160	160
Max. variation between cylinders (psi)	20	20	20

1972 models *

	307	350	350
Displacement (cu in.)	307	350	350
Horsepower (@ rpm)	200 @ 4600	165 @ 4000	200 @ 4400
Torque (lb ft @ rpm)	230 @ 2400	280 @ 2400	300 @ 2800

* Other Engine - general details are as for 1971 models.

1973 models *

	307	350	350
Displacement (cu in.)	307	350	350
Horsepower (@ rpm)	115 @ 4000	145 @ 4000	175 @ 4000
Torque (lb ft @ rpm)	205 @ 2000	255 @ 2400	270 @ 2400

* Other Engine - general details are as for 1971 models.

1974 models *

	350	350	350
Displacement (cu in.)	350	350	350
Horsepower (@ rpm)	145 @ 3600	160 @ 3800	185 @ 4000
Torque (lb ft @ rpm)	250 @ 2200	245 @ 2400	270 @ 2600

* Other Engine - general details are as for 1971 models.

1975 models

Displacement (cu in.)	262 (4.3 liter)	350	350
Horsepower	110	145	155
Bore	3.67	14.0	4.0
Stroke	3.10	3.48	3:48
Compression ratio	8.5: 1	8.5 : 1	8.5 : 1

Cylinder head warpage limit (all) .... 0.003 in per 6 inches

1969 models

Cylinder bore

	307 cu in.	350 cu in.	396 cu in.
Diameter (in.)	3.8745/3.8775	3.9995/4.0025	4.0925/4.0995
Out of round in. (max.)	0.002	0.002	0.002
Taper, in. (max.)	0.005	0.005	0.005
Oversize bore	+ 0.030	+ 0.030	+ 0.030

Pistons

	307 cu in.	350 cu in.	396 cu in.
Clearance in. (max.)	0.0025	0.0025	0.0025 (350 HP) 0.0065 (375 HP)
Oversize piston availability (in.)	+ 0.001 + 0.030	+0.001 + 0.030	+0.001 + 0.030

Piston rings

	307 cu in.	350 cu in.	396 cu in.
Compression:			
Top (groove clearance in.)	0.001/0.0027	0.001/0.0032	0.001/0.0032
Second (groove clearance in.)	0.001/0.0032	0.001/0.0027	0.001/0.0032
Compression:			
Top (gap in.)	0.01/0.02	0.01/0.023	0.01/0.02
Second (gap in.)	0.01/0.02	0.01/0.025	0.01/0.02

	307 cu in.	350 cu in.	396 cu in.
Oil control:			
Groove clearance (in.)	0.000/0.005	0.000/0.005	0.0005/0.0065
Gap(in.)...	0.01/0.055	0.01/0.055	0.01/0.030

Piston pins

	307 cu in.	350 cu in.	396 cu in.
Diameter (in.)	0.927/0.9273	0.927/0.9273	0.9895/0.9898
Clearance in. (max.)..	0.001	0.001	0.001
Fit in connecting rod	0.0008/0.0016 Interference	0.0008/0.0016 Interference	0.0008/0.0016 Interference

Crankshaft

	307 cu in.	350 cu in.	396 cu in.
Main journal diameter (in.):			
No. 1, No. 2..	2.4484/2.4493	2.4484/2.4493	2.7484/2.7493 (No. 1 journal only on 375 HP)
No. 3, No. 4..	2.4484/2.4493	2.4484/2.4493	2.7481/2.7490 (Including No. 2 journal on 375 HP)
No. 5	2.4479/2.4488	2.4479/2.4488	2.7478/2.7488
Taper in. (max.)..	0.001	0.001	0.001
Out of round in. (max.)	0.001	0.001	0.001
Main bearing clearance in. (max.)	0.004	0.004	0.004
Replacement bearing availability (in.)......................................	-0.001 -0.002 -0.009 -0.01 0 -0.020 -0.030	-0 001 -0.002 -0.009 -0.01 0 -0 020 -0.030	-0 001 -0.002 -0.009 -0.010 -0.020 -0.030
Crankshaft endplay	0.003/0.011	0.003/0.011	0.006/0.010
Crankpin diameter (in.)	2.099/2.100	2.099/2.100	2.199/2.200 (350 HP) 2.1985/2.1995 (375 HP)
Taper in. (max.)..	0.001	0.001	0.001
Out of round in. (max.)	0.001	0.001	0.001
Rod bearing clearance in. (max.)..	0.004	0.004	0.004
Replacement bearing availability (in.)......................................	-0.001 -0.002 -0.010 -0.020	-0.001 -0.002 -0.01 0 -0.020	-0.001 -0.002 -0.010 -0.020
Rod side-clearance (in.)	0.009/0.013	0.009/0.013	0.015/0.021 (350 HP) 0.019/0.025 (375 HP)

Camshaft

	307 cu in.	350 cu in.	396 cu in.
Lobe lift in. (+/- 0.002):			
Intake	0.260	0.26	0.2714 (350 HP) 0.3057 (375 HP)
Exhaust	0.2733	0.2733	0.2714 (350 HP) 0.3057 (375 HP)
Journal diameter (in.)..	1.8682/1.8692	1.8682/1.8692	1.9482/1.9492
Runout in. (max.)	0.0015	0.0015	0.0015

Valve system

	307 cu in.	350 cu in.	396 cu in.
Lifter	Hydraulic	Hydraulic	Hydraulic (350 HP) Mechanical (375 HP)
Valve lash (in.):			
Intake	3/4 turn down from zero lash	3/4 turn down from zero lash	3/4 turn down from zero lash (350 HP) 0.024 hot (375 HP)

Valve lash (continued)

	307 cu in.	350 cu in.	396 cu in.
Exhaust	3/4 turn down from zero lash	3/4 turn down from zero lash	3/4 turn down from zero lash (350 HP) 0.028 hot (375 HP)
Face angle	45°	45°	45°
Seat angle:			
Iron head	46°	46°	46°
Aluminum head	45°	45°	45°
Seat width (in.)			
Intake	1/32 / 1/16	1/32 / 1/16	1/32 / 1/16
Exhaust	1/16 / 3/32	1/16 / 3/32	1/16 / 3/32
Stem clearance (in.):			
Intake	0.001/0.0027	0.001/0.0027	0.001/0.0025
Exhaust	0.002/0.0027	0.002/0.0027	0.002/0.0027
Valve spring free-length (in.)	2.03	2.03	2.09
Spring load (@ in.)	76/84 @ 1.7 194/206 @ 1.25	76/84 @ 1.7 194/206 @ 1.25	94/106 @ 1.88 303/327 @ 1.38
Installed height in. (+/- 1/32)	1 5/32	1 5/32	1 7/8
Damper free-length (in.)	1.94	1.94	1.94/2.00

1970 & 1971 models

The following Section lists the differences between 1970/71 models and 1969 models as listed in the previous Section (396 cu in. applicable to 1970 models only).

Cylinder bore

Diameter (in.):	
396 cu in.	4.1246/4.1274

Pistons

Clearance (in.):	
350 cu in	0.0027 max.
396 cu in. & 350 HP	0.0038 max.
396 cu in. & 375 HP	0.0065 max.

Piston rings

Clearance:	
Top in. 350 cu in, 250 HP	0.001/0.0032
2nd in. 350 cu in, 250 HP	0.001/0.0032
Top in. 350 cu in, 300 HP	0.001/0.0032
2nd in. 350 cu in, 300 HP	0.001/0.0027
Gap:	
Top in. 350 cu in.	0.01/0.02
2nd in. 350 cu in.	0.01/0.025
Top and 2nd in. 396 cu in.	0.01/0.020

Crankshaft

Main journal diameter (in.):	
307 and 350 cu in.:	
No. 1, No. 2	2.4484/2.4493
No. 3, No. 4	2.4484/2.4493
No. 5	2.4479/2.4488
396 cu in.:	
No. 1, No. 2	2.7487/2.7496 1350 HP), 2.7481/2.7490 (375 HP)
No. 3, No. 4	2.7481 /2.7490 (350 and 375 HP)
No. 5	2.7478/2.7488 (350 HP), 2.7473/2.7483 (375 HP)
Main bearing clearance in. (max.):	
No. 1	0.002
All others	0.0035
Crankshaft endplay (in.) 307 and 350 cu in.	0.002/0.006
Rod bearing clearance in. (max.)	0.0035
Rod side-clearance (in.) 307 and 350 cu in.	0.008/0.014

Camshaft

Lobe lift (in.), Exhaust 396 cu in. 350 HP	0.2824 +/- 0.002
Journal diameter (in.) 396 cu in.	1.9487/1.9497

Valve system

Stem clearance (in):	
Intake, 396 cu in.	0.002/0.0027
Exhaust, 307 and 350 cu in.	0.001/0.0029
Valve spring, free-length (in.) 396 cu in.	2.12
Spring load (@ in.) 396 cu in..	69/81 @ 1.88
	228/252 @ 1.38
Installed height (in.), 307 and 350 cu in...	1 23/32 +/- 1 /32
Valve spring (inner), free-length (in.) 396 cu in.	2.06
Spring load (@ in.) 396 cu in...	26/34 @ 1.78
	81/99 @ 1.28
Damper free-length (in.) 396 cu in..	No longer applicable

1972 models

The following Section lists the differences between 307 and 350 cu in. 1972 models and 1970/71 models as listed in the previous Section.

Crankshaft

Main journal diameter (in.):	
No. 1	2.4484/2.4493
No. 2, No. 3, No. 4	2.4481/2.4490
No. 5	2.4479/2.4498

1973 through 1975 models

The following Section lists the differences between 1973 models and 1972 models as listed in the previous Section (307 cu in. not applicable to 1974 models).

Valve system

Valve spring, free-length (in.)	1.91
Spring load (@ in.)	76/84 @ 1.81
	183/195 @ 1.20
Installed height (in.)	1 5/8 + 1/32

Camshaft (California only)

Lobe lift (in.):	
Intake	0.2671 +/- 0.002
Exhaust	0.2733 +/- 0.002

Engine lubrication

Pump type..	Gear, driven by distributor shaft from camshaft helical gear
Oil pressure	40 psi
Crankcase capacity...	5 US quarts (including 1 quart filter capacity)

Torque specifications

Ft-lbs (unless otherwise indicated)

Note: One foot-pound (ft-lb) of torque is equivalent to 12 inch-pounds (in-lbs) of torque. Torque values below approximately 15 ft-lbs are expressed in inch-pounds, because most foot-pound torque wrenches are not accurate at these smaller values.

Crankcase front cover	80 in-lbs
Flywheel housing pan	80 in-lbs
Oil filter bypass valve...	80 in-lbs
Oil pan to crankcase (except 396 cu in.)	80 in-lbs
Oil pan to front cover (396 cu in.)	80 in-lbs
Oil pump cover	80 in-lbs
Rocker arm cover (except 396 cu in.)..	45 in-lbs
Rocker arm cover (396 cu in.)..	50 in-lbs
Camshaft sprocket	20
Oil pan to crankcase (except 396 cu in.)	65 in-lbs
Oil pan to crankcase (396 cu in.)	135 in-lbs
Clutch pressure plate	35
Distributor clamp	20
Flywheel housing	30
Exhaust manifold, outer bolts...	20
Exhaust manifold, inside bolts (except 396 cu in.)	30
Exhaust manifold, inside bolts (396 cu in.)	20
Intake manifold	30
Water outlet	20
Water pump	30

1973 through 1975 models (continued)

Torque specifications

	Ft-lbs (unless otherwise indicated)
Connecting rod cap (except 396 cu in.)	45
Connecting rod cap (396 cu in.) 3/8 - 24	50
Cylinder head (except 396 cu in.)	65
Cylinder head (396 cu in. iron)	80
Cylinder head (396 cu in. aluminum, short bolt)	65
Cylinder head (396 cu in. aluminum, long bolt)	75
Main bearing cap (except outer bolts with 4 bolt caps)	75
Main bearing cap (outer bolts with 4 bolt caps	65
Oil pump	65
Rocker arm stud (396 cu in.)	50
Flywheel (except 396 cu in.)	60
Flywheel (396 cu in)	65
Torsional damper (7/16- 14)	60
Connecting rod cap (7/16 - 20)	70
Main bearing cap (1/2- 13)	105
Temperature sender	20
Torsional damper (1/2 - 20)	85
Oil filter	25
Oil pan drain plug	20
Spark plug	25

1976 and later models

Engine - general

Type	V8 water-cooled, overhead valve
Firing order	1-8-4-3-6-5-7-2

Engine availability and performance

1976 models

Displacement (cu in)	Carburetor type	Horsepower @ rpm	Torque lbf ft @ rpm	Compression ratio
305	2GC	140 @ 3800	245 @ 2000	8.5 : 1
350	2GC	145@ 3800	50 @ 2200	8.5 : 1
350	M4MC	165 @ 3800	260 @ 2400	8.5 : 1

1977 model

Displacement (cu in)	Carburetor type	Horsepower @ rpm	Torque lbf ft @ rpm	Compression ratio
305	2GC	145 @ 3800	245 @ 240	8.5 :1
350	M4MC	165 @ 3800	260 @ 2400	8.5 : 1

1978 models

Displacement (cu in)	Carburetor type	Horsepower @ rpm	Torque lbf ft @ rpm	Compression ratio
305	2GC	140 @ 3800	240 @ 2000	8.5 : 1
350	M4MC	170 @ 3800	270@2400	8.5 : 1

1979 models

Displacement (cu in)	Carburetor type	Horsepower @ rpm	Torque lbf ft @ rpm	Compression ratio
305	M2ME	130 @ 3200	245 @ 2000	8.5 : 1
350	M4MC	170 @ 3800	270 @ 2400	8.2 : 1

General engine dimensions

Engine	305 cu in	350 cu in
Displacement		
Bore (in)	3.736	4.00
Stroke (in)		3.48 3.48
Cylinder bore (diameter) (in)	3.7350 - 3.7385	3.9995 - 4.0025
Out of round (max) (in)	0.002	0.002
Taper (max) (in)	0.005	0.005

*Piston oversizes available 0.001 and 0.030 in

1976 models

All dimensions in inches

Pistons and piston rings

Piston clearance in bore

    305 and 350 engines .... 0.0027 max

Piston ring clearance in groove

    305 and 350 engines Top .... 0.0012 - 0 0032

    305 and 350 engines 2nd .... 0.0012 - 0 0027

    305 and 350 (2 bbl carb) oil control .... 0.005 max

    350 (4 bbl carb) oil control .... 0.002 to 0.007

Piston ring end gap

    All engines Top .... 0.010 - 0.020

    305 engine 2nd .... 0.010 - 0 025

    350 (4 bbl carb) engine .... 0.013 - 0 025

    All engines oil control .... 0.015 - 0.055

Piston pin diameter

    All engines .... 0.9270 - 0.9273

Clearance in piston .... 0.001 max

Interference fit in rod .... 0.0008 - 0.0016

Crankshaft

Main journal diameters

    305 and 350 engines

        Journal 1 .... 2.4484 - 2.4493

        Journals 2-3-4 .... 2.4481 - 2.4490

        Journal 5 .... 2.4479 - 2.4488

Journal taper .... 0.001 max

Journal out of round .... 0.001 max

Main bearing running clearance

    Journal 1 .... 0.002 max

    All others .... 0.0035 max

Crankshaft, end play

    All engines 0.002-0.006

Crankpin diameter

    305 and 350 engines .... 2.099 - 2.100

Crankpin taper .... 0.001 max

Crankpin out of round .... 0.001 max

Rod bearing running clearance .... 0.0035 max

Rod side clearance

    All engines .... 0.008 - 0.014

Camshaft

Lobe lift [intake)

    305 engine 0.2485

    350 engine 0.2600

Lobe lift (exhaust)

    All engines 0.2733

Camshaft journal diameter

    305 and 350 engines .... 1.8682 - 1.8692

Camshaft runout .... 0.0015 max

Valve system

Lifter Hydraulic

Rocker arm ratio

    305 and 350 engines 1.50: 1

Valve lash .... 3/4 turn down from zero

Valve face angle .... 45°

Valve seat angle .... 46°

Valve seat width

    Intake .... 1/32 - 1/16

    Exhaust .... 1/16 - 3/32

Valve stem clearance

    Intake .... 0.0010 - 0.0027

    Exhaust .... 0.0010 - 0.0027

Valve spring free length

    All engines 2.03

Valve spring installed height

    305 and 350 engines 1 23/32

Damper free length

    All engines 1.94

1977 models

All dimensions in inches

Pistons and piston rings

Piston clearance in bore	
305 and 350 engines..	0.0027 max
piston ring clearance in groove	
305 and 350 engines Top..	0.0012 - 0.0032
305 and 350 engines 2nd..	0.0012 - 0.0027
All engines oil control..	0.002 - 0.007
Piston ring end gap	
All engines Top..	0.010 - 0.020
305 engine 2nd..	0.010 - 0.025
350 (4 bbl Carter) engine..	0.013 -0.025
All engines oil control..	0.015 -0.055
Piston pin diameter	
All engines..	0.9270 - 0.9273
Clearance in piston..	0.001
Interference fit in rod	0.0008 - 0.0016

Crankshaft

Main journal diameters	
305 and 350 engines	
Journal 1	2.4484 - 2.4493
Journals 2 - 3 - 4..	2.4481 - 2.4490
Journal 5	2.4479 - 2.4488
Journal taper..	0.001 max
Journal out of round	0.001 max
Main bearing running clearance	
Journal 1...	0.002
All others...	0.0035 max
Crankshaft end play	
All engines..	0.002 - 0.006
Crankpin diameter	
305 and 350 engines...	2.099 - 2.100
Crankpin taper	0.001 max
Crankpin out of round...	0.001 max
Rod bearing running clearance	0.0035 max
Rod side clearance	
All engines..	0.008 - 0.014

Camshaft

Lobe lift (intake)	
305 engine..	0.2485
350 engine..	0.2600
Lobe lift (exhaust)	
All engines..	0.2733
Camshaft journal diameter	
305 and 350 engines..	1.8682 - 1.8692
Camshaft runout..	0.0015 max

Valve system

Lifter...	Hydraulic
Rocker arm ratio	
305 and 350 engines..	1.50 : 1
Valve lash...	3/4 turn down from zero
Valve face angle..	45°
Valve seat angle..	46°
Valve seat width	
Intake	1/32 - 1/16
Exhaust	1/16 - 3/32
Valve stem clearance	
Intake	0.0010 - 0.0027
Exhaust	0.0010 - 0.0027
Valve spring free length	
All engines..	2.03
Valve spring installed height	
Intake	1 23/32
Exhaust	1 19/32
Damper free length	
All engines..	1.86

1978 models

All dimensions in inches

Pistons and piston rings

Piston clearance in bore	
305 and 350 engines	0.0027 max
Piston ring clearance in groove	
305 and 350 engines Top	0.0012 - 0.0032
305 and 350 engines 2nd	0.0012- 0.0032
All engines oil control	0.002-0.007
Piston ring end gap	
All engines	
Top	0.010 - 0.020
2nd	0.010 - 0.025
Oil control	0.015 - 0.055
Piston pin diameter	
All engines	0.9270 - 0.9273
Clearance in piston	0.001
Interference fit in rod	0.0008 - 0.0016
Crankshaft	
Main journal diameters	
305 and 350 engines	
Journal 1	2.4484 - 2.4493
Journals 2 - 3 - 4	2.4481 - 2.4490
Journal 5	2.4479 - 2.4488
Journal taper	0.001 max
Journal out of round	0.001 max
Main bearing running clearance (max)	
Journal 1	0.001 - 0.0015
Journal 2 - 3 - 4	0.001 - 0.0025
Journal 5	0.0025 - 0.0035
Crankshaft end play	
All engines	0.002 - 0.006
Crankpin diameter	
305 and 350 engines	2.099 - 2.100
Crankpin taper	0.001 max
Crankpin out of round	0.001 max
Rod bearing running clearance	0.0030 max
Rod side clearance	
All engines	0.008 - 0.014

Camshaft

Lobe lift (intake)	
305 engine	0.2485
350 engine	0.2600
Lobe lift (exhaust)	
305 engine	0.2667
350 engine	0.2733
Camshaft journal diameter	
305 and 350 engines	1.8682 - 1.8692
Camshaft runout	0.0015 max

Valve system

Lifter	Hydraulic
Rocker arm ratio	
305 and 350 engines	1.50: 1
Valve lash	3/4 turn down from zero
Valve face angle	45°
Valve seat angle	46°
Valve seat width	
Intake	1/32 - 1/16
Exhaust	1/16 - 3/32
Valve stem clearance	
Intake	0.0010 - 0.0027
Exhaust	0.0010 - 0.0027
Valve spring free length	
All engines	2.03
Valve spring installed height	
305 and 350 engines	1 23/32
Damper free length	
All engines	1.86

1979 models

Pistons and piston rings   *All dimensions in inches*

Piston clearance in bore
    305 and 350 engines.. 0.0027 max
Piston ring clearance in groove
    305 and 350 engines Top.. 0.0012 - 0.0032
    305 and 350 engines 2nd.. 0.0012 - 0.0032
    All engines oil control .... 0.002 - 0.007
Piston ring end gap
    All engines
        Top.. 0.010 - 0.020
        2nd.. 0.010 - 0.025
        Oil control .... 0.015 - 0.055
Piston pin diameter
    All engines .... 0.9270 - 0.9273
Clearance in piston.. 0.001 max
Interference fit in rod .... 0.0008 - 0.0016

Crankshaft

Main journal diameters
    305 and 350 engines
        Journal 1 .... 2.4484 - 2.4493
        Journals 2 - 3 - 4 .... 2.4481 - 2.4490
        Journal 5 .... 2.4479 - 2.4488
Journal taper.. 0.001 max
Journal out of round .... 0.001 max
Main bearing running clearance
    Journal 1.. 0.001 - 0.0015
    Journal 2 - 3 - 4 .... 0.001 - 0.0025
    Journal 5.. 0.0025 - 0.0035
Crankshaft end play
    All engines .... 0.002 - 0.006
Crankpin diameter
    305 and 350 engines.. 2.099 - 2.100
Crankpin taper... 0.001 max
Crankpin out of round.. 0.001 max
Rod bearing running clearance .... 0.0030 max
Rod side clearance
    All engines .... 0.006 - 0.014

Camshaft

Lobe lift (intake)
    305 engine.. 0.2484
    350 engine.. 0.2600
Lobe lift (exhaust)
    305 engine.. 0.2667
    350 engine.. 0.2733
Camshaft journal diameter
    305 and 350 engines.. 1.8682 - 1.8692
Camshaft runout.. 0.0015 max

Valve system

Lifter... Hydraulic
Rocker arm ratio
    305 and 350 engine.. 1.50: 1
Valve lash.. 3/4 turn down from zero
Valve face angle.. 45°
Valve seat angle.. 46°
Valve seat width
    Intake .... 1/32 - 1/16
    Exhaust .... 1/16 - 3/32
Valve stem clearance
    Intake .... 0.0010 - 0.0027
    Exhaust .... 0.0010 - 0.0027
Valve spring free length
    All engines 2.03
Valve spring installed height
    305 and 350 engines.. 1 23/32
Damper free length
    All engines 1.86

Crankcase oil capacity

Without oil filter change	4 US qts
With oil filter change	5 US qts

Torque specifications

Ft-lbs (unless otherwise indicated)

Note: One foot-pound (ft-lb) of torque is equivalent to 12 inch-pounds (in-lbs) of torque. Torque values below approximately 15 ft-lbs are expressed in inch-pounds, because most foot-pound torque wrenches are not accurate at these smaller values.

Crankcase front cover	120 in-lbs
Flywheel housing cover	120 in-lbs
Oil filter by-pass valve	120 in-lbs
Oil pan screws	
small	120 in-lbs
large	30
Oil pump cover screws..	120 in-lbs
Rocker arm cover screws..	48 in-lbs
Camshaft sprocket bolt	20
Clutch pressure plate bolts...	35
Distributor clamp bolts	20
Flywheel housing bolts	30
Exhaust manifold bolts	20
Inner bolts 350 engine...	30
Intake manifold bolts	30
Water outlet bolts	30
Water pump bolts	30
Connecting rod cap bolts	45
Cylinder head bolts...	65
Main bearing cap bolts*..	70
Oil pump bolts	65
Rocker arm stud	50
Flywheel bolts	60
Torsional damper bolt..	60
Temperature sender unit	20
Oil pan drain plug	20
Spark plug	15
*Outer bolts on 4 bolt caps..	65

1 General description

The engines used on the Nova models dealt with in this manual are of V8, overhead valve type. Chevrolet classify all the engines, with the exception of the 396 (400) cu in. version, as small-block V8s; the 396 cu in, is classified as a big-block V8.

The cylinders are numbered 1, 3, 5, 7 (left-bank) and 2, 4, 6, 8, working from front to rear. Crankshaft rotation is clockwise viewed from the radiator end.

A five main bearing crankshaft is used and is drilled to feed oil to the main and connecting rod bearings. The camshaft also has five bearings, and operates the valve lifters, hollow pushrods, valve rockers and valves.

Full pressure lubrication is provided from a gear type oil pump driven from the distributor shaft.

2 Engine repair operations - general notes

1   The following repair operations can be carried out with the engine installed:

Removal of the intake and exhaust manifolds
Removal of the valve mechanism
Removal of the cylinder head
Removal of the oil pan and oil pump
Removal of the rear main oil seal
Removal of the torsional damper, crankcase front cover, oil seal, and timing chain and sprockets
Removal of the camshaft
Removal of the pistons, connecting rods and associated bearings
Removal of the flywheel
Removal of the engine mounts

The repair operations given in this Chapter generally assume the engine is removed. Where applicable, separate instructions or cautionary notes are given to enable operations to be performed with the engine in the car.

2   The following repair operations can only be carried out after removal of the engine:

Removal of the crankshaft and main bearings (see Section 12, paragraph 7).

3 Engine - removal and installation

Warning 1: *The air conditioning system is under high pressure! Have a dealer service department or service station discharge the system before disconnecting any A/C system hoses or fittings.*

Warning 2: *Gasoline is extremely flammable, so take extra precautions when you work on any part of the fuel system. Don't smoke or allow open flames or bare light bulbs near the work area, and don't work in a garage where a natural gas-type appliance (such as a water heater or a clothes dryer) with a pilot light is present. Since gasoline is carcinogenic, wear latex gloves when there's a possibility of being exposed to fuel, and, if you spill any fuel on your skin, rinse it off immediately with soap and water. Mop up any spills immediately and do not store fuel-soaked rags where they could ignite. When you perform any kind of work on the fuel system, wear safety glasses and have a Class B type fire extinguisher on hand.*

Removal

Refer to illustrations 3.5, 3.20 and 3.22

1   Disconnect the cable from the negative terminal of the battery.
2   Cover the fenders and cowl and remove the hood (see Chapter 12). Special pads are available to protect the fenders, but an old bedspread or blanket will also work.
3   Remove the air cleaner assembly.
4   Drain the cooling system (see Chapter 2).
5   Label the vacuum lines, emissions sys-

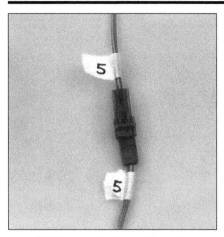

3.5 Flag each wire and hose to ease reassembly

tem hoses, wiring connectors, and ground straps to ensure correct reinstallation, then detach them. Pieces of masking tape with numbers or letters written on them work well **(see illustration)**. If there's any possibility of confusion, make a sketch of the engine compartment and clearly label the lines, hoses and wires.

6   Label and detach all coolant hoses from the engine.

7   Remove the cooling fan, shroud and radiator (see Chapter 2).

8   Remove the drivebelts.

9   Disconnect the fuel lines running from the engine to the chassis. Plug or cap all open fittings/lines.

10   Disconnect the throttle linkage (and TV linkage/speed control cable, if equipped) from the engine.

11   On power steering equipped vehicles, unbolt the power steering pump (see Chapter 11). Leave the lines/hoses attached and make sure the pump is kept in an upright position in the engine compartment (use wire or rope to restrain it out of the way).

12   On A/C equipped vehicles, unbolt the compressor and set it aside. Do not disconnect the hoses.

13   Drain the engine oil and remove the filter.

14   Remove the starter motor (see Chapter 10).

15   Remove the alternator (see Chapter 10).

16   Unbolt the exhaust system from the engine (see Chapter 3).

17   If you're working on a vehicle with an automatic transmission, refer to Chapter 6B and remove the torque converter-to-drive-plate fasteners.

18   Support the transmission with a floor jack. Position a block of wood between the jack and the transmission to prevent damage to the transmission. Special transmission jacks with safety chains are available - use one if possible.

19   Attach an engine sling or a length of chain to the lifting brackets on the engine.

20   Roll the hoist into position and connect the sling or chain to it **(see illustration)**. Take up the slack in the sling or chain, but don't lift the engine. **Warning:** *DO NOT place any part of your body under the engine when it's supported only by a hoist or other lifting device.*

21   Remove the transmission-to-engine block bolts.

22   Remove the engine mount through-bolts **(see illustration)**.

23   Recheck to be sure nothing is still connecting the engine to the transmission or vehicle. Disconnect anything still remaining.

24   Raise the engine slightly. Carefully work it forward to separate it from the transmission. If you're working on a vehicle with an automatic transmission, be sure the torque converter stays in the transmission (clamp a pair of vise-grips to the housing to keep the converter from sliding out). If you're working on a vehicle with a manual transmission, the input shaft must be completely disengaged from the clutch. Slowly raise the engine out of the engine compartment. Check carefully to make sure nothing is hanging up.

25   Remove the flywheel/driveplate and mount the engine on an engine stand.

Installation

26   Check the engine and transmission mounts. If they're worn or damaged, replace them.

27   If you're working on a manual transmission equipped vehicle, install the clutch and

pressure plate (Chapter 6A). Now is a good time to install a new clutch.

28   Carefully lower the engine into the engine compartment - make sure the engine mounts line up. Install the engine mount through-bolts.

29   If you're working on an automatic transmission equipped vehicle, guide the torque converter into the crankshaft following the procedure outlined in Chapter 6B.

30   If you're working on a manual transmission equipped vehicle, apply a dab of high-temperature grease to the input shaft and guide it into the crankshaft pilot bearing until the bellhousing is flush with the engine block.

31   Install the transmission-to-engine bolts and tighten them securely. **Caution:** *DO NOT use the bolts to force the transmission and engine together!*

32   Reinstall the remaining components in the reverse order of removal.

33   Add coolant, oil, power steering and transmission fluid as needed.

34   Run the engine and check for leaks and proper operation of all accessories, then install the hood and test drive the vehicle.

35   Have the A/C system recharged and leak tested if any lines were disconnected.

---

4   Engine - disassembly (general)

1   After the engine has been removed, attach it to an engine stand. These can be obtained at many auto parts stores and are also available at most equipment rental yards.

2   During the disassembly process the greatest care should be taken to keep the exposed parts free from dirt. As an aid to achieving this, it is absolutely necessary to thoroughly clean down the outside of the engine, removing all traces of oil and dirt.

3   Use a water soluble grease solvent. This will make the job much easier, as, after the solvent has been applied and allowed to stand for a time, a vigorous jet of water will wash off the solvent and all the grease and filth. If the dirt is thick and deeply embed-

3.20 Properly attached lifting chain

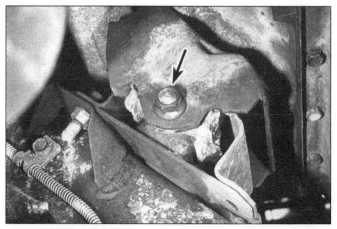

3.22 Remove the engine mount through-bolts

5.10 Removing the exhaust manifold

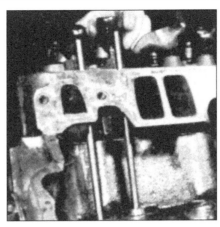

5.14a Removing a pushrod . . .

5.14b . . . and a valve lifter

ded, work the solvent into it with a wire brush.

4   Finally wipe down the exterior of the engine with a rag and only then, when it is quite clean should the disassembly process begin. As the engine is stripped, clean each part with solvent.

5   Clean the oilways of the crankshaft with a pipe cleaner and if an air line is available, blow the oilways through to clean them.

6   Discard all old gaskets; use new ones on reassembly.

7   Wherever possible, replace nuts, bolts and washers finger-tight from wherever they were removed. If they cannot be replaced then lay them out in such a fashion that it is clear from where they came.

---

5   Cylinder head - removal

Refer to illustrations 5.10, 5.14a and 5.14b

Intake manifold

1   Drain the radiator and remove the air cleaner housing.

2   Disconnect the battery cables, upper radiator and heater hose at the manifold, the accelerator linkage at the pedal, the fuel line at the carburetor, the coil and temperature sender switch wires, the power brake hose, the distributor spark advance hose and the appropriate emission control and crankcase ventilation hoses.

3   Remove the distributor cap and mark the distributor body and rotor position (where applicable). Remove the distributor clamp and distributor, then position the cap rearwards, clear of the manifold.

4   Remove the alternator upper bracket, and the coil and its bracket.

5   Remove the manifold-to-head attaching bolts, and lift off the manifold complete with carburetor.

6   Remove the carburetor (and choke tube assembly, where applicable), the water outlet and thermostat, heater hose adapter, choke coil and EGR valve, if the manifold is to be replaced.

Exhaust manifold

7   Remove the air injection reactor (A.I.R.) air manifold and tubes (refer to Chapter 3), if considered necessary.

8   Remove the battery ground cable and the air cleaner pre-heater air stove.

9   Remove the manifold to exhaust flange nuts, then lower the pipe assembly. Hang it from the frame to prevent undue loading.

10   Remove the end mounting bolts followed by the center ones and lift the manifold away from the engine **(see illustration)**.

Head assembly

11   If the engine is installed, remove the appropriate crankcase ventilation hoses.

12   Disconnect the wiring harness from the clips on the rocker cover(s), remove the cover retaining screws and lift off the cover(s).

13   If the engine is installed, drain the coolant from the cylinder block (refer to Chapter 2).

14   Working on each valve in turn, loosen the rocker arm nut until the rocker can be pivoted, then remove the pushrod and valve lifter. Place the pushrods and valve lifters in a rack so that they may be installed in the same location during engine assembly **(see illustrations)**.

15   Remove the valve rocker arm nuts, balls and rocker arms.

16   Unscrew the cylinder head bolts, one turn at a time and remove them.

17   With the aid of an assistant lift the cylinder head(s) from the block. If they are stuck, do not attempt to pry them off, but tap upwards using a block of wood and a hammer at each end. Place the head(s) on a clean workbench for further disassembly.

---

6   Cylinder head - disassembly

1   Using a suitable valve spring compressor, compress each spring in turn to permit the locks to be removed. Release the compressor and remove the spring cap or rotator, shield (where applicable), spring and damper, then remove the oil seals and spring shims.

2   Remove the valves and place them in the rack in their proper sequence with their associated pushrods etc.

3   Unscrew and remove each spark plug, if not already done.

---

7   Oil pan and oil pump - removal

Refer to illustration 7.14

1   Disconnect the battery ground cable, then for safety's sake remove the distributor cap.

2   Remove the fan shroud retaining bolts,

3   On big-block V8 engines, place a piece of heavy cardboard between the radiator and fan to prevent contact when the engine is raised.

4   Disconnect the exhaust pipes or crossover pipes.

5   Drain the engine oil.

6   Where automatic transmission is fitted, remove the converter housing underpan and splash shield.

7   On pre-1975 models, disconnect the steering idler lever at the frame and swing the linkage down clear of the oil pan.

8   Rotate the crankshaft until the timing mark on the torsional damper is at the 6 o'clock position.

9   On all vehicles, except the 350 cu in. and big-block V8 with automatic transmission, disconnect the starter brace at the starter. Remove the inboard starter bolt and loosen the outboard one, then swing the starter outboard.

10   Remove both front engine mount through-bolts.

11   Using suitable jacks and wooden blocks positioned beneath the torsional damper, raise the engine until 3 inch wooden spacers can be inserted at the engine mounts. Lower the engine onto the 3 inch spacers,

12   Remove the oil pan retaining bolts and the oil pan. If stuck, tap it sharply with a soft-faced mallet or cut around the joint with a thin sharp knife, Do not pry against the crankcase or irreparable distortion may occur.

13   Where applicable, remove the oil pan baffle.

7.14 Remove the oil pump and screen assembly

8.3 Remove the accessory drive pulley

8.4 Using a three-jaw puller to draw off the torsional damper

14   Remove the oil pump and screen assembly and the extension shaft **(see illustration)**.

8   Timing chain, sprockets and camshaft - removal

Refer to illustrations 8.3, 8.4, 8.8 and 8.13
Note: *Where the engine has been removed, only the procedure given in paragraph 3 onwards, excluding paragraphs 10 and 11, is applicable. On pre-1974 small-block V8 models, initially remove the oil pan (Section 7).*
1   Remove the fan belt, fan and pulley (refer to Chapter 2).
2   Remove the fan shroud. If the camshaft is to be removed, remove the radiator (Chapter 2).
3   Remove the accessory drive pulley, then remove the torsional damper retaining bolt **(see illustration)**.
4   Using a suitable puller, draw off the torsional damper **(see illustration)**. **Caution:**

Use only a puller that pulls on the hub of the damper.
5   Remove the water pump (refer to Chapter 2).
6   **Small-block V8:** Unscrew the crankcase front cover attaching bolts, then remove the cover.
7   **Big-block V8:** Remove the front cover to block attaching screws, pull the cover slightly forwards to permit the oil pan front seal to be cut with a sharp knife flush with the block at both sides of the cover. Remove the front cover.
8   Rotate the crankshaft until the marks on the camshaft and crankshaft sprockets are aligned, then remove the three camshaft sprocket retaining bolts. Using a soft-faced mallet, tap off the camshaft sprocket and remove the timing chain.
9   If it is found necessary to replace the crankshaft sprocket, it can be removed using a suitable puller.
10   If the engine is installed, remove both front mount through-bolts and remove the right engine mount. Lower the engine on the frame.

11   Remove the two center bolts and one lower bolt securing the hood catch support to the grille and radiator support to provide clearance for removal of the camshaft.
12   If not previously removed, remove the fuel pump and pushrod (refer to Chapter 3).
13   Taking a great deal of care, in order to prevent damage to the camshaft journals and cam peaks, withdraw the camshaft **(see illustration)**. This will be made easier if two 5/16 inch - 18 x 4 inch bolts are installed through the sprocket flange holes and tightened evenly, to start it moving.

9   Pistons, connecting rods and bearings - removal

Refer to illustrations 9.3 and 9.4
1   Initially remove the cylinder heads, oil pan and oil pump, as described previously in this Chapter.
2   Ensure that identification marks are present on each connecting rod and bearing cap

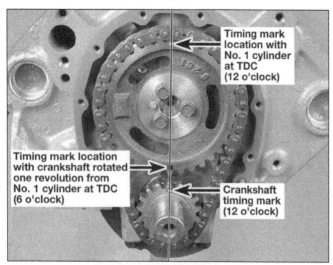

8.8 Align the timing marks as shown - the cam gear mark should be in the six o'clock position and the crankshaft gear mark should be in the12 o'clock position

8.13 Remove the camshaft slowly, supporting it in two places, and make sure the cam lobes and distributor gear don't gouge the camshaft bearings in the block

9.3 A ridge reamer is required to remove the ridge from the top of each cylinder - do this before removing the pistons!

9.4 Use a feeler gauge to check the connecting rod side clearance (end play)

to enable them to be installed in their original positions. If no marks are present, small punch indentations will be satisfactory (left-bank - 1,3,5,7, right-bank - 2,4,6,8).

3 Turn the crankshaft so that the relevant piston is at the lowest point of its stroke. Place a rag on top of the piston and then carefully remove the wear ridge from the top of the cylinder bore **(see illustration)**. Remove the rag and the metal scrapings.

4 The connecting rod side clearance should be measured with a feeler gauge between the connecting rod caps. If the side clearance is outside the specified tolerance, replace the rod(s) **(see illustration)**.

5 Remove the connecting rod bearing cap, push a piece of rubber or plastic tubing onto each of the connecting rod studs to prevent them scratching the cylinder bores as the rods are removed.

6 Push the piston/connecting rod assembly out of the top of the cylinder block.

7 Repeat the operations on the remaining seven cylinders turning the crankshaft as necessary, to gain access to the connecting

rod cap bolts and to bring the piston to the bottom or top of its stroke as required.

10 Flywheel-removal

1 The flywheel may be unbolted from the crankshaft rear flange after removal of the transmission unit and in the case of vehicles with a manual gearbox, unbolting the clutch housing and clutch (refer to Chapters 5 and 6).

11 Engine ancillary components - removal

Refer to illustration 11.1

1 Having removed an engine for major overhaul as described in the following Sections, it is important to remove the various ancillary components. This is essential where the cylinder block is to be serviced or replaced. Some components will have been removed already, whether the engine has been removed or is still installed, but at this

stage check, and remove where applicable, those which remain; Typically these will be:
a) *Oil filter (unscrew)*
b) *Distributor (refer to Chapter 4)*
c) *Engine ground strap*
d) *Oil dipstick and dipstick tube*
e) *Coolant hoses*
f) *Water temperature sender (from left-hand cylinder head)*
g) *Oil pressure sender (from the rear of the block* **(see illustration)***)*
h) *The appropriate emission control items and the associated pipes and wiring*

12 Crankshaft and main bearings - removal

Refer to illustration 12.2

1 The crankshaft can be removed only after the engine has been removed from the vehicle and completely disassembled as described in the earlier Sections of this Chapter.

2 Check that the main bearing caps are marked in respect of their location in the

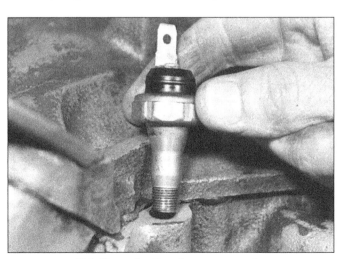

11.1 Removing the oil pressure sender

12.2 A main bearing cap showing its number

crankcase (and their orientation), as carried out for the connecting rod bearing caps (Section 9) **(see illustration)**.

3   Unbolt the main bearing caps, keeping their shell bearings together with their respective caps (if they are to be used again).

4   Lift the crankshaft from the crankcase, with help from an assistant.

5   If the upper half shell bearings are to be removed and it is intended to use them again, (not advised) keep them identified with (but not interchanged with) their respective lower half shell bearings.

6   Remove the two halves of the crankshaft rear oil seal.

Replacement of main bearings - engine in vehicle

7   This operation can be carried out, but it is considered that a certain amount of risk is involved since it is impossible to ensure that there are no deposits of dirt etc., on the bearing shell faces as they are being installed. However, the owner may consider it worth the risk, particularly in view of the work saving.

8   Remove the engine oil pan, oil pump and spark plugs.

9   Identify the main bearing caps (see paragraph 2).

10   To replace the main bearing shells with the crankshaft in position, first loosen (but do not remove) each of the main bearing caps.

11   Remove No. 1 main bearing cap, extract the shell bearing from the cap and fit the new one, lubricating it with engine oil.

12   Insert a cotter pin in each of the crankshaft journal oil holes and then turn the crankshaft in a clockwise direction until the head of the cotter pin slides the upper shell from its seat in the crankcase. Fit the new upper shell in the same way, oiling it liberally and inserting the plain (unnotched) end first.

13   Remove the cotter pin, fit the bearing cap ('F' mark towards the front of the engine). Tighten the cap bolts finger-tight.

14   Repeat the operations on the remaining main bearings and then tighten all the cap bolts to the specified torque.

13   Examination and renovation - general

1   With the engine completely stripped, clean every component (except the cylinder bores) in solvent and dry off. Make sure that all oilways are then thoroughly cleaned out to remove all trace of kerosene.

2   Pay particular attention to the engine block. Scrape off old pieces of gasket or jointing compound, probe the oilways and waterways, examine the casting for cracks, and check the freeze plugs for security.

3   Never clean the cylinder bores with solvent, but use hot water and detergent and when dry, apply clean engine oil.

4   The individual components should be carefully checked for wear or distortion, as described in the following Sections.

14.5 Checking the crankshaft endplay by inserting a feeler gauge between the crankshaft and the face of the thrust bearing

14   Crankshaft and bearings- examination and renovation

Refer to illustration 14.5

1   Examine the crankpin and main journal surfaces for scoring; scratches or corrosion. If evident, then the crankshaft will have to be reground professionally.

2   Using a micrometer, test each journal and crankpin at several different points for ovality. If this is found to be more than 0.001 inch then the crankshaft must be reground. Undersize bearings are available as listed in Specifications to suit the recommended reground diameter, but normally your parts salesperson will supply the correct matching bearings with the reconditioned crankshaft.

3   After a high mileage, the main bearings and the connecting rod bearings may have worn to give an excessive running clearance. The correct running clearance for the different journals is given in the Specifications.

 a) *The clearance is best checked using a proprietary product such as 'Plastigage' having refitted the original bearings and caps and tightened the cap bolts to the torque settings listed in the Specifications. Always fit new shell bearings, having first checked the crankshaft journals and crankpins for ovality and to establish whether their diameters are of standard or reground sizes.*

 b) *Do not turn the crankshaft while the 'Plastigage' material is in position.*

4   Checking the connecting rod bearings is carried out in a similar manner to that described for the main bearings. The correct running clearance is given in the Specifications.

5   The crankshaft endplay should be checked by forcing the crankshaft to the extreme front position, then using a feeler gauge at the front end of the rear main bearing. Refer to the Specifications for the permissible clearance **(see illustration)**.

15   Cylinder block - examination and renovation

1   The cylinder bores must be examined for taper, ovality, scoring and scratches. Start by carefully examining the top of the bores If they are worn a ridge will be found on the thrust side. The bottom of the ridge marks the upper limit of piston ring travel and the thickness of the ridge will be a guide to the amount of bore wear.

2   Another indication of cylinder bore wear will be evident before engine disassembly takes place by the emission of blue smoke from the exhaust and the frequent need for topping-up the engine oil.

3   Using an internal type dial gauge, measure each bore at three different points in both the thrust and axial directions. Carryout this operation near the top of the bore and then near the bottom of the bore. From the readings obtained, establish the out-of-round which must not exceed 0.002 inch, and the taper which must not exceed 0.005 inch.

4   Where the cylinder bores are worn beyond the permitted tolerance then they must be honed or bored to the next oversize. This is a specialist operation and must be carried out in a properly equipped workshop.

5   New pistons are available in standard and oversizes as listed in the Specifications Section and they will be supplied to match the new bore diameters of the cylinder block. Keep each piston identified in respect of its cylinder. The maximum permissible piston to bore clearance is given in the Specifications.

16   Pistons, connecting rods and rings - servicing

Piston/connecting rod assemblies - inspection Refer to illustrations 16.4a, 16.4b, 16.10 and 16.11

1   Before the inspection process can be carried out, the piston/connecting rod assemblies must be cleaned and the original

16.4a The piston ring grooves can be cleaned with a special tool, as shown here . . .

16.4b . . . or with a section of broken ring

piston rings removed from the pistons. **Note:** *Always use new piston rings when the engine is reassembled.*

2   Using a piston ring installation tool, carefully remove the rings from the pistons. Be careful not to nick or gouge the pistons in the process.

3   Scrape all traces of carbon from the top of the piston. A hand-held wire brush or a piece of fine emery cloth can be used once the majority of the deposits have been scraped away. Do not, under any circumstances, use a wire brush mounted in a drill motor to remove deposits from the pistons. The piston material is soft and may be eroded away by the wire brush.

4   Use a piston ring groove cleaning tool to remove carbon deposits from the ring grooves. If a tool isn't available, a piece broken off the old ring will do the job. Be very careful to remove only the carbon deposits - don't remove any metal and do not nick or scratch the sides of the ring grooves **(see illustrations)**.

5   Once the deposits have been removed, clean the piston/rod assemblies with solvent and dry them with compressed air (if available). Make sure the oil return holes in the back sides of the ring grooves are clear.

6   If the pistons and cylinder walls aren't damaged or worn excessively, and if the engine block is not rebored, new pistons won't be necessary. Normal piston wear appears as even vertical wear on the piston thrust surfaces and slight looseness of the top ring in its groove. New piston rings, however, should always be used when an engine is rebuilt.

7   Carefully inspect each piston for cracks around the skirt, at the pin bosses and at the ring lands.

8   Look for scoring and scuffing on the thrust faces of the skirt, holes in the piston crown and burned areas at the edge of the crown. If the skirt is scored or scuffed, the engine may have been suffering from overheating and/or abnormal combustion, which caused excessively high operating temperatures. The cooling and lubrication systems should be checked thoroughly. A hole in the piston crown is an indication that abnormal combustion (preignition) was occurring.

Burned areas at the edge of the piston crown are usually evidence of spark knock (detonation). If any of the above problems exist, the causes must be corrected or the damage will occur again. The causes may include intake air leaks, incorrect fuel/air mixture, incorrect ignition timing and EGR system malfunctions.

9   Corrosion of the piston, in the form of small pits, indicates that coolant is leaking into the combustion chamber and/or the crankcase. Again, the cause must be corrected or the problem may persist in the rebuilt engine.

10   Measure the piston ring side clearance by laying a new piston ring in each ring groove and slipping a feeler gauge in beside it **(see illustration)**. Check the clearance at three or four locations around each groove. Be sure to use the correct ring for each groove - they are different. If the side clearance is greater than specified, new pistons will have to be used.

11   Check the piston-to-bore clearance by measuring the bore and the piston diameter. Make sure the pistons and bores are correctly matched. Measure the piston across

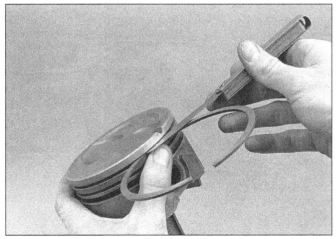

16.10 Check the ring side clearance with a feeler gauge at several points around the groove

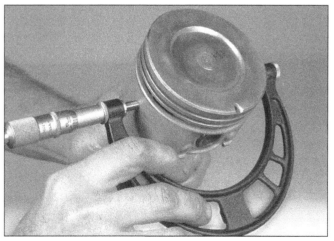

16.11 Measure the piston diameter at a 90-degree angle to the piston pin, and also in line with it

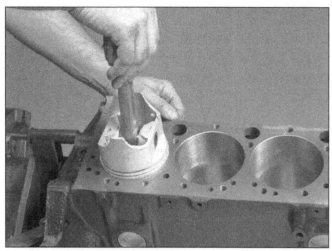

16.17 When checking piston ring end gap, the ring must be square in the cylinder bore (this is done be pushing the ring down with the top of a piston as shown)

16.18 With the ring square in the cylinder, measure the end gap with a feeler gauge

the skirt, at a 90-degree angle to and in line with the piston pin **(see illustration)**. Subtract the piston diameter from the bore diameter to obtain the clearance. If it's greater than specified, the block will have to be rebored and new pistons and rings installed.

12   Check the piston-to-rod clearance by twisting the piston and rod in opposite directions. Any noticeable play indicates excessive wear, which must be corrected. The piston/connecting rod assemblies should be taken to an automotive machine shop to have the pistons and rods resized and new pins installed.

13   If the pistons must be removed from the connecting rods for any reason, they should be taken to an automotive machine shop. While they are there have the connecting rods checked for bend and twist, since automotive machine shops have special equipment for this purpose. **Note:** *Unless new pistons and/or connecting rods must be installed, do not disassemble the pistons and connecting rods.*

14   Check the connecting rods for cracks and other damage. Temporarily remove the rod caps, lift out the old bearing inserts, wipe the rod and cap bearing surfaces clean and inspect them for nicks, gouges and scratches. After checking the rods, replace the old bearings, slip the caps into place and tighten the nuts finger tight. **Note:** *If the engine is being rebuilt because of a connecting rod knock, be sure to install new or remanufactured rods.*

Piston rings - installation

Refer to illustrations 16.17, 16.18, 16.19, 16.23a, 16.23b and 16.26

15   Before installing the new piston rings, the ring end gaps must be checked. It's assumed that the piston ring side clearance has been checked and verified correct.

16   Lay out the piston/connecting rod assemblies and the new ring sets so the ring sets will be matched with the same piston

16.19 If the end gap is too small, clamp a file in a vise and file the ring ends (from the outside in only) to enlarge the gap slightly

and cylinder during the end gap measurement and engine assembly.

17   Insert the top (number one) ring into the first cylinder and square it up with the cylinder walls by pushing it in with the top of the piston **(see illustration)**. The ring should be near the bottom of the cylinder, at the lower limit of ring travel.

18   To measure the end gap, slip feeler gauges between the ends of the ring until a gauge equal to the gap width is found **(see illustration)**. The feeler gauge should slide between the ring ends with a slight amount of drag. Compare the measurement to the Specifications. If the gap is larger or smaller than specified, double-check to make sure you have the correct rings before proceeding.

19   If the gap is too small, it must be enlarged or the ring ends may come in contact with each other during engine operation, which can cause serious damage to the engine. The end gap can be increased by filing the ring ends very carefully with a fine file. Mount the file in a vise equipped with soft jaws, slip the ring over the file with the ends contacting the file face and slowly move the

ring to remove material from the ends. When performing this operation, file only from the outside in **(see illustration)**.

20   Excess end gap isn't critical unless it's greater than 0.040-inch. Again, double-check to make sure you have the correct rings for your engine.

21   Repeat the procedure for each ring that will be installed in the first cylinder and for each ring in the remaining cylinders. Remember to keep rings, pistons and cylinders matched up.

22   Once the ring end gaps have been checked/corrected, the rings can be installed on the pistons.

23   The oil control ring (lowest one on the piston) is usually installed first. It's composed of three separate components. Slip the spacer/expander into the groove **(see illustration)**. If an anti-rotation tang is used, make sure it's inserted into the drilled hole in the ring groove. Next, install the lower side rail. Don't use a piston ring installation tool on the oil ring side rails, as they may be damaged. Instead, place one end of the side rail into the groove between the spacer/expander and

16.23a Installing the spacer/expander in the oil control ring groove

16.23b DO NOT use a piston ring installation tool when installing the oil ring side rails

16.26 Installing the compression rings with a ring expander - the mark (arrow) must face up

18.5 Inspect the gear housing for scoring, pitting and wear - this housing shows normal wear

the ring land, hold it firmly in place and slide a finger around the piston while pushing the rail into the groove **(see illustration)**. Next, install the upper side rail in the same manner.

24   After the three oil ring components have been installed, check to make sure that both the upper and lower side rails can be turned smoothly in the ring groove.

25   The number two (middle) ring is installed next. It's usually stamped with a mark which must face up, toward the top of the piston. **Note:** *Always follow the instructions printed on the ring package or box - different manufacturers may require different approaches. Do not mix up the top and middle rings, as they have different cross sections.*

26   Use a piston ring installation tool and make sure the identification mark is facing the top of the piston **(see illustration)**, then slip the ring into the middle groove on the piston. Don't expand the ring any more than necessary to slide it over the piston.

27   Install the number one (top) ring in the same manner. Make sure the mark is facing up. Be careful not to confuse the number one and number two rings.

28   Repeat the procedure for the remaining pistons and rings.

17 Camshaft and bearings- examination and renovation

1   Mount the camshaft on V-blocks and use a dial gauge to measure lobe lift and run out. Reject a camshaft which does not meet the specified limits.

2   Measure the journal diameters using a micrometer. Reject a camshaft which does not meet the specified limits.

3   Examine the bearing surfaces and the surfaces of the cam lobes. If the lobes and bearing surfaces are in good condition, the camshaft can be reused. If the journals are damaged, the bearing inserts in the block are probably damaged as well.

4   Camshaft bearing replacement requires special tools and expertise that place it outside the scope of the home mechanic. Take the block to an automotive machine shop to ensure that the job is done correctly.

18 Oil pump - disassembly, examination and reassembly

Refer to illustrations 18.5 and 18.6

Note: *Oil pump gears and the body are not serviced separately. If wear or damage is evident on the gears or body, the complete pump assembly must be replaced.*

1   Remove the pump cover retaining screws and the pump cover. Index mark the gear teeth to permit reassembly in the same position.

2   Remove the idler gear, drive gear and shaft from the body.

3   Remove the pressure regulator valve retaining pin, the regulator valve and the related parts.

4   If necessary, the pick-up screen and pipe assembly can be extracted from the pump body.

5   Wash all the parts in solvent and thoroughly dry them. Inspect the body for cracks, wear or other damage **(see illustration)**. Similarly inspect the gears.

18.6 Inspect the pump cover for scoring and pitting - this cover shows wear; the pump should be replaced

6   Check the drive gear shaft for looseness in the pump body, and the inside of the pump cover for wear that would permit oil leakage past the ends of the gears **(see illustration)**.

7   Inspect the pick-up screen and pipe assembly for damage to the screen, pipe or relief grommet.

8   Apply a gasket sealant to the end of the pipe (pick-up screen and pipe assembly) and tap it into the pump body taking care that no damage occurs. If the original press-fit cannot be obtained a rear assembly must be used to prevent air leaks and loss of pressure.

9   Install the pressure regulator valve and related parts.

10   Install the drive gear and shaft in the pump body, followed by the idler gear with the smooth side towards the pump cover opening. Pack all cavities in the pump and between the gear with petroleum jelly (this will ensure the pump is primed).

11   Install the cover and torque tighten the screws.

12   Turn the driveshaft to ensure that the pump operates freely.

19 Oil filter- servicing

1   Remove the element from the filter casing and discard it.

2   Clean all the component parts in kerosene or gasoline, shake off the surplus and dry with a lint-free cloth.

3   Check the oil filter fiber bypass valve and spring for operation (mounted in the block). Inspect for a broken valve; if evident the oil filter adapter and the bypass valve assembly must be replaced.

4   Clean the valve chamber in the block and torque tighten the screws to the specified value.

20 Positive Crankcase Ventilation (PCV) system - description and servicing

Refer to illustration 20.1

1   The engine has a positive crankcase ventilation system which utilizes intake manifold vacuum to draw fumes and 'blow-by' vapors from the crankcase into the combustion chambers where they are burned **(see illustration)**.

2   Fresh air enters the engine via a filter on the periphery of the air cleaner, circulates through the engine, mixes with the blow-by vapors, passes out through the PCV control valve and mixes with the combustible mixture entering the intake manifold.

3   Servicing comprises flushing (or replacement) of the hoses, and replacement of the PCV valve and filter, at the intervals stated in routine maintenance.

21 Cylinder head - decarbonizing, examination and servicing

Decarbonizing

1   This can be carried out with the engine either in or out of the car. With the cylinder head off carefully remove with a blunt scraper or wire brush all traces of carbon deposits from the combustion spaces and the ports. The valve head stems and valve guides should also be freed from any carbon deposits. Wash the combustion spaces and ports down with solvent and scrape the cylinder head surface free of any foreign matter with a gasket.

2   Clean the pistons and top of the cylinder bores. If the pistons are still in the block then it is essential that great care is taken to ensure that no carbon gets into the cylinder bores as this could scratch the cylinder walls or cause damage to the piston and rings, To ensure this does not happen, first turn the crankshaft so that two of the pistons are at the top of their bores. Stuff rag into the other bores or seal them off 1 with paper and masking tape. The waterways should also be covered with small pieces of masking tape to prevent particles of carbon entering the cooling system and damaging the water pump.

3   Press a little grease into the gap between the cylinder walls and the two pistons which are to be worked on.

4   With a blunt scraper carefully scrape away the carbon from the piston crown, taking great care not to scratch the aluminum. Also scrape away the carbon from the surrounding lip of the cylinder wall, When all carbon has been removed, scrape away the grease which will now be contaminated with carbon particles, taking care not to press any into the bores.

5   Rotate the crankshaft until the next two pistons are at the top of their bores. Repeat the foregoing operation until all eight cylinders are decarbonized.

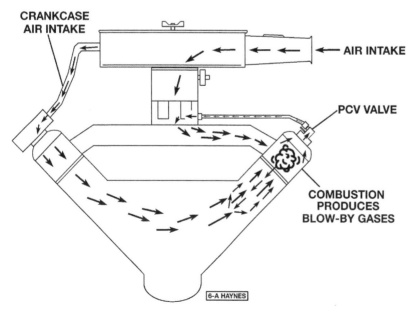

20.1 Gas flow in a typical V8 PCV system

CRANKCASE AIR INTAKE

AIR INTAKE

PCV VALVE

COMBUSTION PRODUCES BLOW-BY GASES

6-A HAYNES

21.7 Check the cylinder head gasket surface for warpage by trying to slip a feeler gauge under the straightedge (see the Specifications for the maximum warpage allowed and use a feeler gauge of that thickness)

23.1 A dial indicator can be used to determine the valve stem-to-guide clearance (move the valve stem as indicated by the arrows)

Examination

Refer to illustration 21.7

6   Inspect the cylinder head for cracks in the exhaust ports and combustion chambers, or external cracks into the water chambers.

7   Using a straightedge and feeler gauge, check the head gasket mating surface for warpage **(see illustration)**. If the warpage exceeds the specified limit, it must be resurfaced at an automotive machine shop.

8   Inspect the rocker arm studs for wear and damage. On big-block V8 cylinder heads inspect the pushrod guides for wear and damage.

Servicing

9   On big-block V8 and some high performance small-block V8 heads the pushrod guides are retained by nuts on the rocker arm studs. These studs can be unscrewed for replacement of the guides. When assembling, coat the replacement stud with a gasket sealant and torque tighten.

10   On small-block V8 engines the studs are pressed in, but replacement is considered a specialist operation involving the reaming of the stud holes 0.003 or 0.013 inch oversize, the new studs being lubricated with hypoid axle oil and pressed in to their original depth.

22   Valves - servicing

1   Because of the complex nature of the job and the special tools and equipment needed, servicing of the valves, the valve seats and the valve guides, commonly known as a valve job, should be done by a professional.

2   The home mechanic can remove and disassemble the heads, do the initial cleaning and inspection, then reassemble and deliver them to a dealer service department or an

automotive machine shop for the actual service work. Doing the inspection will enable you to see what condition the heads and valvetrain components are in and will ensure that you know what work and new parts are required when dealing with an automotive machine shop.

3   The dealer service department, or automotive machine shop, will remove the valves and springs, recondition or replace the valves and valve seats, recondition the valve guides, check and replace the valve springs, spring retainers or rotators and keepers (as necessary), replace the valve seals with new ones, reassemble the valve components and make sure the installed spring height is correct. The cylinder head gasket surfaces will also be resurfaced if they're warped.

4   After the valve job has been performed by a professional, the heads will be in like-new condition. When the heads are returned, be sure to clean them again before installation on the engine to remove any metal particles and abrasive grit that may still be present from the valve service or head resurfacing operations. Use compressed air, if available, to blow out all the oil holes and coolant passages.

23   Valve guides and springs - examination and renovation

Refer to illustration 23.1

1   Thoroughly clean out each valve guide and then insert the appropriate valve, Using a dial gauge test the movement of the valve (at 90-degrees to the center line of the cylinder head) making sure that the valve is held from its seat by about 1/16 inch **(see illustration)**. The total valve stem movement indicated by the gauge needle must be divided by two to obtain the actual stem clearance.

2   If the stem clearance exceeds that given

in the Specifications it will be necessary to ream the valve guide and f it oversize valve stems. Check the availability with your Chevrolet dealer or engine repair specialist for this servicing operation.

3   Each valve spring and damper should be compared with the specified free-length and renewed if it is shorter. In any event it is recommended that new springs and dampers are installed if the engine has covered more than 30,000 miles since they were new.

24   Flywheel and starter ring gear - examination

1   Examine the starter ring gear for broken or chipped teeth. If evident, the flywheel must be replaced.

2   On manual transmission versions, examine for glazing and scoring on the clutch friction face. Light glazing may be dressed out using emery cloth, but where there is scoring the flywheel must be refaced or replaced or clutch damage will soon occur. It's a good idea to have the flywheel refaced as a matter of course.

3   On automatic transmission models, examine the converter securing bolt holes for elongation. Replace the driveplate if elongation is present.

25   Valve lifters - disassembly, cleaning and reassembly

Refer to illustration 25.2

1   Clean the lifters with solvent and dry them thoroughly without mixing them up.

2   Check each lifter wall, pushrod seat and foot for scuffing, score marks and uneven wear. Each lifter foot (the surface that rides on the cam lobe) must be slightly convex, although this can be difficult to determine by

25.2 The foot of each lifter should be slightly convex - the side of another lifter can be used as a straightedge to check it; if it appears flat, it is worn and must not be reused

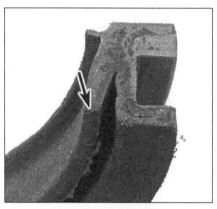

27.1 The rear main oil seal may have two lips - the seal lip (arrow) must point to the front of the engine, which means that the dust seal will face out, toward the rear of the engine

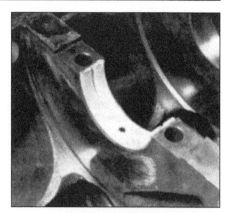

27.3a A main bearing shell properly installed

27.3b The rear main bearing cap contains the thrust bearing

27.4 Lubricate the crankshaft journals with engine oil

eye **(see illustration)**. If the base of the lifter is concave, the lifters and camshaft must be replaced. If the lifter walls are damaged or worn (which is not very likely), inspect the lifter bores in the engine block as well. If the pushrod seats are worn, check the pushrod ends.

3   If new lifters are being installed, a new camshaft must also be installed. If a new camshaft is installed, use new lifters as well. Never install used lifters unless the original camshaft is used and the lifters can be installed in their original locations.

26   Engine reassembly - general

1   To ensure maximum life with minimum trouble from a rebuilt engine, not only must everything be correctly assembled, but everything must be spotlessly clean, all the oilways must be clear, locking washers and spring washers must always be fitted where indicated and all bearing and other working surfaces must be thoroughly lubricated during assembly.

2   Before assembly begins replace any bolts or studs the threads of which are in any way damaged.

3   Apart from your normal tools, a supply of clean rags, an oil can filled with engine oil, a supply of assorted spring washers, a set of new gaskets, and a torque wrench, should be collected together.

27   Crankshaft and main bearings- installation

Refer to illustrations 27.1, 27.3a, 27.3b, 27.4, 27.6, 27.7a and 27.7b

1   Install the rear main bearing oil seal in the cylinder block and rear main bearing cap grooves, with the seal lip towards the front of the engine **(see illustration)**. (Where a seal has two lips, the lip with the helix is towards the front of the engine).

2   Lubricate the seal lips with engine oil.

3   Install the main bearings in the cylinder block and main bearing caps and lubricate the bearing surface with engine oil **(see illustrations)**.

4   With the aid of an assistant, install the crankshaft, taking care not to damage the bearing surfaces. Lubricate the crankshaft journals with engine oil **(see illustration)**.

5   Apply a thin film of RTV sealant to the mating surface of the rear main bearing cap **(see illustration)**. Do not allow the sealant to contact the crankshaft or seal.

6   Install the main bearing caps in their

27.5 Apply a thin film of RTV sealant to the ends of the seal and the adjacent areas on the block (arrows) before installing the rear main bearing cap

27.6 Install the main bearing caps

27.7a Driving the crankshaft rearwards

27.7b Torque-tighten the main
bearing caps

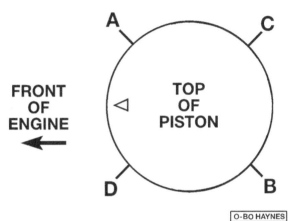

FRONT
OF
ENGINE
←

TOP
OF
PISTON

28.1 Ring end gap positions

A   Oil ring side rail gap - lower
B   Oil ring side rail gap - upper
C   Top compression ring gap
D   Second compression ring
     gap and oil ring spacer gap

O-BO HAYNES

28.3a Install the connecting rod bearing
cap shell

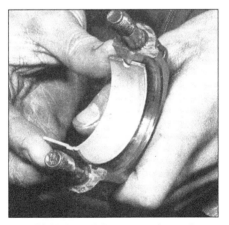

28.3b Install the connecting rod
bearing shell

28.4a Identification numbers on the
connecting rod and bearing cap

28.4b The notch or arrow in the top of
each piston must face the front of the
engine as the pistons are installed

correct positions, arrows towards the front of
the engine **(see illustration)**.

7   Torque tighten the main bearing caps
with the exception of the rear cap bolts.
Tighten these to 12 ft-lbs, then tap the end of
the crankshaft rearwards then forwards with
a hammer interposed with a block of wood to
align the rear main bearing thrust faces **(see
illustration)**. Retorque all the bearing caps to
the specified value **(see illustration)**.

28   Pistons and connecting rods - installation

*Refer to illustrations 28.1, 28.3a, 28.3b,
28.4a, 28.4b, 28.5 and 28.6*

1   Stagger the piston ring end gaps around
the piston **(see illustration)**. also that the piston is correctly aligned to the connecting rod.
Apply engine oil to the cylinder bores.

2   Fit a piston ring compressor over the
piston rings which should have been well
lubricated.

3   Fit pieces of plastic or rubber tube to the
threads of the connecting rod bolts. Fit and
lubricate the bearing shells **(see illustrations)**.

4   Insert the connecting rod/piston assembly into its respective bore, ensuring that it is
the correct way round **(see illustrations)**.

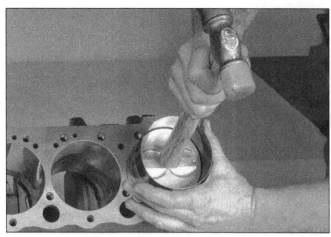

28.5 The piston can be driven (gently) into the cylinder bore with the end of a wooden hammer handle

28.6 Fitting the bearing cap to a connecting rod

5   With the base of the piston ring compressor resting on the top face of the cylinder block, tap the assembly into the bore using the wooden handle of a hammer **(see illustration)**.

6   Carefully guide the connecting rod to engage with the crankpin and fit the bearing cap so that the numbers on the rod and cap are adjacent **(see illustration)**.

7   Tighten the cap bolts to the specified torque,

8   Repeat the operations on the remaining pistons. The crankshaft will have to be rotated to facilitate connection of the connecting rods and bearing caps.

29 Flywheel- installation

Refer to illustration 29.1
Note: *To permit the crankshaft to be rotated more easily, the flywheel is fitted at this stage. However, if it is required to keep the engine weight to a minimum to aid maneuverability during rebuild, the operation can be left until later.*

1   Install the flywheel with the dowel hole aligned with the dowel hole in the crankshaft. On vehicles with automatic transmission, the converter attaching pads and flange collar should face towards the transmission **(see illustration)**.

2   Torque tighten the attaching bolts. A clean wooden block wedged between the crankshaft and cylinder block will prevent rotation while the bolts are being tightened. Tighten the bolts in a criss-cross pattern.

3   Where this operation is being carried out with the engine installed, install the clutch, clutch housing and manual transmission, or the automatic transmission (refer to Chapters 5 and 6).

30 Camshaft, timing chain and   sprockets - installation

Refer to illustrations 30.3, 30.22a and 30.22b
Note: *Where a new camshaft or valve lifters are being fitted, the cam lobes must be coated with camshaft installation lubricant.*

1   Install two 5/16 inch - 18 x 4 inch bolts in the camshaft bolt holes, then lubricate the journals with engine oil.

2   Carefully install the camshaft, taking care to feed the journals and cam lobes through the bearings to prevent damage.

When the camshaft is fully home, remove the two 5/16 inch bolts.

3   Place the timing chain on the camshaft sprocket then align the marks on the camshaft and crankshaft sprockets. Connect the chain to the crankshaft sprocket, align the camshaft dowel with the dowel hole in the sprocket and install the sprocket on the camshaft **(see illustration)**.

4   Draw the sprocket onto the camshaft using the attaching bolts. Do not drive the sprocket on, or the rear plug may be loosened. Torque tighten the sprocket attaching bolts.

5   Lubricate the timing chain with engine oil.

6   If not already carried out, pry out the old seal from the front of the front cover using a screwdriver.

7   Ensure that the seal housing is clean then install a new seal so that the open end is towards the inside of the cover. The seal must be carefully pressed in with the cover supported and care must be taken to prevent damage to the seal lips.

8   Smear the front cover oil seal lips with engine oil.

Small-block V8

9   Ensure that the block and crankcase front cover are clean.

10   Use a sharp knife to remove any oil pan gasket material protruding at the oil pan to engine block junction (1974 models onwards).

11   Apply a 1/8 inch bead of RTV sealant to the joint formed at the oil pan and block, as well as the front lip of the oil pan (1974 models onwards).

12   Coat the cover gasket with a non-setting sealant, position it on the cover then loosely install the cover. First install the top four bolts loosely then install two 1/inch - 20 x 1/2 inch screws at the lower cover holes. Apply a bead of the RTV sealant on the bottom of the cover then install the cover, tightening the screws alternately and evenly while using a suitable tool to align the dowel pins.

29.1 The flywheel dowel hole correctly aligned

30.3 The camshaft and crankshaft sprocket marks aligned

30.22a Install the torsional damper

30.22b Using a special tool to install the damper - the box-end wrench holds the main shaft stationary while the adjustable wrench rotates the large nut that pushes the damper onto the crankshaft

31.5 When installing the oil pan, note the clips which may be used

13   Remove the two 1/4 inch - 20 x 1/2 inch screws and install the remaining cover screws.

Big-block V8

14   Ensure that the block and crankcase front cover are clean.
15   Cut the tabs from the new oil pan front seal, using a sharp knife.
16   Install the seal to the front cover, pressing the tips into the holes provided in the cover.
17   Coat the gasket with a non-setting sealant and position it on the cover.
18   Apply a 1/8 inch bead of RTV sealant to the joint formed at the oil pan and block.
19   Install the cover attaching screws and torque tighten to the specified value.

All versions

20   Install the fuel pump pushrod and the pump (refer to Chapter 3).
21   Install the water pump (refer to Chapter 2).
22   Coat the front cover seal area of the torsional damper with engine oil, place the damper in position, then use a damper installer or a suitable bolt and spacers to draw the damper into position **(see illustra-**

tions). Install and torque tighten the damper retaining bolt. Take care that the damper is not damaged during this operation.
23   Install the accessory drive pulley.
24   Install the fan shroud (and radiator, where applicable).
25   Install the fan pulley, fan and fan belt. Adjust the fan belt tension (refer to Chapter 2).

31   Oil pump and oil pan - installation

Refer to illustrations 31.5
Note: *On pre-1974 models, the crankcase front cover must have been installed prior to installation of the oil pan.*
1   Assemble the pump and extension shaft to the rear main bearing cap. Align the slot at the top end of the extension shaft with the drive tang on the lower end of the distributor shaft, if the distributor is installed.
2   Install the pump to the rear bearing cap bolt and torque tighten the bolts. Where applicable install the oil baffle.
3   Use a non-setting gasket sealant and install the side gaskets on the cylinder block.

Apply the sealant at the intersection of the side gaskets and end seals.
4   Install the oil pan rear seal in the groove in the rear main bearing cap, with the ends butting the side gaskets.
5   Install the oil pan and torque tighten the bolts, making sure any clips are in their proper positions **(see illustration)**.
6   Where applicable, install the torsional damper and pulley (paragraphs 22 and 23 of Section 30).
7   Where the engine is installed, the remainder of the oil pan installation procedure is the reverse of that given in Section 7, paragraphs 1 thru 11.

32   Cylinder head - assembly

Refer to illustrations 32.1, 32.2a, 32.2b, 32.2c, 32.3 and 32.9
1   Apply engine oil to the stem of No. 1 valve and insert it in its correct port **(see illustration)**.

Small-block V8

2   Set the valve spring shim, spring, damper, valve shield and cap (intake valve) or rotator (exhaust valve) in place **(see illustrations)**.

32.1 Inserting No. 1 valve

32.2a The intake valve shield and cap

32.2b The exhaust valve shield and rotator

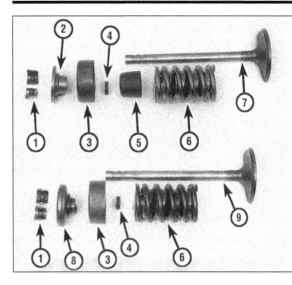

32.2c Valve spring components - exploded view

1 Locks
2 Retainer
3 Oil shield
4 O-ring oil seal
5 Umbrella seal
6 Spring and damper
7 Intake valve
8 Retainer/rotator
9 Exhaust valve

32.3 Using a spring compressor on one of the valves

32.9 Measuring valve spring installed height

33.3 The cylinder head gasket in position

3   Using a suitable valve spring compressor, compress the spring and install the oil seal in the lower groove of the valve stem, ensuring that it is seated squarely **(see illustration)**.
4   Insert the valve locks and release the spring compressor, ensuring that the locks seat properly in the stem upper groove.

Big-block V8

5   Install the valve spring shim on the valve spring seat and then install a new valve stem oil seal over the valve and valve guide.
6   Set the valve spring, damper and valve cap in place, then use a suitable compressor to compress the spring. **Note:** *The valve springs are wound closer at one end, install the more closely wound end next to the head.*
7   Insert the valve locks and release the spring compressor, ensuring that the locks seat properly in the stem groove.

All versions

8   Install the remaining valves in their correct positions following the same procedure.
9   Using a soft-faced mallet, tap the end of each valve stem sharply to settle the valve

components, then measure the installed height of the valve spring. To do this, use a narrow scale passed down beside the spring to rest on the valve spring seat. Now measure the height of the upper surface of the spring and compare the measurement with that given in the Specifications **(see illustration)**. If the specified height is exceeded, install a valve spring seat shim of approximately 1/16 inch thickness. At no time should the spring, installed height be less than that specified.

---

33 Cylinder head - installation

Refer to illustrations 33.3, 33.4, 33.5a, 33.5b, 33.7, 33.8a, 33.8b, 33.16a, 33.16b, 33.17a, 33.17b, 33.17c and 33.17d

Head assembly

1   Scrupulously clean the mating surfaces of the head and block and the cylinder head bolt threads.
2   Where a steel gasket is used, coat both sides thinly and evenly with a copper coat-type gasket sealant. Where a steel/asbestos

composition gasket is used, no gasket sealant is permitted.
3   Place the gasket over the dowel pins with the bead upwards **(see illustration)**.
4   Guide the cylinder head into position over the dowel pins and gasket **(see illustration)**.
5   Apply gasket sealant to the bolt threads and install them finger-tight. Tighten them progressively in the sequence shown **(see illustrations)**, to the specified torque.
6   Install the valve lifters and pushrods in their original positions. If new lifters are being used, coat the foot of each one with camshaft installation lubricant or an equivalent anti-scuffing compound.
7   Install the valve rocker arms, rocker arm balls and nuts, and tighten the rocker arm nuts until all endplay has *just* been taken up. Do not install the rocker cover. **Note:** Where new rocker arms and balls are being used, the contact surfaces should be coated with moly-based grease or an equivalent anti-scuffing compound **(see illustration)**.

Exhaust manifold

8   Clean the mating surfaces of the mani-

33.4 Installing the cylinder head

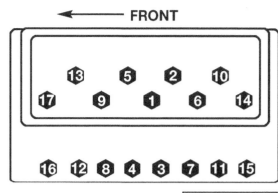

FRONT

33.5a Cylinder head bolt tightening sequence -
small-block V8 engines

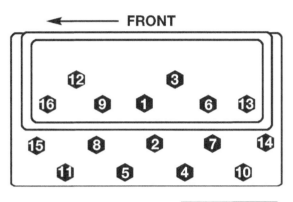

FRONT

33.5b Cylinder head
bolt tightening
sequence - big-block
V8 engines

33.7 Install the rocker arms

33.8a An exhaust manifold gasket
in position

33.8b A spark plug heat shield in position

33.16a Make sure the intake manifold
gaskets are installed right side up or all
the passages and bolt holes may not line
up properly

fold and head, then install the manifold in position with the center bolts, using new gaskets. Where applicable, at this stage, it may be found easier to install the spark plug heat shields **(see illustrations)**.

9   Install the end bolts then lightly tighten all the bolts.

10   Torque tighten the two center bolts to the specified value, then torque-tighten the four end bolts.

Engine installed

11   Using a new flange gasket, fit the exhaust pipe to the manifold flange.

12   Connect the battery ground cable.

13   Install the air cleaner pre-heater.

14   Connect the air injection reactor (A.I.R.) air manifold and tubes (refer to Chapter 3).

Intake manifold

15   Thoroughly clean the gasket and sealing

surfaces of the manifold, cylinder heads and block.

16   Install the manifold end seals on the block and the side gaskets on the cylinder heads **(see illustrations)**. Use a gasket sealant around the water passages.

17   Install the manifold and carburetor, and torque tighten the bolts to the specified value

33.16b Install the intake manifold end seals

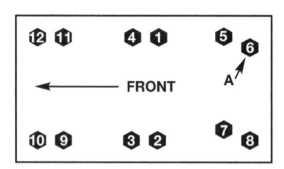

33.17a Intake manifold bolt tightening sequence - small-block V8 engines

A - Stud bolt

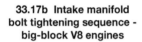

33.17b Intake manifold bolt tightening sequence - big-block V8 engines

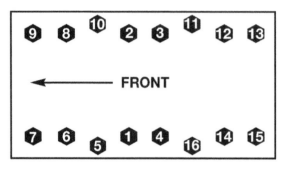

33.17c The spark plug wire clips on the manifold

in the proper sequence **(see illustrations)**. Note the spark plug wire clips **(see illustrations)**.

Engine installed

18   Install the ignition coil and distributor. Refer to Chapter 4 for ignition timing details where the crankshaft has been rotated. If the crankshaft has not been rotated, align the marks made during removal.

19   Install the alternator upper bracket and adjust the belt tension (refer to Chapter 2). Torque tighten the bolts.

20   Connect the battery cables, coolant hoses, accelerator pedal linkage, fuel line at the carburetor, ignition Spark plug wires, power brake hose, spark advance and crankcase ventilation hoses, as applicable.

21   Fill the cooling system with the correct amount of coolant (refer to Chapter 2).

34   Valve lash- adjustment

Refer to illustration 34.2

Note: *For this operation to be carried out, the air cleaner, rocker cover ventilation hoses, rocker cover wiring harness clips and rocker covers must be removed. Where applicable, refer to Chapter 3 (air cleaner) and para-*

graphs 11 and 12 of Section 5 for further information.

Hydraulic valve lifters

1   Rotate the crankshaft until the mark on the torsional damper aligns with the center or "0" marking on the timing indicator. If No. 1 cylinder valves are moving, the engine is in No. 6 cylinder firing position and the crankshaft must be rotated 360 degrees. If No. 1 cylinder valves are not moving, the piston is at top-dead-center (TDC) which is correct.

2   Back-off the rocker arm stud adjusting nut on No. 1 intake and exhaust valves in

33.17d The spark plug wire clips on the manifold

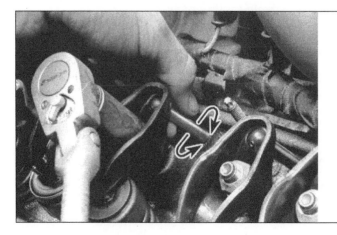

34.2 Rotate each pushrod as the rocker arm nut is tightened; tighten the nut until a slight drag is felt - this is the point at which all play is removed. Tighten the nut an additional 3/4-turn

turn, until there is play in the pushrod; tighten the nut to *just* eliminate play then rotate the nut 3/4-turn **(see illustration)**.

3 With the engine in the No. 1 firing position, as determined in paragraph 1, also adjust the exhaust valves of cylinders 1, 3, 4 and 8 and the intake valves of cylinders 1, 2, 5 and 7.

4 Rotate the crankshaft through 360 degrees to align the torsional damper mark once more, then repeat paragraph 2 for exhaust valves 2, 5, 6 and 7, and intake valves 3, 4, 6 and 8.

Mechanical valve lifters

Note: *Where mechanical lifters are used, adjustment must be carried out with the engine at normal operating temperature. After major repair or overhaul, initially set the lash as described below, install the engine (where applicable), run the engine carefully to attain the normal operating temperature, then repeat the procedure with a hot engine.*

5 Initially proceed as for hydraulic lifters in paragraph 1.

6 Rotate the rocker arm stud adjusting nuts on exhaust valves 4 and 8, and intake valves 2 and 7, until a feeler gauge of the specified clearance is a firm sliding fit between the rocker and the end of the valve stem.

7 Rotate the crankshaft through 180 degrees clockwise and repeat paragraph 6 for exhaust valves 3 and 6, and intake valves 1 and 8.

8 Rotate the crankshaft through a further 180-degrees clockwise (torsional damper and timing indicator marks aligned) and repeat paragraph 6 for exhaust valves 5 and 7, and intake valves 3 and 4.

9 Rotate the crankshaft through a further 180-degrees clockwise and repeat paragraph 6 for exhaust valves 1 and 2, and intake valves 5 and 6.

All versions

10 Clean the gasket surfaces of the cylinder head and rocker arm cover with gasoline and wipe dry with a lint-free cloth.

11 Using a new gasket, install the rocker arm cover and torque tighten the bolts to the specified value. Note the spark plug wire clips.

12 Where applicable, connect the crankcase ventilation hoses and the electrical wiring harness at the rocker arm cover clips, Install the air cleaner.

35 Engine ancillary components - installation

1 After an engine repair or overhaul it will now be necessary to install the various ancillary components. The items to be installed will obviously be those which were removed during disassembly, and will typically be as follows:

a) *Spark plugs. Ensure that they are of the correct type, preferably new (but at least properly cleaned) with correctly set gaps (refer to Chapter 4).*
b) *Oil filter (new).*
c) *Distributor (refer to Chapter 4).*
d) *Engine ground strap.*
e) *Oil dipstick and filler tube.*
f) *The appropriate emission control items, and the associated pipes and wiring.*
g) *Water temperature sender.*
h) *Oil pressure sender.*

36 Engine start-up after major repair or overhaul

1 Refill the cooling system (refer to Chapter 2)

2 Refill the crankcase with the correct grade and quantity of engine oil.

3 Make a final check to ensure that all cables and pipes have been connected and that no tools or rags have been left in the engine compartment.

4 Start the engine and check for water and oil leaks. **Note:** *The engine may not start readily since there may be condensation*

inside; also it may take a little while for the fuel pump to deliver fuel to the carburetor.

5 Check that the instruments are indicating satisfactory readings.

6 Run the vehicle until normal engine operating temperature is reached and check the valve lash (mechanical lifters only).

7 Check the carburetor and emission control settings, as described in Chapter 3.

8 Check the ignition timing (Chapter 4).

9 Top-up the engine oil level to make up for the oil absorbed by the new filter element.

10 Run the vehicle for between 500 and 1000 miles and with the engine cold, check the torque settings of all engine nuts and bolts, particularly the cylinder head bolts (only if the gasket manufacturer requires it). Also change the engine oil and filter at this time.

37 Engine mounts- replacement

Refer to illustrations 37.2a and 37.2b

1 If on inspection, the flexible mounts have become hard or are split or separated from their metal backing, they must be replaced. This operation may be carried out with the engine/transmission still in the vehicle.

Front mounts

2 Remove the through-bolt and nut **(see illustrations)**.

3 Raise the engine slightly using a hoist, then remove the mount and frame bracket assembly from the crossmember.

4 Install the new mount, install the through-bolt and nut, then tighten all the bolts to the specified torque.

Rear mount

5 Remove the crossmember to mount bolts then raise the transmission slightly.

6 Remove the mount to transmission bolts, followed by the mount.

7 Install the new mount, lower the transmission and align the crossmember to mount bolts.

8 Tighten all the bolts to the specified torque.

38 Crankshaft oil seals (front and rear) - replacement in-vehicle

Front oil seal

1 Remove the torsional damper, as described in paragraphs 1 thru 4 of Section 8.

2 Pry out the old seal from the front cover with a large screwdriver taking care not to mark the crankshaft.

3 Install a new seal, open end towards the inside of the cover. Using a suitable tubular spacer, carefully drive it into position.

4 Install the torsional damper, as described in paragraphs 22 thru 25 of Section 30.

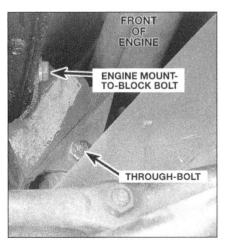

37.2a Typical engine mount details - left side

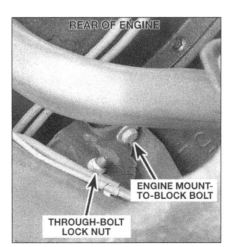

37.2b Typical engine mount details - right side

38.7 Tap the seal end with a brass punch and hammer, until it can be gripped with a pair of pliers and pulled out

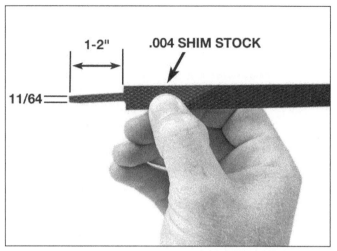

38.10 If the new seal did not include an installation tool, make one from a piece of brass or plastic shim stock

Rear oil seal

Refer to illustrations 38.7 and 38.10

5   Initially remove the oil pan and pump, as described in Section 7.

6   Remove the rear main bearing cap and pry out the oil seal with a screwdriver.

7   Use a hammer and small aluminum or brass drift to drive the upper half of the seal from around the crankshaft **(see illustration)**. Once it has started to move, pliers can be used to draw it out.

8   Clean all foreign material from the crankshaft and bearing cap, and inspect all the sealing faces for nicks. scratches and burrs.

9   Coat the seal lips and seal bead with a light grade of engine oil, keeping it off the seal mating ends.

10   Make up a tool as shown in the accompanying illustration, if one was not included with the new seal **(see illustration)**.

11   Position the tip of the tool between the crankshaft and seal seat in the crankcase.

12   Position the seal between the crankshaft and tip of the tool, so that the seal bead is contacting the tool tip. Ensure that oil seal lip is towards the front of the engine.

13   Roll the seal round the crankshaft using the tool as a 'shoe-horn' to protect the seal

bead from the sharp corner of the seal seat surface in the crankcase.

14   As soon as both ends of the tool are flush with the ends of the block, remove it, taking care not to withdraw the seal.

15   Install the seal half in the bearing cap using the same method and by using light thumb and finger pressure.

16   Apply gasket sealant to the cap-to-case interface, being careful to keep it off the seal split line. Install the bearing cap and torque tighten the bolts to the specified value.

17   Install the oil pump and oil pan, as described in Section 31.

39   Troubleshooting - engine

Symptom	Reason
Engine fails to start	Discharged battery
	Loose battery connection
	Disconnected or broken ignition leads
	Moisture on spark plugs, distributor or leads
	Incorrect contact points gap, cracked distributor cap or rotor
	Incorrect spark plug gap
	Dirt or water in carburetor jets
	Empty fuel tank
	Faulty fuel pump
	Faulty starter motor
	Transmission park or neutral switch inoperative
	Faulty carburetor choke mechanism
Engine idles erratically	Air leak at intake manifold
	Leaking cylinder head gasket
	Worn timing sprockets
	Worn camshaft lobes
	Overheating
	Choked PCV valve
	Faulty fuel pump
	Leaking EGR valve (emission control)

Engine misses at idling speed	Incorrect spark plug gap
	Uneven compression between cylinders
	Faulty coil or condenser
	Faulty contact points
	Poor connections or condition of ignition leads
	Dirt in carburetor jets
	Incorrectly adjusted carburetor
	Worn distributor cam
	Air leak at carburetor flange gasket
	Faulty ignition advance mechanism
	Sticking valves
	Incorrect valve lash
	Low cylinder compression
	Leaky EGR valve (emission control)
Engine misses throughout speed range	Dirt or water in carburetor or fuel lines
	Incorrect ignition timing
	Contact points incorrectly gapped
	Worn distributor
	Faulty coil or condenser
	Spark plug gaps incorrect
	Weak valve spring
	Overheating
	Leaking EGR valve (emission control)
Engine stalls	Incorrectly adjusted carburetor
	Dirt or water in fuel
	Ignition system incorrectly adjusted
	Sticking choke mechanism
	Faulty spark plugs or incorrectly gapped
	Faulty coil or condenser
	Incorrect contact points gap
	Exhaust system clogged
	Distributor advance inoperative
	Air leak at intake manifold
	Air leak at carburetor mounting flange
	Incorrect valve lash
	Sticking valve
	Overheating
	Low compression
	Poor electrical connections on ignition system
	Leaking EGR valve (emission control)
Engine lacks power	Incorrect ignition timing
	Faulty coil or condenser
	Worn distributor
	Dirt in carburetor
	Spark plugs incorrectly gapped
	Incorrectly adjusted carburetor
	Faulty fuel pump
	Weak valve springs
	Sticking valve
	Incorrect valve timing
	Incorrect valve lash
	Blown cylinder head gasket
	Low compression
	Brakes dragging
	Clutch slipping
	Overheating
	Transmission regulator valve sticking
	(Hydra-Matic automatic transmission)

Note: *In addition to the foregoing, reference should also be made to the fault finding chart for emission control equipment which is to be found at the end of Chapter 3. Such a fault can have an immediate effect upon engine performance.*

Chapter 2   Cooling system

Contents

Specifications

Pressure cap rating	15 psi
Thermostat type	Wax pellet
Thermostat rating	195-degrees F (180-degrees F for 396 cu in. 375 HP)
Water pump	Impeller type
Radiator type	Crossflow
Cooling fan	Belt driven from engine
	Automatic viscous fan clutch fitted on some models
Coolant capacity	16 quarts + 1 quart with air conditioning (23 quarts + 2 quarts with air conditioning for 396 cu in. engine)

Torque specifications   Ft-lbs (unless otherwise indicated)

Note: One foot-pound (ft-lb) of torque is equivalent to 12 inch-pounds (in-lbs) of torque. Torque values below approximately 15 ft-lbs are expressed in inch-pounds, because most foot-pound torque wrenches are not accurate at these smaller values.

Water outlet	30
Water pump	30
Fan shroud	50 in-lbs
Coolant recovery tank bracket	50 in-lbs
Radiator mounting to support	18
Thermostatic fan clutch hub bolts	25

1   General description

Refer to illustrations 1.2a and 1.2b

All vehicles covered by this manual employ a pressurized engine cooling system with thermostatically controlled coolant cir-culation. An impeller type water pump mounted on the front of the block pumps coolant through the engine. The coolant flows around each cylinder and toward the rear of the engine. Cast-in coolant passages direct coolant around the intake and exhaust ports, near the spark plug areas and in close proximity to the exhaust valve guides.

A wax pellet type thermostat is located in the thermostat housing near the front of the engine. During warm up, the closed ther-mostat prevents coolant from circulating through the radiator. When the engine reaches normal operating temperature, the

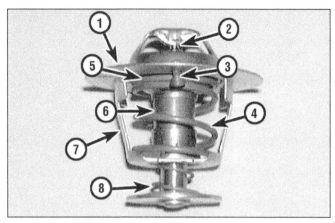

1.2a Typical thermostat

1 Flange	4 Main coil	7 Frame
2 Piston	spring	8 Secondary
3 Jiggle valve	5 Valve seat	coil spring
	6 Valve	

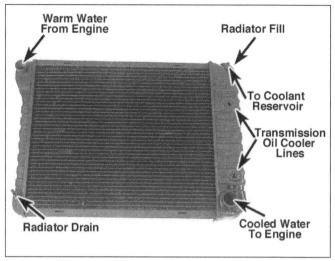

1.2b Typical crossflow radiator

thermostat opens and allows hot coolant to travel through the radiator, where it is cooled before returning to the engine **(see illustrations)**.

The cooling system is sealed by a pressure type radiator cap. This raises the boiling point of the coolant, and the higher boiling point of the coolant increases the cooling efficiency of the radiator. If the system pressure exceeds the cap pressure relief value, the excess pressure in the system forces the spring-loaded valve inside the cap off its seat and allows the coolant to escape through the overflow tube. On models built after 1972, the coolant escapes into a coolant reservoir. When the system cools, the excess coolant is automatically drawn from the reservoir back into the radiator.

The coolant reservoir, on models so equipped, serves as both the point at which fresh coolant is added to the cooling system to maintain the proper fluid level and as a holding tank for overheated coolant.

2 Antifreeze - general information

Warning: *Do not allow antifreeze to come in contact with your skin or painted surfaces of the vehicle. Rinse off spills immediately with plenty of water. Antifreeze is highly toxic if ingested. Never leave antifreeze lying around in an open container or in puddles on the floor; children and pets are attracted by it's sweet smell and may drink it. Check with local authorities about disposing of used antifreeze. Many communities have collection centers which will see that antifreeze is disposed of safely. Never dump used anti-freeze on the ground or into drains.*
Note: *Non-toxic coolant is available at most auto parts stores. Although the coolant is non-toxic, proper disposal is still required.*

The cooling system should be filled with a water/ethylene glycol based antifreeze

solution, which will prevent freezing down to at least -20-degrees F, or lower if local climate requires it. It also provides protection against corrosion and increases the coolant boiling point.

The cooling system should be drained, flushed and refilled at least every other year. The use of antifreeze solutions for periods of longer than two years is likely to cause damage and encourage the formation of rust and scale in the system. If your tap water is "hard", use distilled water with the antifreeze.

Before adding antifreeze to the system, check all hose connections, because antifreeze tends to leak through very minute openings. Engines do not normally consume coolant. Therefore, if the level goes down find the cause and correct it.

The exact mixture of antifreeze-to-water which you should use depends on the relative weather conditions. The mixture should contain at least 50 percent antifreeze, but should never contain more than 70 percent antifreeze. Consult the mixture ratio chart on the antifreeze container before adding coolant. Hydrometers are available at most auto parts stores to test the ratio of antifreeze to water. Use antifreeze which meets the vehicle manufacturer's specifications.

3 Coolant level

Warning: *Do not allow antifreeze to come in contact with your skin or painted surfaces of the vehicle. Flush contaminated areas immediately with plenty of water. Don't store new coolant or leave old coolant lying around where it's accessible to children or pets - they're attracted by its sweet taste. Ingestion of even a small amount of coolant can be fatal! Wipe up garage floor and drip pan coolant spills immediately. Keep antifreeze containers covered, and repair leaks in your cooling system immediately. Refer to illustration 3.1*

1   Most vehicles covered by this manual are equipped with a pressurized coolant recovery system. A white plastic coolant reservoir located in the engine compartment is connected by a hose to the radiator filler neck **(see illustration)**. If the engine overheats, coolant escapes through a valve in the radiator cap and travels through the hose into the reservoir. As the engine cools, the coolant is automatically drawn back into the cooling system to maintain the correct level.

2   If your particular vehicle is not equipped with a coolant recovery system, the level should be checked by removing the radiator cap. However, the cap should not under any circumstances be removed while the system is hot, as escaping stream could cause serious injury. Wait until the engine has completely cooled, then wrap a thick cloth around the cap and turn it to its first stop. If any steam escapes from the cap, allow the engine to cool further. Then remove the cap and check the level in the radiator. It should be about two to three inches below the bottom of the filler neck.

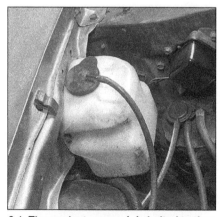

3.1 The coolant reservoir is bolted to thee side of the engine compartment

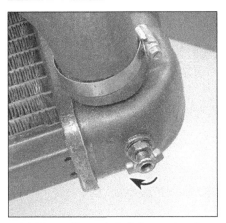

4.6 The radiator drain is located at the lower side of the radiator

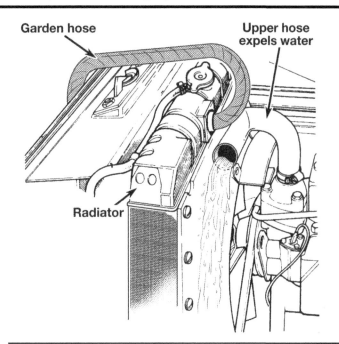

4.13 With the thermostat removed, disconnect the upper radiator hose and flush the radiator and engine block with a garden hose

3   The coolant level in the reservoir should be checked regularly. **Warning:** *Do not remove the radiator cap to check the coolant level when the engine is warm. The level in the reservoir varies with the temperature of the engine. When the engine is cold, the coolant level should be at or slightly above the FULL COLD mark on the reservoir. Once the engine has warmed up, the level should be at or near the FULL HOT mark. If it isn't, allow the engine to cool, then remove the cap from the reservoir and add a 50/50 mixture of ethylene glycol based antifreeze and water.*

4   Drive the vehicle and recheck the coolant level. If only a small amount of coolant is required to bring the system up to the proper level, water can be used. However, repeated additions of water will dilute the antifreeze and water solution. In order to maintain the proper ratio of antifreeze and water, always top up the coolant level with the correct mixture. An empty plastic milk jug or bleach bottle makes an excellent container for mixing coolant. Do not use rust inhibitors or additives.

5   If the coolant level drops consistently, there may be a leak in the system. Inspect the radiator, hoses, filler cap, drain plugs and water pump. If no leaks are noted, have the radiator cap pressure tested by a service station.

6   If you have to remove the radiator cap, wait until the engine has cooled, then wrap a thick cloth around the cap and turn it to the first stop. If coolant or steam escapes, let the engine cool down longer, then remove the cap.

7   Check the condition of the coolant as well. It should be relatively clear. If it's brown or rust colored, the system should be drained, flushed and refilled. Even if the coolant appears to be normal, the corrosion inhibitors wear out, so it must be replaced at the specified intervals.

4   Cooling system - draining, flushing and refilling

Warning 1: *Wait until the engine is completely cool before beginning this procedure.*

Warning 2: *Do not allow antifreeze to come in contact with your skin or painted surfaces of the vehicle. Rinse off spills immediately with plenty of water. Antifreeze is highly toxic if ingested. Never leave antifreeze lying around in an open container or in puddles on the floor; children and pets are attracted by its sweet smell and may drink it. Check with local authorities about disposing of used antifreeze. Many communities have collection centers that will see that antifreeze is disposed of safely.*
Note: *Non-toxic antifreeze solutions are now widely available, but even these should be disposed of properly.*

1   Every two years or 24,000 miles, whichever occurs first, the cooling system should be drained, flushed and refilled to replenish the antifreeze mixture and prevent formation of rust and corrosion, which can impair the performance of the cooling system and cause engine damage.

2   At the same time the cooling system is serviced, all hoses and the radiator cap should be inspected and replaced if defective.

3   Since antifreeze is a corrosive and poisonous solution, be careful not to spill any of the coolant mixture on the vehicle's paint or your skin. If this happens, rinse it off immediately with plenty of clean water. Consult local authorities about where to recycle or dispose of antifreeze before draining the cooling system. In many areas, reclamation centers have been set up to collect automobile oil and drained antifreeze/water mixtures, rather than allowing them to be added to the sewage system.

Draining

Refer to illustration 4.6

4   Apply the parking brake and block the wheels. If the vehicle has just been driven, wait several hours to allow the engine to cool down before beginning this procedure.

5   Once the engine is completely cool, remove the radiator cap and the coolant reservoir cap.

6   Place a drain pan under the radiator. Drain the radiator by opening the drain plug at the bottom of the radiator **(see illustration)**. If the drain plug is corroded and can't be turned easily, or if the radiator isn't equipped with a plug, disconnect the lower radiator hose to allow the coolant to drain. Be careful not to get antifreeze on your skin or in your eyes.

7   After the coolant stops flowing out of the radiator, remove the lower radiator hose and allow the remaining fluid in the upper half of the engine block to drain.

8   While the coolant is draining from the engine block, disconnect the hose from the coolant reservoir (if equipped) and remove the reservoir. Flush the reservoir out with water until it's clean, and if necessary, wash the inside with soapy water and a brush to make reading the fluid level easier.

9   While the coolant is draining, check the condition of the radiator hoses, heater hoses and clamps.

10   Replace any damaged clamps or hoses.

Flushing

Refer to illustration 4.13

11   Once the system is been completely drained, remove the thermostat from the engine (see Section 6). Then reinstall the thermostat housing without the thermostat. This will allow the system to be flushed.

12   Reinstall the lower radiator hose and tighten the radiator drain plug. Turn your heating system controls to Hot, so that the heater core will be flushed at the same time as the rest of the cooling system.

13   Disconnect the upper radiator hose, then place a garden hose in the upper radiator inlet and flush the system until the water runs clear at the upper radiator hose **(see illustration)**.

5.4 Removing the radiator upper panel

5.6 Removing the radiator. The shroud is pushed back over the fan

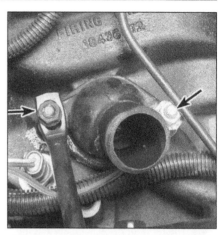

6.9 Remove the fasteners and lift off the thermostat housing

14   In severe cases of contamination or clogging of the radiator, remove the radiator and have a radiator repair facility clean and repair it if necessary.

15   Many deposits can be removed by the chemical action of a cleaner available at auto parts stores. Follow the procedure outlined in the manufacturer's instructions. **Note:** *When the coolant is regularly drained and the system refilled with the correct antifreeze/water mixture, there should be no need to use chemical cleaners or descalers.*

Refilling

16   To refill the system, install the thermostat, reconnect any radiator hoses and install the reservoir and the overflow hose.

17   Place the heater temperature control in the maximum heat position.

18   Be sure to use the proper coolant mixture listed in this Chapter's Specifications. Slowly fill the radiator with the recommended mixture of antifreeze and water to the base of the filler neck. Add coolant to the reservoir (if equipped) until it reaches the FULL COLD mark. Wait five minutes and recheck the coolant level in the radiator, adding if necessary.

19   Leave the radiator cap off and run the engine in a well-ventilated area until the thermostat opens (coolant will begin flowing through the radiator and the upper radiator hose will become hot).

20   Turn the engine off and let it cool. Add more coolant mixture to bring the level back up to the base of the filler neck.

21   Squeeze the upper radiator hose to expel air, then add more coolant mixture if necessary. Reinstall the radiator cap.

22   Start the engine, allow it to reach normal operating temperature and check for leaks.

5   Radiator - removal, inspection and installation

Refer to illustrations 5.4 and 5.6
Warning: *Wait until the engine is completely cool before beginning this procedure.*

1   The method of fitment of the radiator (and shroud, where applicable) has changed slightly throughout the production years of the Nova, but the following is a typical procedure.

2   Disconnect the negative battery cable. Drain the cooling system (see Section 3).

3   Detach the radiator hoses from the radiator. If equipped with an automatic transmission, unscrew the cooler line fittings at the side tank of the radiator.

4   Remove the radiator upper panel **(see illustration)**.

5   Remove the shroud from the radiator (where applicable) and hang it over the fan.

6   Carefully lift out the radiator **(see illustration)**.

7   The radiator should be cleared of dirt and bugs by blowing from the rear with compressed air and the use of a soft brush. Take care that the fins are not bent.

8   Installation of the radiator is the reverse of the removal procedure. Make sure the rubber mounting insulators at the bottom of the radiator are in place, and are in good shape.

9   Refer to Section 4 for filling instructions.

6   Thermostat - check and replacement

Refer to illustrations 6.9 and 6.10
Warning: *The engine must be completely cool when this procedure is performed.*

Check

1   Before assuming the thermostat is to blame for a cooling system problem, check coolant level, drivebelt tension and temperature gauge (or light) operation.

2   If the engine takes a long time to warm up, the thermostat is probably stuck open. Replace the thermostat.

3   If the engine runs hot, use your hand to check the temperature of the upper radiator hose. If the hose is not hot, but the engine is,

the thermostat is probably stuck in the closed position, preventing the coolant inside the engine from escaping to the radiator. Replace the thermostat.

4   If the upper radiator hose is hot, it means that the coolant is flowing and the thermostat is open.

Replacement

5   Disconnect the negative battery cable.

6   Drain the cooling system.

7   Remove the upper radiator hose from the thermostat housing.

9   Remove the bolts from the thermostat housing and detach the housing **(see illustration)**. Be prepared for some coolant to spill as the gasket seal is broken.

10   Remove the thermostat, noting the way it was installed **(see illustration)**.

11   Remove all traces of gasket material from the sealing surfaces.

12   Apply gasket sealer to both sides of a new gasket and position it on the engine.

13   Install the thermostat (spring end first), housing and bolts. Tighten the bolts to the specified torque.

14   Refill the cooling system.

6.10 Lift out the thermostat, noting how it is installed

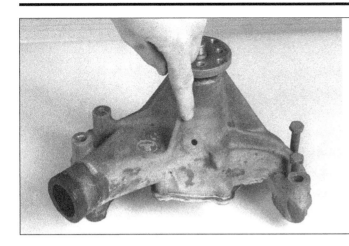

7.2 The water pump weep hole is located on the pump snout; some pumps have two holes

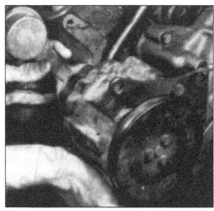

7.10 Remove the AIR pump

7.12a After the accessories and brackets have been removed, remove the remaining water pump bolts (arrows)

7.12b Grasp the water pump and rock it back-and-forth to break the gasket seal

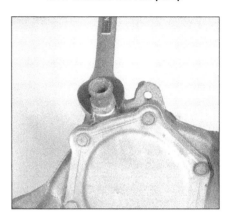

7.13 Unscrew the heater hose pipe so it can be transferred to the new pump

7 Water pump - check, removal and installation

Check

Refer to illustration 7.2

1   Water pump failure can cause overheating of and serious damage to the engine. There are two ways to check the operation of the water pump while it's installed on the engine. If either of the following quick checks indicates water pump failure, it should be replaced immediately.

2   A seal protects the water pump impeller shaft bearing from contamination by engine coolant. If the seal fails, a "weep hole" in the water pump snout **(see illustration)** will leak coolant under the vehicle. If the weep hole is leaking, shaft bearing failure will follow. Replace the water pump immediately.

3   Besides contamination by coolant after a seal failure, the water pump impeller shaft bearing can also be prematurely worn out by an improperly tensioned drivebelt. When the bearing wears out, it emits a high pitched squealing sound. If noise is coming from the water pump during engine operation, the shaft bearing has failed. Replace the water pump immediately.

4   To identify excessive bearing wear before the bearing actually fails, grasp the water pump pulley and try to force it up and

down or from side to side. If the pulley can be moved either horizontally or vertically, the bearing is nearing the end of its service life. Replace the water pump.

Removal and installation

Removal

Refer to illustrations 7.10, 7.12a and 7.12b

5   Disconnect the negative battery cable.

6   Drain the radiator, referring to Chapter 1 if necessary.

7   Remove the engine cooling fan.

8   Remove the radiator shroud for better access to the water pump.

9   Remove the drivebelts from the water pump pulley. The number of belts will depend upon model, year, and equipment. Loosen the adjusting and pivot bolts of each component (air pump, alternator, air conditioning compressor, power steering pump) and push the component inward to loosen the belt enough to be removed from the water pump pulley.

10   Disconnect and remove all mounting brackets which are attached to the water pump. These may include the air pump, the alternator, air conditioning compressor, and power steering brackets **(see illustrations)**.

11   Disconnect the lower radiator hose, heater hose, and bypass hose at the water pump.

12   Remove the remaining bolts which secure the water pump to the engine or front cover **(see illustration)**. Rock the pump back-and-

forth or lightly tap the water pump housing with a hammer to break the gasket seal and remove it from the engine **(see illustration)**.

Installation

Refer to illustrations 7.13, 7.14 and 7.18

13   If installing a new or rebuilt water pump, transfer all fittings and studs to the new water pump **(see illustration)**.

14   Clean the gasket surfaces of the front cover completely using a gasket scraper or putty knife **(see illustration)**.

7.14 Remove all traces of old gasket material from the water pump mating surfaces on the engine block

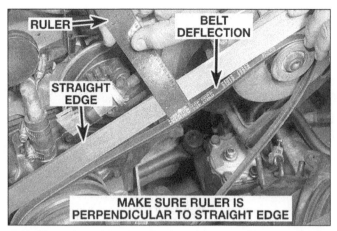

7.18 Measuring drivebelt deflection with a straightedge and ruler

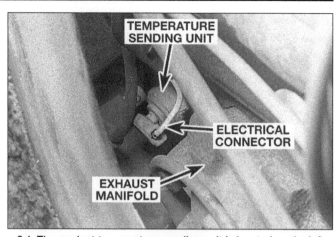

8.1 The coolant temperature sending unit is located on the left side between the number one and number three spark plugs

15   Use a thin coat of gasket sealant on the new gaskets and install the new water pump. Place the pump into position on the front cover and secure loosely with the bolts. Do not tighten the bolts until all brackets have been installed to their original position on the water pump.

16   Tighten the water pump bolts to the torque listed in this Chapter's Specifications.

17   Install the engine components in the reverse order of removal, tightening the appropriate fasteners securely.

18   Adjust all drivebelts to the proper tension. The tension of each belt is checked by pushing the belt at a distance halfway between the pulleys. Push firmly with your thumb and see how much the belt moves (deflects) **(see illustration)**. As a rule of thumb, if the distance from pulley center-to-pulley center is between 7 and 11 inches, the belt should deflect 1/4-inch. If the belt travels between pulleys spaced 12 to 16-inches apart, the belt should deflect 1/2-inch.

19   Connect the negative battery cable and fill the radiator with the proper mixture of antifreeze and water. Start the engine and allow it to idle until the upper radiator hose gets hot. Check for leaks. With the engine hot, fill the coolant reservoir with coolant mixture to the correct level. Check coolant level periodically over the next few miles of driving.

---

8   Coolant temperature sending unit - check and replacement

Temperature warning light system check

Refer to illustration 8.1

1   If the light doesn't come on when the ignition switch is turned on, check the bulb. If the light stays on even with the engine cold, unplug the wire at the sending unit **(see illustration)**. If the light goes off, replace the sending unit. If the light stays on, the wire is grounded somewhere in the harness.

Temperature gauge system check

2   If the gauge is inoperative, check the fuse (Chapter 11).

3   If the fuse is OK, unplug the wire connected to the sending unit and ground it with a jumper wire. Turn on the ignition switch momentarily. The gauge should register at maximum. If it does, replace the sending unit. If it's still inoperative, the gauge or wiring may be faulty.

Sending unit replacement

Warning: *Wait until the engine is completely cool before beginning this procedure.*

4   Unplug the wire connected to the sending unit.

5   Unscrew the sending unit and quickly install the new unit to prevent loss of coolant. Tighten the sending unit securely.

6   Connect the wire and check indicator operation.

---

9   Troubleshooting - cooling system

Symptom	Reason
Overheating	Low coolant level
	Faulty radiator pressure cap
	Thermostat stuck shut
	Drive belt slipping or incorrectly tensioned
	Clogged radiator core
	Incorrect ignition timing
	Corroded system
	Incorrect proportion of antifreeze to water in cooling system
	Incorrect operation of viscous fan clutch
Cool running and/or slow warm up	Incorrect type thermostat
	Thermostat stuck open
	Incorrect operation of viscous fan clutch
Coolant loss	Faulty radiator pressure cap
	Split hose
	Leaking water pump to block joint
	Leaking freeze plug
	Blown cylinder head gasket
	Cracked cylinder head or block
	Leaking water pump
Excessive noise from thermostatic fan clutch	Clutch bearing fault
Fluid leakage from bearing assembly	Clutch bearing/seal fault

Chapter 3
Carburetion, fuel, exhaust and emission control systems

Contents

Specifications

Fuel pump .... Diaphragm type, pushrod operated from camshaft

Fuel tank capacity (approx.)
Pre 1971 .... 18 US gallons
1971-1972 .... 16 US gallons
1973 onwards .... 21 US gallons

Carburetor

1969 & 1970; 307 cu in. 200 HP models

Type	Rochester 2GV (1 1/4)
Part No. and transmission:	
7029101	Manual
7029103	Manual with air conditioning
7029110	Automatic
7029112	Automatic with air conditioning
Float level	27/32 in
Float drop	1%
Accelerator pump	1-1/8 in
Idle vent	0.020 in
Fast idle:	
Mechanical	One turn
Running	2200/2400 rpm
Choke rod	0.060 in
Choke vacuum break	0.1 in
Choke unloader	0.215 in
Main metering jet	0.052 in Manual (1969)
	0.051 in All other models
Throttle bore (primary)	1-7/16 in

Bottom of thermostat choke rod should be even with top of hole at closed choke.

1969; 350 & 396 cu in. 255, 300 & 350 HP models

Type	Rochester 4MV
Part No. and transmission:	
7029203	Manual (255 and 300 HP)
7029202	Automatic (255 and 300 HP)
7029215	Manual (350 HP)
7029204	Automatic (350 HP)
Float level	7/32 in (255 and 300 HP) 1/4 in (350 HP)
Accelerator pump	5/16 in
Idle vent	3/8 in
Fast idle:	
Mechanical	Two turns
Running	2400 rpm
Choke rod	0.1 in
Choke vacuum break	0.245 in Manual, 0.180 in Automatic
Choke unloader	0.45 in
Air valve spring	7/16 in (255 and 300 HP) 13/16 in (350 HP)
Secondary closing	0.02 in
Secondary opening	0.07 in
Secondary lockout	0.015 in
Main metering jet	0.067 in (255 and 300 HP), 0.071 in (350 HP)
Throttle bore:	
Primary	1-3/8 in
Secondary	2-1/4 in
Air valve dashpot	0.015 in

Top of thermostat choke rod should be even with bottom of hole.

1969; 396 cu in. 375 HP models

Type	Holley 4150
Part No. and transmission	
3959164	Manual and Automatic
Float level:	
Mechanical:	
Primary	0.350 in
Secondary	0.50 in
Final	Use sight plug

Accelerator pump	0.015 in
Fast idle:	
Mechanical	0.025 in
Final	2200 rpm
Choke vacuum break	0.3 in
Choke unloader	0.35 in
Secondary stop	1/2 turn open
Main metering jet:	
Primary	No. 68
Secondary	No. 76
Throttle bore (primary and secondary)	1-9/16 in

Top of thermostat choke rod should be even with bottom of hole.

1970; 350 cu in. 250 HP models

Type	Rochester 2GV (1-1/2 in)
Part No. and transmission:	
7040114	Automatic
7040116	Automatic with air conditioning
7040113	Manual
7040115	Manual with air conditioning
Float level	23/32 in
Float drop	1-3/8 in
Accelerator pump	1-17/32 in
Idle vent	0.020 in
Fast idle:	
Mechanical	One turn
Running	2200/2400 rpm
Choke rod	0.085 in
Choke vacuum break	0.2 in Automatic, 0.215 in Manual
Choke unloader	0.325 Automatic, 0.275 in Manual
Main metering jet (primary)	0.060 in Automatic, 0.059 in Manual
Throttle bore (primary)	1-11/16 in

Bottom of thermostat choke rod should be even with top of hole at closed choke.

1970; 350 & 396 cu in. 300 and 350 HP models

Type	Rochester 4MV
Part No. and transmission:	
7040202	Automatic (300 HP)
7040203	Manual (300 HP)
7040204	Automatic (350 HP)
7040205	Manual (350 HP)
Float level	1/4 in
Accelerator pump	5/16 in
Fast idle:	
Mechanical	Two turns
Running	2400 rpm
Choke rod	0.1 in
Choke vacuum break	6.245 in Automatic, 0.275 in Manual
Choke unloader	0.45 in
Air valve spring	7/16 in (300 HP) 13/16 in (350 HP)
Main metering jet (primary)	0.076 in (300 HP), 0.078 in (350 HP)
Throttle bore:	
Primary	1-3/8 in
Secondary	2-1/4 in
Air valve dashpot	0.02 in

Top of thermostat choke rod should be even with bottom of hole.

1970; 396 cu in. 375 HP models

This carburetor is similar to that fitted for 1969 396 cu in., 375 HP models, except as follows:

Part No. and transmission:	
3969898	Automatic
3967477	Manual
Choke vacuum break	0.035 in
Main metering jet:	
Primary	No. 70

Bottom of thermostat choke rod should be even with top of hole at closed choke.

Carburetor (continued)

1971; 307 cu in. 200 HP models

Type	Rochester 2GV(1-1/4 in)
Part No. and transmission:	
7041101	Manual
7041110	Automatic
Float level	13/16 in
Float drop	1-3/4 in
Pump rod	1-3/64 in
Choke rod...	0.075 in Manual, 0.040 in Automatic
Vacuum break	0.11 in Manual, 0.08 in Automatic
Unloader...	0.215 in
Idle speed:	
Initial...	700 rpm, Manual (in neutral), 580 rpm, Automatic (in Drive)
Final	600 rpm, Manual (in neutral), 550 rpm, Automatic (in Drive)
CO % at idle...	0.5 in
CEC valve engine speed:	
Manual...	900 rpm
Automatic	650 rpm

1971; 350 cu in. 245 HP models

Type	Rochester 2GV (1-1/2 in)
Part No. and transmission:	
7041113	Manual
7041114	Automatic
Float level	
Manual	23/32 in
Automatic...	25/32 in
Float drop...	1-3/8 in
Pump rod	1-5/32 in
Choke rod...	0.1 in
Vacuum break	0.18 in Manual, 0.17 in Automatic
Unloader..	0.325 in

All idle settings are as for 1971 307 cu in. 200 HP.

1971; 350 cu in. 270 HP models

Type	Rochester 4MV
Part No. and transmission:	
7041203	Manual
7041202	Automatic
Float level	1/4 in
Choke rod..	0.1 in
Air valve dashpot..	0.02 in
Vacuum break	0.275 in Manual, 0.26 in Automatic
Idle speed (initial)..	675 in Manual (in neutral)
CO % at idle...	1.0 in (Manual only)

All other idle settings are as for 1971 307 cu in., 200 HP.

1972; 307 cu in. models

Type	Rochester 2GV (1-1/4 in)
Part No. and transmission:	
7042101	Manual
7042100	Automatic
7042821	Manual (California)
7042820	Automatic (California)
Float level	25/32 in
Float drop...	1-31/32 in
Pump rod	1-5/16 in
Choke rod (fast idle cam)	0.075 in Manual, 0.04 in Automatic
Vacuum break	0.11 in Manual, 0.08 in Automatic
Unloader..	0.215 in
Fast idle (without vacuum advance)...	1850 rpm
Curb idle speed:	
Initial...	1000 rpm Manual (in neutral)
	650 rpm Automatic (in Drive)
Final	900 rpm Manual (in neutral)
	600 rpm Automatic (in Drive)

* *For vehicles with A.I.R., idle adjustment is 1/4 turn rich from lean roll*

** *Set low idle at 450 rpm.*

1972; 350 cu in. RPO L-65 models

Type	Rochester 2GV (1-1/2 in)
Part No. and transmission:	
7042113	Manual (Federal)
7042114	Automatic (Federal)
7042833	Manual (California)
7042834	Automatic (California)
Float level	23/32 in
Float drop	1-9/32 in
Pump rod	1-1/2 in
Choke rod (fast idle cam)	0.1 in
Vacuum break	0.18 in Manual, 0.17 in Automatic
Unloader...	0.325 in
Fast idle (without vacuum advance)......................	2200 rpm on high step
Curb idle speed (initial).......................................	1050 rpm Manual (in neutral)

All other idle settings are as for 1972 307 cu in.

1972; 350 cu in. RPO L-48 models

Type	Rochester 4MV
Part No. and transmission:	
7042203	Manual
7042202	Automatic
7042903	Manual (California)
7042902	Automatic (California)
Float level	1/4 in
Pump rod	3/8 in
Choke rod (fast idle cam)	0.02 in
Vacuum break	0.215 in
Unloader...	0.45 in
Fast idle (without vacuum advance)......................	1350 rpm Manual (on second step)
Curb idle speed (final)	1500 rpm Automatic (on second step)
	630 rpm Automatic (in Drive)

All other idle settings are as for 1972, 307 cu in.

1973; 307 cu in. models

Type	Rochester 2GV (1-1/4 in)
Part No. and transmission:	
7043101	Manual
7043100	Automatic
Float level	21/32 in
Float drop	19/32 in
Pump rod	15/16 in
Choke rod (fast idle cam)	0.15 in
Vacuum break	0.08 in
Unloader...	0.215 in
Fast idle (with vacuum advance).........................	1600 rpm
Curb idle speed:	
Initial...	950 rpm Manual (in neutral)
	630 rpm Automatic (in Drive)
Final *	900 rpm Manual (in neutral)
	600 rpm Automatic (in Drive)

Set low idle at 500 rpm.

1973; 350 cu in. RPO L-65 models

Type	Rochester 2GV (1-1/2 in)
Part No. and transmission:	
7043113	Manual
7043114	Automatic
Float level	19/32 in
Float drop	19/32 in
Pump rod	17/16 in
Choke rod (fast idle cam)	0.2 in Manual, 0.245 in Automatic
Vacuum break	0.14 in Manual, 0.13 in Automatic
Unloader...	0.25 in Manual, 0.325 in Automatic
Fast idle...	1600 rpm on high step
Curb idle speed (initial).......................................	1000 rpm

All other idle settings are as for 1973 307 cu in. except low idle set at 400 rpm.

Carburetor (continued)

1973; 350 cu in. RPO L-48

Type	Rochester 4MV
Part No. and transmission:	
7043203	Manual
7043202	Automatic
Float level	7/32 in
Pump rod (inner location)	13/32 in
Choke rod (fast idle cam)	0.43 in
Air valve wind-up	1/2 turn
Vacuum break	0.25 in
Unloader	0.45 in
Fast idle:	
With vacuum advance	1600 rpm Automatic (on high step)
Without vacuum advance	1300 rpm Manual (on high step)
Curb idle speed (initial)	920 rpm Manual (in neutral)
	620 rpm Automatic (in Drive)

All other idle settings are as for 1973 307 cu in.

1974 350 cu in. RPO L-65 models

Type	Rochester 2GV (1-1/2 in)
Part No. and transmission:	
7044115	Manual
7044116	Automatic
Float level	19/32 in
Float drop	19/32 in
Pump rod	19/32 in Manual, 13/16 in Automatic
Choke rod (fast idle cam)	0.2 in Manual, 0.245 in Automatic
Vacuum break	0.14 in Manual, 0.13 in Automatic
Unloader	0.25 in Manual, 0.325 in Automatic
Fast idle (with vacuum advance and without EGR signal)	1600 rpm, on high step
Curb idle, solenoid screw energized	900 rpm (N) Manual
	600 rpm (DR) Automatic
Low idle, de-energized	500 rpm (N) Manual
	500 rpm (DR) Automatic
Lean drop idle mixture	1000/900 (H) Manual
	650/600 (DR) Automatic
Max. acceptable CO level	0.5%

All settings are with the throttle lever ball stud in the lower hole.

1974; 350 cu in. RPO L-48 & RPO LM-1 models

Type	Rochester 4MV
Part No. and transmission:	
7044207	Manual (Federal)
7044206	Automatic (Federal)
7044507	Manual (California)
7044506	Automatic (California)
Float level	1/4 in
Pump rod (inner location)	13/32 in
Choke rod (fast idle cam)	0.43 in
Air valve wind-up	7/8 turn
Vacuum break	0.23 in
Unloader	0.45 in
Fast idle:	
Without vacuum advance	1300 rpm Manual (on high step)
With vacuum advance and without EGR signal	1600 rpm Automatic (on high step)
Curb idle, solenoid screw energized	900 rpm (N) Manual
	600 rpm (DR) Automatic
Low idle, de-energized	500 rpm (N) Manual
	500 rpm (DR) Automatic
Lean drop idle mixture	950/900 rpm (N) Manual
	650/600 (Federal), 630/600 (California) rpm (DR) Automatic
Max. acceptable CO level	0.5%

1975; 262 cu in./4-3 liter, RPO LV-1 models

Type	Rochester 2GC
Part No. and transmission:	
7045105	Manual (Federal)
7045106	Automatic (Federal)

7045405	Manual (California)
7045406	Automatic (California)
Float level	19/32 in Federal, 21/32 in California
Float drop	17/32 in
Pump rod	1-19/32 in
Choke rod (fast idle cam)	0.375 in Manual (Federal), 0.38 in Other types
Vacuum break	0.13 in
Unloader	0.35 in
Idle speed	800 rpm (N) Manual
	600 rpm (DR) Automatic

1975; 350 cu in. RPO L-65 models

Type	Rochester 2GC
Part No. and transmission:	
7045123	Manual
7045124	Automatic
Float level	21/32 in
Float drop	31/32 in
Pump rod	1 5/8 in
Choke rod (fast idle cam)	0.4 in
Vacuum break	0.13 in
Unloader	0.35 in
Idle speed	800 rpm (N) Manual
	600 rpm (DR) Automatic

1975; 350 cu in. RPO LM-1 models

Type	Rochester M4MC (Federal)
	M4MCA (California)
Part No. and transmission:	
7045207	Manual (Federal)
7045206	Automatic (Federal)
7045507	Manual (California)
7045506	Automatic (California)
Float level	15/32 in
Pump rod (inner location)	0.275 in
Choke coil lever	0.12 in
Choke rod (fast idle cam)	0.3 in
Air valve dashpot	0.015 in
Vacuum break:	
Front	0.18 in
Rear	0.17 in
Spring wind-up	7/8 turn
Unloader	0.325 in
Low idle speed	800 rpm (N) Manual (Federal)
	600 rpm (N or DR) (other types)
Curb idle speed	1600 rpm (N) (all types)

Note: *The choke valve setting is at the top of the valve.*

1976 models w/ 305 engine

Type	2GC series
Part number	
Federal models w/ manual trans.	17056113
Federal models w/ automatic trans.	17056110
California models w/ automatic trans.	17056410

1976 models w/ 350 engine (2 bbl carb)

Type	2GC
Part number	
California models w/ automatic trans.	17056420

1976 models w/ 350 engine (4 bbl carb)

Type	M4MC
Part number	
Federal models w/ manual trans.	17056207
Federal models w/ automatic trans.	17056206
California models w/ manual trans.	17056507
California models w/ automatic trans.	17056506

Carburetor (continued)

1976 models w/ optional high performance 350 engine (4 bbl carb)
Type M4MC
Part number
   Federal models w/ manual trans. 17056203
   Federal models w/ automatic trans. 17056202
   California models w/ manual trans. 17056503
   California models w/ automatic trans. 17056502

1977 models w/ 305 engine
Type 2GC
Part number
   Federal models (all)
      With air conditioning .... 17057123 or 17057110
      Without air conditioning .... 17057113 or 17057108
   California models w/ automatic trans.
      With air conditioning 17057410
      Without air conditioning 17057408

1977 models w/ 350 engine
Type M4MC
Part number
   Federal models w/ manual trans. 17057203
   Federal models w/ automatic trans.
      With air conditioning 17057204
      Without air conditioning 17057202
   California models w/ automatic trans.
      With air conditioning 17057504
      Without air conditioning 17057502
   High altitude models w/ automatic trans.
      With air conditioning 17057584
      Without air conditioning 17057582

1978 models w/ 305 engine
Type 2GC
Part number
   Federal models w/ manual trans.
      With air conditioning 17058123
      Without air conditioning 17058113
   Federal models w/ automatic trans.
      With air conditioning 17058110
      Without air conditioning 17058108
   California models w/ automatic trans.
      With air conditioning 17058410
      Without air conditioning 17058408

1978 models w/ 350 engine
Type M4MC
Part number
   California models w/ automatic trans.
      With air conditioning 17058504
      Without air conditioning 17058502
   High altitude models w/ automatic trans.
      With air conditioning 17058584
      Without air conditioning 17058582

1979 models w/ 305 engine
Type M2MC
Part number
   Federal models w/ manual trans.
      With air conditioning 17059137
      Without air conditioning 17059135
   Federal models w/ automatic trans.
      With air-conditioning 17059136
      Without air conditioning 17059134
   California models w/ automatic trans.
      With air conditioning 17059436
      Without air conditioning 17059434

1979 models w/ 350 engine

Type M4MC
Part number
    California models w/ automatic trans.
        With air conditioning 17059504
        Without air conditioning 17059502
    High altitude models w/ automatic trans.
        With air conditioning 17059584
        Without air conditioning 17059582

Idle speed and mixture adjustments (rpm) - 1976 and later models

	Curb idle speed	Speed with A/C solenoid on	Fast idle speed	Idle speed before lean drop	Idle speed after lean drop
1976 models					
305 engine					
Automatic trans.	600	—	—	630	600
Manual trans.	800	—	—	900	800
350 engine (2 bbl carb)					
Automatic trans. (Fed)	600	—	—	650	600
350 engine (4 bbl carb)					
Automatic trans. (Fed)	600	—	1600	650	600
Manual trans. (Fed)	800	—	1600	900	800
Automatic trans. (Ca)	600	—	1600	650	600
1977 models					
305 engine					
Manual trans.	600	700	—	650	600
Automatic trans.	500	650	—	550	500
350 engine					
Manual trans.	700	—	1300	800	700
Automatic trans.	500	650	1600	550	500
Automatic trans. (high aft)	600	650	1600	650	600
1978 models					
305 engine					
Manual trans. (Fed)	600	—	—	—	—
Automatic trans. (Ca)	500	600	—	—	—
Automatic trans. (Fed)	500				
350 engine					
Automatic trans. (Ca)	500	600	1600	—	—
Automatic trans. (high aft)	500	650	1600	—	—
1979 models					
305 engine					
Manual trans. (Fed)	600	700	1300	—	—
Automatic trans. (Fed)	500	600	1600	—	—
Automatic trans. (Ca)	600	650	1950	—	—
350 engine					
Automatic trans. (Ca)	500	600	1600	—	—
Automatic trans. (high aft)	600	650	1750	—	—

Carburetor adjustments (2GC)

Float level
    1976 models (all) .... 9/16 in
    1977 Federal models .... 19/32 in
    1977 California models .... 21/32 in
    1978 Federal models .... 15/32 in
    1978 California models .... 1/2 in
Float drop (all models) .... 1-9/32 in
Accelerator pump rod
    All except 17056113 .... 1-21/32 in
    17056113 .... 1-11/16 in
Choke rod (fast idle cam) (all models) .... 0.260 in
Vacuum break
    All 1976 models except 17056113 .... 0.140 in
    17056113 .... 0.130 in

Carburetor adjustments (2GC) (continued)

Vacuum break (continued)

1977 Federal models..	0.130 in*
1977 California models	0.140 in*
1978 Federal models..	0.130 in**
1978 California models	0.140 in*

* 0.160 in if engine has over 22 500 miles on it
** reset to 0.160 in after first tune-up

Unloader (all models)...	0.325 in
Automatic choke coil	
All except below	Index mark
1977 and 1978 California models	1/2 notch lean
Choke coil lever (all models)..	0.120 in

Carburetor adjustments (M4MC and M2MC)

Float level adjustment

1976 models (all)	13/32 in
1977 thru 1979 models w/ 350 engines...	15/32 in
1979 models w/ 305 engine	13/32 in
Accelerator pump rod	
1976 and 1978 models (ail)	9/32 in
1977 models (all except high altitude)...	9/32 in (inner hole)
1977 high altitude models	9/32 in (outer hole)
1979 California models w/ 350 engine...	1/4 in
1979 high altitude models w/ 350 engine	11/32 in (outer hole)
1979 models w/ 305 engine (all)..	1/4 in (inner hole)
Choke coil lever (all models)...	0.120 in
Choke rod (fast idle cam)	
1976 and 1977 models (all)	0.325 in
1978 models (all)	46°
1979 models (all)	38°
Air valve rod	
1976 thru 1979 models w/ 350 engine...	0.015 in
Front vacuum break	
1976 models (all)	0.185
1977 Federal models...	27°
1977 California models	28.5°
1977 high altitude models	30°
1978 California models	28°*
1978 high altitude models	30°**
1979 California models w/ 350 engine...	28°
1979 high altitude models w/ 350 engine	33°
1979 Federal models w/ 305 engine..	27°
1979 California models w/ 305 engine..	29°

*31° after 30000 miles
** 33° after 22 500 miles

Unloader	
1976 models (all)	0.325 in
1977 and 1978 models (all)	42°
1979 high altitude models w/350 engine	46°
1979 California models w/ 350 engine...	38°
1979 models w/ 305 engine	38°
Automatic choke coil	
All 1976 and 1977 models(except Fed models w/ manual trans.)	2 notches lean
1976 and 1977 Federal models w/ manual trans.............................	3 notches lean
1978 models (all)	2 notches lean
1979 California models w/ 350 engine...	2 notches lean
1979 high altitude models w/ 350 engine	1 notch lean
1979 models w/ 305 engine	1 notch lean
Secondary lock out	
1976 thru 1979 models w/ 350 engine...	0.015 in
Secondary closing	
1976 thru 1979 models w/ 350 engine...	0.020 in
Air valve spring	
1976 thru 1979 models w/ 350 engine...	7/8 turn

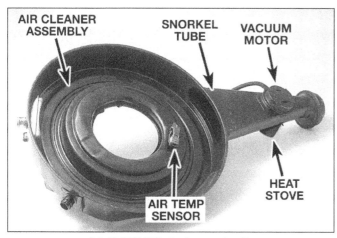

2.2 THERMAC air cleaner details

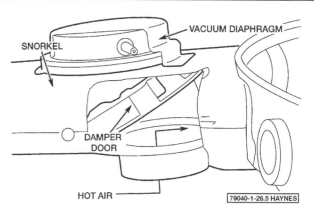

3.5 When the engine is cold, the damper door closes off the snorkel passage, allowing air warmed by the exhaust to enter the carburetor

1   General description

Warning: *Gasoline is extremely flammable, so take extra precautions when you work on any part of the fuel system. Don't smoke or allow open flames or bare light bulbs near the work area, and don't work in a garage where a natural gas-type appliance (such as a water heater or a clothes dryer) with a pilot light is present. Since gasoline is carcinogenic, wear latex gloves when there's a possibility of being exposed to fuel, and, if you spill any fuel on your skin, rinse it off immediately with soap and water. Mop up any spills immediately and do not store fuel-soaked rags where they could ignite. When you perform any kind of work on the fuel system, wear safety glasses and have a Class B type fire extinguisher on hand.*

1   The fuel system comprises a rear-mounted fuel tank, an engine-mounted fuel pump driven from the camshaft by a pushrod, a carburetor and the necessary interconnecting pipes.

2   Emission control systems are fitted to the Nova models but these vary according to the year of manufacture and the region in which the vehicle is sold. These systems are dealt with separately in this Chapter, with the exception of Positive Crankcase Ventilation (PCV) which is dealt with in Chapter 1.

3   Vehicles manufactured from 1975 require the use of unleaded fuels and incorporate a smaller diameter inlet pipe to the fuel tank designed to accept the special nozzles or pumps dispensing unleaded fuel. A catalytic converter is incorporated in the exhaust system of these vehicles to reduce hydrocarbon and carbon monoxide pollutants in the exhaust gases.

2   Air cleaner element - replacement

1   Remove the air cleaner cover and take out the element.
2   Disconnect the hose(s). Discard the element and gasket.

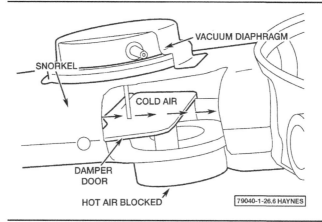

3.6 As the engine warms up, the damper door moves to close off the heat stove passage and open the snorkel passage to outside air can enter the carburetor

3   Clean the bottom of the air cleaner, the gasket surfaces and the cover.
4   Install the new air cleaner using new gaskets, and align the heat stove tube.
5   Install the air cleaner cover and tighten the fender nut.

3   Thermostatically controlled air cleaner - description, check and component replacement

Description

Refer to illustration 3.2

1   The thermostatic air cleaner (THERMAC) system improves engine efficiency and drive-ability under varying climatic conditions by controlling the temperature of the air coming into the air cleaner. A uniform incoming air temperature allows leaner air/fuel ratios during warm-up, which reduces hydrocarbon emissions.

2   The system uses a damper assembly, located in the snorkel of the air cleaner housing, to control the ratio of cold and warm air directed into the carburetor or throttle body. This damper is controlled by a vacuum motor which is, in turn, modulated by a temperature sensor in the air cleaner **(see illustration)**. On some engines a check valve is used in the sensor, which delays the opening of the damper

when the engine is cold and the vacuum signal is low.

Check

Refer to illustrations 3.5 and 3.6

3   This is a visual check. If access is limited, a small mirror may have to be used. Locate the damper door inside the air cleaner assembly. It's inside the long snorkel of the metal air cleaner housing.

4   If there is a flexible air duct attached to the end of the snorkel, leading to an area behind the grille, disconnect it at the snorkel. This will enable you to look through the end of the snorkel and see the damper inside.

5   The check should be done when the engine is cold. Start the engine and look through the snorkel at the damper, which should move to a closed position. With the damper closed, air cannot enter through the end of the snorkel, but instead enters the air cleaner through the flexible duct attached to the exhaust manifold and the heat stove passage **(see illustration)**.

6   As the engine warms up to operating temperature, the damper should open to allow air through the snorkel end **(see illustration)**. Depending on outside temperature, this may take 10 to 15 minutes. To speed up this check you can reconnect the snorkel air duct, drive the vehicle, then check to see if the damper is completely open.

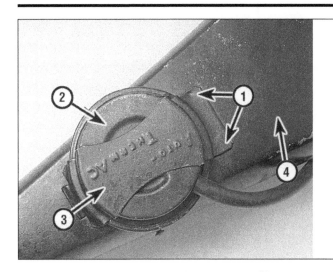

3.10 To remove the vacuum motor, drill out the spot welds and detach the retaining strap

1 *Spot welds*
2 *vacuum diaphragm*
3 *Retaining strap*
4 *Snorkel*

3.17 The sensor retainer must be pried off with a screwdriver

7   If the damper does not close off snorkel air when the cold engine is started, disconnect the vacuum hose at the snorkel vacuum motor and place your thumb over the hose end, checking for vacuum. Replace the vacuum motor if the hose routing is correct and the damper moves freely.

8   If there was no vacuum going to the motor in the above test, check the hoses for cracks, crimped areas and proper connection. If the hoses are clear and in good condition, replace the temperature sensor inside the air cleaner housing.

Component replacement

Note: *Check on parts availability before attempting component replacement. Some components may no longer be available for all models, requiring replacement of the complete air cleaner assembly.*

Air cleaner vacuum motor

Refer to illustration 3.10

9   Remove the air cleaner assembly from the engine and disconnect the vacuum hose from the motor.

10   Drill out the two spot welds which secure the vacuum motor retaining strap to the snorkel tube **(see illustration)**. Remove the motor retaining strap.

11   Lift up the motor, cocking it to one side to unhook the motor linkage at the damper assembly.

12   To install, drill a 7/64-inch hole in the snorkel tube at the center of the vacuum motor retaining strap.

13   Insert the vacuum motor linkage into the damper assembly.

14   Using the sheet metal screw supplied with the motor service kit, attach the motor and retaining strap to the snorkel. Make sure the sheet metal screw does not interfere with the operation of the damper door. Shorten the screw if necessary.

Air cleaner temperature sensor

Refer to illustration 3.17

15   Remove the air cleaner from the engine

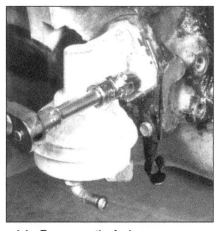

4.4a To remove the fuel pump, remove the two mounting bolts . . .

and disconnect the vacuum hoses at the sensor.

16   Carefully note the position of the sensor. The new sensor must be installed in exactly the same position.

17   Pry up the tabs on the sensor retaining clip and remove the sensor and clip from the air cleaner **(see illustration)**.

18   Install the new sensor with a new gasket in the same position as the old one.

19   Press the retaining clip onto the sensor. Do not damage the control mechanism in the center of the sensor.

20   Connect the vacuum hoses and attach the air cleaner to the engine.

---

4   Fuel pump - check and replacement

Refer to illustrations 4.4a, 4.4b and 4.6

Warning: *Gasoline is extremely flammable, so take extra precautions when you work on any part of the fuel system. See the* **Warning** *in Section 1.*

1   The fuel pump is a sealed unit and cannot be serviced in the event of malfunction.

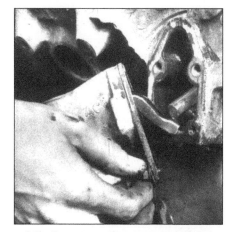

4.4b the remove the pump and gasket from the block (be sure to remove all old gasket material from the pump and block mating surfaces)

Check

2   If a faulty pump is suspected, Detach the ignition coil primary leads and detach the fuel line at the carburetor. Position a container to catch any fuel and have an assistant crank the engine over. If little or no fuel flows out, the pump is defective or the pipe is clogged.

Replacement

3   Detach the fuel inlet and outlet lines from the fuel pump. If threaded fittings are used, unscrew them using two wrenches to prevent twisting the line.

4   Remove the fuel pump mounting bolts, the pump, and the gasket **(see illustrations)**.

5   If the pushrod is to be removed, first remove the pipe plug or the pump adapter and gasket, as appropriate.

6   Installation is the reverse of removal. When installing the rod (if removed), make sure it's positioned with the correct end out. Heavy grease can be used to hold it in place while the fuel pump is installed **(see illustration)**.

4.6 When installing the fuel pump rod, use heavy grease to hold it in place

7.1a Adjusting an idle mixture screw on a Holley 4150 carburetor

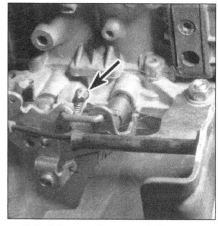

7.1b Idle speed screw - Holley 4150

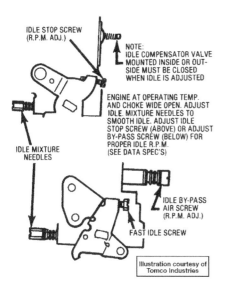

7.1c Idle speed and mixture adjustment details - Rochester 2GC/2GV carburetors

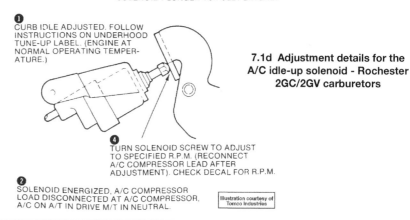

7.1d Adjustment details for the A/C idle-up solenoid - Rochester 2GC/2GV carburetors

7   Install the pump using a new gasket. Use gasket sealant on the screw threads.
8   Connect the fuel lines (including the one at the carburetor, if it was disconnected for the checking procedure), start the engine and check for leaks.

5   Carburetors - basic description

1   Carburetors used on the Nova models covered by this manual include Rochester 2GV (1-1/4 inch), 2GV (1-1/2 inch), Rochester 4MV, Rochester 2GC, Rochester M4MC or M4MCA, or Holley 4150 types.
2   The Rochester 2GV and 2GC carburetors are 2-barrel down draft types. The Rochester 4MV and Holley 4150 are 4-barrel down draft types.
3   The units fitted on automatic and manual transmission models are generally similar for any particular year of manufacture. However, they are calibrated differently and must not be interchanged or substituted for different types.

4   The main metering jets are of fixed type, calibration being accomplished through a system of air bleeds.
5   A power enrichment valve assembly is incorporated by which power mixtures are controlled by air velocity past the boost venturi according to engine demands.
6   An internal fuel filter is fitted and the float bowl is vented internally in the interest of reducing vapor emissions.
7   Some carburetors incorporate a combined emission control (CEC) solenoid which controls distributor vacuum and increases the idle speed during top gear deceleration (refer to Section 59).
8   Carburetors used for most other years of manufacture incorporate an electrically operated throttle closing solenoid. This is used to ensure that the throttle valve closes fully after the ignition is switched off, to prevent running-on (dieseling).
9   The choke is automatic and is operated by a manifold-heated coil.

6   Throttle linkage - inspection and maintenance

1   Inspect the throttle linkage for damage

and missing parts, worn bushings and for binding and interference when the accelerator pedal is depressed.
2   Make sure the throttle lever on the carburetor opens fully when the accelerator is depressed fully, and make sure it returns to its stop when the accelerator pedal is released.
3   Periodically lubricate the linkage pivot points with engine oil.

7   Carburetors - idle speed adjustment

Refer to illustrations 7.1a, 7.1b, 7.1c, 7.1d, 7.1e, 7.1f and 7.1g

1   Idle speed adjustment must be carried out after the engine has fully warmed up **(see illustrations)**. The air cleaner must be in place, except where otherwise specified, and it is essential that the ignition timing and dwell angle are correctly set. All emission control systems must also be functioning correctly. In order to check engine speed, an external tachometer must be connected, following the tool manufacturer's instructions. **Note**: *If the information given on the underhood label has superseded the information*

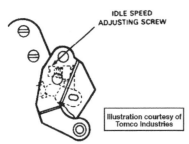

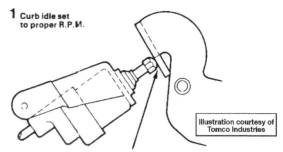

7.1e Idle speed adjustment details - Rochester 4MV/M4MC carburetors (without an idle-up solenoid)

7.1f Adjustment details for the A/C idle-up solenoid - Rochester 4MV/M4MC carburetors

given in this Section, the label should be assumed to be correct.

1969 models

2   Turn the idle mixture screw(s) in until it seats gently, then back it out three full turns.
3   Run the engine and adjust the idle mixture screw to obtain the highest steady idle speed where no idle solenoid is fitted. Adjust the idle speed screw to obtain the recommended idle speed given on the underhood label (the label also states whether the air conditioning should be On or Off).
4   Where an idle solenoid is fitted, adjust the solenoid plunger hex to obtain 600 rpm. Disconnect the solenoid lead then adjust the carburetor idle screw to obtain 400 rpm.
5   Adjust one mixture screw in to obtain a 20 rpm drop (lean roll), then screw out 1/4 turn.
6   Where applicable, repeat paragraph 5 for the second mixture screw.
7   Re-adjust the idle speed if necessary.

1970 models (307 cu in.)

8   Disconnect and plug the distributor vacuum line then turn the mixture screw(s) in, until lightly contacting the seat; back the screw(s) out four full turns.
9 Disconnect the fuel tank line from the vapor canister and turn the air conditioning off.
10   Adjust the idle speed screw to obtain 800 rpm (manual) or 630 rpm (automatic, in Drive).
11   Screw in the mixture screws equally to obtain 700 rpm (manual) or 600 rpm (automatic).
12   On automatic transmission vehicles, disconnect the solenoid lead and set the idle screw to obtain 450 rpm. Reconnect the lead.
13   Reconnect the distributor vacuum line and the fuel tank line to the vapor canister.

1970 models (350 cu in., 250 HP)

14   Follow the procedure of paragraph 9 thru 13, except that manual transmissions are set to 830 rpm (paragraph 10) and 750 rpm (paragraph 11).

1970 models (350 cu in., 300 HP)

15   Follow the procedure of paragraph 9 thru 13, excluding paragraph 12, except that manual transmissions are set to 775 rpm (paragraph 10).

1970 models (396 cu in., 375 HP)

16   Remove the air cleaner then disconnect and plug the distributor vacuum line.
17   Adjust the mixture screws for maximum idle speed.
18   Adjust the idle speed screw to obtain 750 rpm (manual) or 700 rpm (automatic, in Drive).
19   Turn one mixture screw in to obtain a 20 rpm drop, then back it out by 1/4 turn. Repeat for the other mixture screw.
20   Re-adjust the idle speed (paragraph 18), then reconnect the distributor vacuum line and install the air cleaner.

1970 models (396 cu in., 350 HP)

21   Disconnect and plug the distributor vacuum line.
22   Turn the mixture screws in until they are lightly contacting the seats, then back them out by 4 full turns.
23   Adjust the carburetor idle speed screw to obtain 630 rpm (automatic, in Drive) or 700 rpm (manual).
24   *Automatics*: Adjust the mixture screws to obtain 600 rpm then reconnect the distributor vacuum line. This completes the adjustment.
25   Turn one mixture screw in until the speed drops by 40 rpm, then regain 700 rpm by adjusting the idle speed screw. Repeat for the other mixture screw then reconnect the distributor vacuum line.

1971 models (307 cu in. and 350 cu in., 2-barrel carburetor)

26   Disconnect and plug the distributor vacuum line.
27   Disconnect the fuel tank line from the vapor canister.

7.1g Adjusting an idle mixture screw - Rochester 4MV/M4MC carburetors (carburetor removed for clarity)

28   Adjust the carburetor idle speed screw to obtain 600 rpm (manual, air conditioning off) or 550 rpm (automatic, in Drive; air conditioning on). Do not adjust the solenoid screw or a decrease in engine braking may result.
29   Reconnect the distributor vacuum line and the fuel tank line to the vapor canister.

1971 models (350 cu in., 4-barrel carburetor)

30   Initially follow the procedure of paragraphs 26 thru 28, then place the fast idle cam follower on the second step of the fast idle cam. Turn the air conditioning off, and adjust the fast idle to 1350 rpm (manual) or 1500 rpm (automatic, in Drive).
31   Reconnect the distributor vacuum line and the fuel tank line to the canister.

1972 models (307 cu in., and 350 cu in., 2-barrel carburetor)

32   Disconnect and plug the distributor vacuum line.
33   Disconnect the fuel tank line from the vapor canister.
34   With the air conditioning off, adjust the idle stop solenoid screw to obtain 900 rpm (manual) or 600 rpm (automatic, in Drive).

35   With the transmission in 'Park' or 'Neutral', adjust the idle cam screw to obtain 1850 rpm (307 cu in.) or 220 rpm (350 cu in.).
36   Reconnect the distributor vacuum line and the fuel tank line to the vapor canister.

1972 models (350 cu in., 4-barrel carburetor)

37   Initially follow the procedure of paragraphs 32 thru 34, except that the idle stop solenoid screw is set to 800 rpm for manual transmission.
38   Place the fast idle cam follower on the second step of the fast idle cam, turn the air conditioning off and adjust the fast idle to 1350 rpm (manual) or 1500 rpm automatic, in Park).
39   Reconnect the distributor vacuum line and the fuel tank line to the vapor canister.

1973 models (307 cu in., and 350 cu in., 2-barrel carburetor)

40   Disconnect and plug the distributor vacuum line.
41   Disconnect the fuel tank line from the vapor canister.
42   With air conditioning off, adjust the idle stop solenoid screw to obtain 900 rpm (manual) or 600 rpm (automatic, in Drive).
43   Disconnect the idle stop solenoid lead and adjust the carburetor idle cam screw on the low step of the cam to obtain 450 rpm (307 cu in.), 400 rpm (350 cu in. automatic) or 500 rpm (350 cu in. manual).
44   Reconnect the distributor vacuum line and the fuel tank line to the vapor canister.

1973 models (350 cu in., 4-barrel carburetor)

45   Initially follow the procedure of paragraph 40 thru 42, then reconnect the distributor vacuum line.
46   Place the fast idle cam follower on the top step of the fast idle cam, turn the air conditioning off and adjust the fast idle to 1300 rpm (manual) or 1600 rpm (automatic, in Park).
47   Reconnect the fuel tank line to the vapor canister.

1974 models (350 cu in., 2-barrel carburetor)

48   Follow the procedure of paragraph 40 thru 44.

1974 models (350 cu in., 4-barrel carburetor)

49   Follow the procedure for 1973 models, given in paragraph 45 thru 47 (including paragraph 40 thru 42).

1975 models (262 cu in. (4.3 liter) and 350 cu in., 2-barrel carburetor)

50   Disconnect the fuel tank line from the vapor canister and switch the air conditioning off.

51   Turn the idle speed screw to obtain an idle speed of 800 rpm (manual) or 600 rpm (automatic, in Drive).
52   Reconnect the fuel tank line to the vapor canister.

1975 models (350 cu in., 4-barrel carburetor)

53   Disconnect the fuel tank line from the vapor canister and switch the air conditioning off.
54   Disconnect the electrical lead from the idle stop solenoid.
55   Turn the idle speed screw to obtain a low idle speed of 800 rpm (manual - Federal) or 600 rpm (manual - except Federal, and automatic, in Drive).
56   Reconnect the idle stop solenoid lead and open the throttle slightly to extend the solenoid plunger.
57   Adjust the solenoid plunger screw to obtain a curb idle speed of 1600 rpm in Neutral.
58   Reconnect the fuel tank line to the vapor canister.

1976 and later models

2GC carburetors

59   Have the engine at normal operating temperature with ignition settings correct and emissions control systems operating correctly.
60   Set the idle speed screw on the low step of the fast idle cam.
61   Turn the idle speed screw to set the curb (initial idle speed) to Specifications (or refer to the vehicle decal).
62   Where a solenoid is fitted to the carburetor, carry out the operations described in paragraphs 1 and 3 and then with (i) the solenoid energized, (ii) the lead disconnected from the air conditioner compressor, (iii) the air conditioner on, open the throttle to allow the solenoid plunger to extend fully. Turn the solenoid hexagonal-headed bolt until the idle speed is 700 rpm (manual) or 650 rpm (automatic). Reconnect the compressor lead on completion.
63   2GC carburetors are fitted with a solenoid when the vehicle is equipped with automatic transmission or air conditioning.

M2MC and M4MC carburetors

64   Have the engine at normal operating temperature, air cleaner in position and air conditioning off. Connect a reliable tachometer to the engine.
65   Disconnect the fuel tank hose from the vapor canister.
66   Disconnect the lead from the idle stop solenoid if equipped.
67   With automatic transmission in Drive or manual in Neutral, turn the idle speed screw to obtain the curb (final) idle speed shown in the Specifications.
68   Reconnect the solenoid, if equipped, and crack open the throttle slightly to extend the solenoid plunger.

69   Now turn the solenoid plunger screw to set the solenoid idle speed shown in the Specifications.
70   Remove the tachometer and reconnect the fuel tank hose.

8   Carburetors - idle mixture adjustment

Note: *Refer to illustrations 7.1a, 7.1c and 7.1g for mixture adjusting screw locations.*
1   Carburetors used on 1971 models onwards have idle needle limiter caps to prevent adjustment in service. These caps should never be removed except where repairs have been carried out which may effect the idle mixture. For these models, mixture adjustment should ideally be carried out using an exhaust gas analyzer. Where this is not possible, the procedure given in this Section is satisfactory. For 1969/1970 models, idle mixture adjustment is given in the preceding Section. **Note:** *If the information given on the underhood label has superseded the information given in this Section, the label should be assumed to be correct.*

1971 models

2   Initially follow the instructions on the vehicle tune-up decal then turn the mixture screw in until it lightly contacts the seat; now turn the screw back 4 full turns.
3   Adjust the idle speed screw to obtain the specified initial idle speed.
4   Adjust the mixture screws equally in to obtain the specified final idle speed.
5   Install service replacement idle needle limiter caps on the mixture screws.
6   Reconnect the distributor vacuum hose and fuel tank line to the canister.

1972 and 1973 models

7   Initially follow the procedure of paragraphs 2 and 3, then adjust the mixture screws equally in, or 1/4 turn rich from the lean roll, (as specified), to obtain the final curb idle speed. **Note:** *1/4 turn rich from lean roll is obtained by screwing the mixture screws equally in to obtain a drop of 20 rpm then screwing out 1/4 turn.*
8   Follow the procedure of paragraphs 5 and 6.

1974 and 1975 models

9   Disconnect and plug the distributor vacuum line.
10   Disconnect the fuel tank line from the vapor canister.
11   With the engine at normal operating temperature, air conditioning off and transmission in Drive (automatic) or neutral (manual), use needle nose pliers to break off the tab on the mixture screw.
12   Adjust the idle speed to the higher of the two specified speeds with the idle solenoid energized.

9.1 When loosening the fuel line fitting ant the carburetor, always use the correct size wrenches, preferably a "flare-nut" wrench to avoid rounding-off the corners of the tube fitting, and another wrench to keep the carburetor inlet fitting from turning

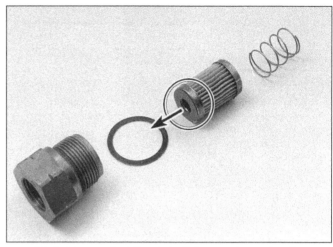

9.2 Carburetor fuel filter, inlet fitting, gasket and spring

13   Equally turn out the mixture screws until the maximum idle speed is obtained, then reset to the higher specified idle speed.

14   Turn in the mixture screws equally until the lower specified idle speed is obtained.

15   Reconnect the distributor vacuum line and the fuel tank line to the canister.

1976 and 1977 models

13   The mixture screws are fitted with limiter caps which restrict their movement between 1/2 and 3/4 turn lean. Any adjustment should be kept to this but where the carburetor has been overhauled or new components fitted then the caps should be broken off and the following operations carried out.

14   Have the engine at normal operating temperature with air conditioning off and a tachometer connected to the engine.

15   Also have the parking brake set. On 1976 models, the air cleaner should be on. On 1977 models, the air cleaner should be removed for access to the carburetor, but leave the vacuum hoses connected to it.

16   On 1976 350 engines, disconnect the fuel tank vent line from the charcoal canister.

17   On 1977 models, refer to the *emission control information label* under the hood and disconnect and plug the appropriate hoses as directed.

18   Disconnect the vacuum advance hose from the distributor and plug it. Check the ignition timing, adjusting it if necessary. Then reconnect the vacuum advance hose to the distributor.

19   Using needle-nosed pliers, carefully break off the caps from the idle mixture screws, being careful not to bend the screws.

20   On 1977 models, lightly seat the screws and then back them out equally just enough so the engine will run.

21   If equipped with a manual transmission, be sure it is in Neutral. If equipped with an automatic transmission, block the tires and have an assistant sit in the driver's seat. While holding the brake firmly depressed, shift it into Drive.

22   Back out both mixture screws 1/8 turn at a time until the maximum idle speed is obtained.

23   Turn the idle speed adjusting screw until the idle speed prior to lean drop (see the Specifications section) is obtained. Again turn the mixture screws a little at a time to be sure you have the maximum idle speed.

24   Again, turning them evenly in 1/8 turn increments, turn the mixture screws in until the proper idle speed after lean drop is obtained (see the Specifications section). Automatic transmission models can now be shifted to Park.

25   On 1976 models with 350 engines, reconnect the fuel tank vent hose to the charcoal canister.

26   On 1977 models, reset the idle speed (see Section 7). Then check and adjust the fast idle, again referring to the label.

27   Reconnect any vacuum hoses that were disconnected, and replace the air cleaner.

1978 and 1979 models

28   In 1978, General Motors changed the design of the carburetor idle mixture screw so that backing out the screw will not appreciably richen the mixture. Idle mixture adjustment is carried out with special propane enrichment equipment, a procedure best left to a dealer or other qualified repair shop.

9   Carburetor fuel filter - replacement

Refer to illustrations 9.1 and 9.2

Warning: *Gasoline is extremely flammable, so take extra precautions when you work on any part of the fuel system. See the* **Warning** *in Section 1.*

1   Disconnect the fuel line connection at the inlet fuel filter nut **(see illustration)**.

2   Unscrew the fuel filter nut, then remove the element and spring **(see illustration)**.

3   Install the element spring and element. On bronze filters (Holley carburetors) the small section of the cone faces outwards.

4   Where applicable, install a new gasket on the inlet fitting nut. Install the nut and tighten it securely (but don't overtighten it, as the threads in the carburetor body are easily stripped).

5   Install the fuel line and tighten the line fitting securely, holding the inlet fitting with another wrench to prevent stripping out the threads in the carburetor body or twisting the fuel line.

10   Combination Emission Control (CEC) valve - adjustment

1   This valve, used on 1971 models, will only need adjustment after replacement of its solenoid, after carburetor overhaul or replacement of the throttle body.

2   Initially adjust the idle speed and mixture as previously described.

3   With the engine running, transmission in Neutral (manual) or Drive (automatic), air conditioning off, distributor vacuum hose removed and plugged, and the fuel tank vapor line disconnected from the vapor canister, manually extend the CEC valve plunger to contact the throttle valve lever.

4   Adjust the plunger length to obtain the specified CEC valve engine speed.

11   Idle stop solenoid - adjustment

1   With the engine at normal operating temperature, the air cleaner installed, choke open and air conditioning off, disconnect the fuel tank line from the vapor canister.

2   Disconnect and plug the distributor vacuum line.

3   Disconnect the lead to the idle stop solenoid.

4   Adjust the carburetor low idle to 450 rpm with the adjustment screw on the low step of the cam.

5   Check the dwell angle and ignition timing and recheck the low idle. Reset if necessary.

12.3 Typical choke coil (arrow)

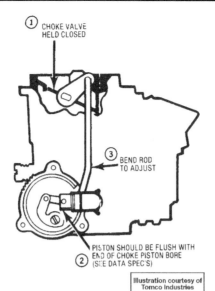

14.3 Choke rod adjustment details - Rochester 2GV carburetors

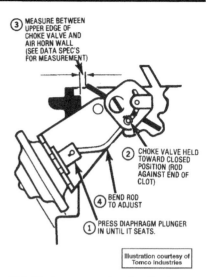

15.5 Choke vacuum break adjustment details - Rochester 2GV carburetors

6   Reconnect the solenoid lead, open the throttle momentarily then adjust the solenoid plunger screw to obtain the specified curb idle speed.

12 Automatic choke coil - replacement

Refer to illustration 12.3
1   Remove the air cleaner and disconnect the choke rod upper clip.
2   Remove the choke coil shield by prying with a screwdriver in the cut out or on the small tangs at the base of the shield, then lifting the shield over the rod.
3   Remove the choke rod, mounting screw and the choke coil assembly **(see illustration)**.
4   Install the replacement coil, ensuring that the locator is in the hole of the intake manifold, then install the mounting screw.
5   Install the choke rod and adjust if necessary, without the choke coil shield installed.
6   Disconnect the choke rod upper end and lower the choke coil shield over the choke rod, install the shield over the choke coil.
7   Check that the valve moves freely over its full range, then start the engine and check for correct operation.
8   Install the air cleaner.

13 Carburetor (all versions) - removal and installation

Warning: *Gasoline is extremely flammable, so take extra precautions when you work on any part of the fuel system. See the* **Warning** *in Section 1.*
1   Remove the air cleaner.
2   Disconnect the fuel line and vacuum hoses from the carburetor **(see illustration 9.1)**. Mark the hoses with pieces of numbered tape to avoid confusion on installation.

3   Disconnect the choke rod.
4   Disconnect the accelerator linkage.
5   Disconnect the throttle valve linkage or downshift cable (automatic transmission).
6   Remove all hoses and electrical connections, making very careful note of where they were removed from.
7   Remove the carburetor attaching nuts and/or bolts.
8   Remove the carburetor from the intake manifold.
9   Remove the gasket and/or insulator.
10   Installation is the reverse of the removal procedure, but the following points should be noted:

a) *By filling the carburetor bowl with fuel, the initial start-up will be easier and less drain on the battery will occur.*
b) *New gaskets must be used.*
c) *Idle speed and mixture settings should be checked, and adjusted if necessary.*

14 Rochester 2GV carburetor - choke rod adjustment

Refer to illustration 14.3
1   Turn the idle stop screw in until it just touches the bottom step of the fast idle cam, then screw it in exactly one full turn.
2   Position the idle screw so that it is on the second step of the fast idle cam against the shoulder of the high step.
3   Hold the choke valve plate towards the closed position (using a rubber band to keep it in place) and check the gap between the upper edge of the choke valve plate and the inside wall of the air horn **(see illustration)**.
4   Adjust to the specified gap, if necessary, by bending the tang on the upper choke lever. The setting should provide the specified fast idle speeds.

15 Rochester 2GV carburetor - choke vacuum break adjustment

Refer to illustration 15.5

Pre 1972 models
1   Remove the air cleaner and plug the Thermac sensor vacuum take-off port.
2   Start the engine or apply suction to the diaphragm vacuum tube.
3   Remove the choke rod from the lever and install a rubber band to the lever to hold the choke towards the closed position.
4   Slowly open the accelerator until the choke is closed as far as the vacuum break link reaction will permit, and the idle is determined by the high step of the fast idle cam. Release the accelerator.
5   With the condition of paragraph 4 maintained, insert a gauge of the specified thickness between the air horn and the choke blade. Bend the rod or tang as necessary to obtain the specified dimension **(see illustration)**.

1972 models onwards
6   Remove the air cleaner and plug the Thermac sensor vacuum take off port.
7   Using an external suction source, apply suction to the vacuum break diaphragm until the plunger is fully seated.
8   With the diaphragm fully seated, push the choke valve towards the closed position and place a gauge of the specified thickness between the air horn and the choke blade.
9   Bend the vacuum break rod if necessary to obtain the specified dimension.

16 Rochester 2GV carburetor - choke unloader adjustment

Refer to illustration 16.3
1   Hold the throttle valve plates in the fully open position.

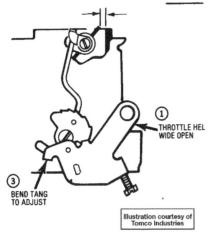

16.3 Choke unloader adjustment details - Rochester 2GV carburetors

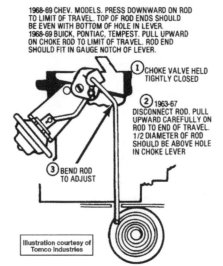

17.2 Choke coil rod adjustment details - Rochester 2GV carburetors

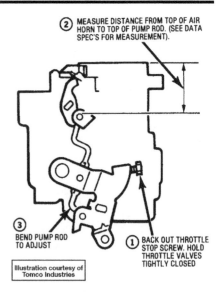

18.2 Accelerator pump rod adjustment details - Rochester 2GV carburetors

2   Hold the choke valve plate towards the closed position using a rubber band to keep it in place.
3   Check the gap between the upper edge of the choke valve plate and the inside wall of the air horn (see illustration).
4   Bend the tang on the throttle lever to adjust the gap to the specified value, if necessary.

17 Rochester 2GV carburetor - choke coil rod adjustment

Refer to illustration 17.2

Pre-1972 models

1   Hold the choke valve open then pull down on the coil rod to the end of its travel.
2   The top of the rod end which slides in the hole in the choke lever should be even with the bottom of the choke lever hole. Bend the rod at the point shown to adjust, if necessary (see illustration).
3   Connect the rod to the choke lever and install the retaining clip. Check that the choke operates freely.

1972 models onwards

4   Hold the choke valve plate fully open.
5   Disconnect the thermostatic coil rod from the upper lever and push down on the rod as far as it will go. The top of the rod should be level with the bottom of the hole in the choke lever.
6   Adjust if necessary by bending the rod.

18 Rochester 2GV carburetor - accelerator pump rod adjustment

Refer to illustration 18.2
1   Unscrew the idle speed screw.

2   Close both throttle valve plates completely and measure from the top surface of the air horn ring to the tip of the pump rod (see illustration).
3   Bend the rod to obtain the specified dimension.

19 Rochester 2GV carburetor - servicing

Refer to illustrations 19.1, 19.25a, 19.25b and 19.25c
Warning: *Gasoline is extremely flammable, so take extra precautions when you work on any part of the fuel system. See the* **Warning** *in Section 1.*
1   When a carburetor develops faults after a considerable mileage, it is usually more economical to replace the complete unit, rather than to completely disassemble it and replace individual components. Where, however, it is decided to strip and rebuild the unit, first obtain a repair kit which will contain all the necessary gaskets and other renewable items, and proceed in the following sequence (see illustration).
2   Bend back the lockwasher tabs then remove the idle stop solenoid (where applicable) from the carburetor.
3   Remove the choke lever from the vacuum break diaphragm link and the vacuum break link from the diaphragm plunger. The diaphragm plunger stem spring need not be removed.
4   Disconnect the vacuum break hose from the tube then remove the diaphragm from the air horn by unscrewing two retaining screws.
5   Remove the fuel inlet filter nut, filter, spring and two gaskets.
6   Remove the pump rod from the throttle lever after removing the retaining clip. Rotate the upper pump lever counterclockwise, then remove the pump rod from the lever by align-

ing the rod squirt with the lever notch.
7   Remove the fast idle cam retaining screw, rotate the cam and remove it from the rod.
8   Hold the choke open, rotate the upper end of the choke rod towards the pump lever and remove the rod from the upper choke lever.
9   Remove the air horn from the float bowl (8 screws).
10   Remove the float hinge pin, float, splash shield and float needle.
11   Unscrew the float needle seat and remove the gasket.
12   Remove the air horn to float bowl gasket.
13   Depress the power piston shaft, and allow the spring to snap sharply and eject the piston from the casting.
14   Remove the inner pump lever retaining screw then remove the outer pump lever and plastic washer from the air horn. Place the plunger in gasoline to prevent the rubber from drying out.
15   Rotate the pump plunger stem out of the hole in the inner lever if it is required to remove it. Do not bend the tang on the inner lever.
16   If the choke shaft or valve need replacement, remove the two staked screws, remove the valve, then remove the shaft and lever from the air horn.
17   Remove the pump plunger return spring from the pump well, followed by the inlet check ball (where applicable).
18   Remove the pump inlet screen from the bottom of the float bowl (where applicable).
19   Unscrew the main jets, power valve and gaskets.
20   Remove the cluster and gasket (3 screws and washers). Note the fiber washer on the center screw.
21   Remove the pump discharge spring retainer, the spring and the check ball.

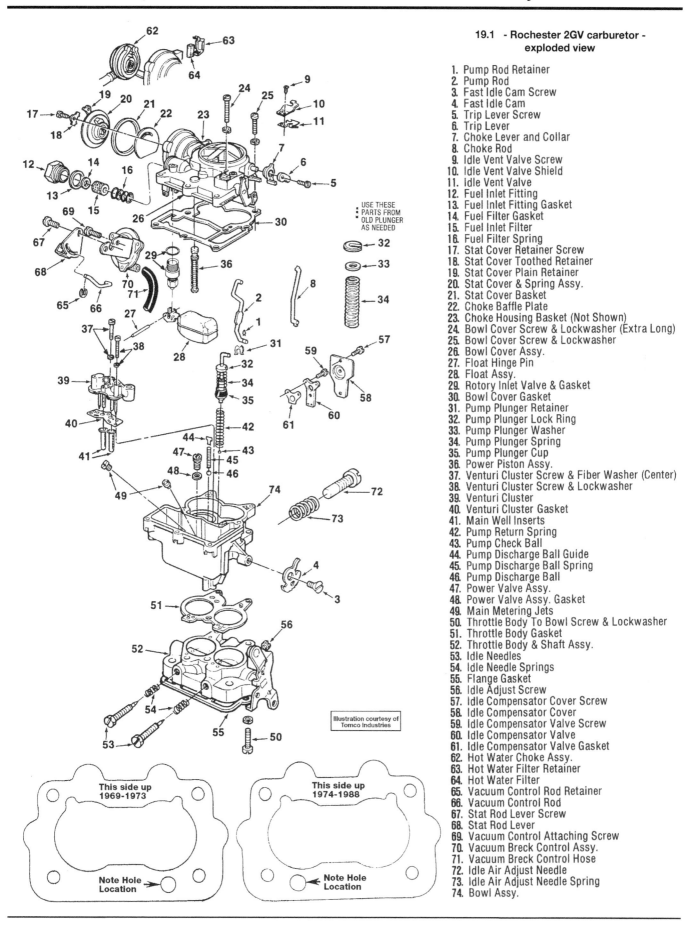

19.1 - Rochester 2GV carburetor - exploded view

1. Pump Rod Retainer
2. Pump Rod
3. Fast Idle Cam Screw
4. Fast Idle Cam
5. Trip Lever Screw
6. Trip Lever
7. Choke Lever and Collar
8. Choke Rod
9. Idle Vent Valve Screw
10. Idle Vent Valve Shield
11. Idle Vent Valve
12. Fuel Inlet Fitting
13. Fuel Inlet Fitting Gasket
14. Fuel Filter Gasket
15. Fuel Inlet Filter
16. Fuel Filter Spring
17. Stat Cover Retainer Screw
18. Stat Cover Toothed Retainer
19. Stat Cover Plain Retainer
20. Stat Cover & Spring Assy.
21. Stat Cover Basket
22. Choke Baffle Plate
23. Choke Housing Basket (Not Shown)
24. Bowl Cover Screw & Lockwasher (Extra Long)
25. Bowl Cover Screw & Lockwasher
26. Bowl Cover Assy.
27. Float Hinge Pin
28. Float Assy.
29. Rotory Inlet Valve & Gasket
30. Bowl Cover Gasket
31. Pump Plunger Retainer
32. Pump Plunger Lock Ring
33. Pump Plunger Washer
34. Pump Plunger Spring
35. Pump Plunger Cup
36. Power Piston Assy.
37. Venturi Cluster Screw & Fiber Washer (Center)
38. Venturi Cluster Screw & Lockwasher
39. Venturi Cluster
40. Venturi Cluster Gasket
41. Main Well Inserts
42. Pump Return Spring
43. Pump Check Ball
44. Pump Discharge Ball Guide
45. Pump Discharge Ball Spring
46. Pump Discharge Ball
47. Power Valve Assy.
48. Power Valve Assy. Gasket
49. Main Metering Jets
50. Throttle Body To Bowl Screw & Lockwasher
51. Throttle Body Gasket
52. Throttle Body & Shaft Assy.
53. Idle Needles
54. Idle Needle Springs
55. Flange Gasket
56. Idle Adjust Screw
57. Idle Compensator Cover Screw
58. Idle Compensator Cover
59. Idle Compensator Valve Screw
60. Idle Compensator Valve
61. Idle Compensator Valve Gasket
62. Hot Water Choke Assy.
63. Hot Water Filter Retainer
64. Hot Water Filter
65. Vacuum Control Rod Retainer
66. Vacuum Control Rod
67. Stat Rod Lever Screw
68. Stat Rod Lever
69. Vacuum Control Attaching Screw
70. Vacuum Breck Control Assy.
71. Vacuum Breck Control Hose
72. Idle Air Adjust Needle
73. Idle Air Adjust Needle Spring
74. Bowl Assy.

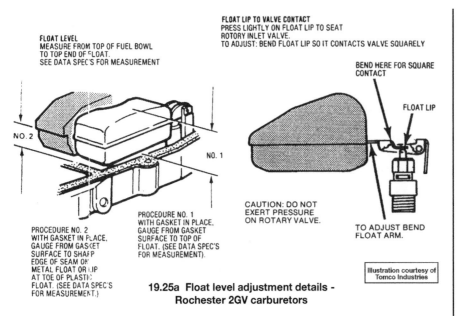

FLOAT LEVEL
MEASURE FROM TOP OF FUEL BOWL
TO TOP END OF FLOAT.
SEE DATA SPEC'S FOR MEASUREMENT

NO. 2

NO. 1

PROCEDURE NO. 2
WITH GASKET IN PLACE,
GAUGE FROM GASKET
SURFACE TO SHARP
EDGE OF SEAM ON
METAL FLOAT OR LIP
AT TOE OF PLASTIC
FLOAT. (SEE DATA SPEC'S
FOR MEASUREMENT.)

PROCEDURE NO. 1
WITH GASKET IN PLACE,
GAUGE FROM GASKET
SURFACE TO TOP OF
FLOAT. (SEE DATA SPEC'S
FOR MEASUREMENT).

FLOAT LIP TO VALVE CONTACT
PRESS LIGHTLY ON FLOAT LIP TO SEAT
ROTARY INLET VALVE.
TO ADJUST: BEND FLOAT LIP SO IT CONTACTS VALVE SQUARELY

BEND HERE FOR SQUARE
CONTACT

FLOAT LIP

CAUTION: DO NOT
EXERT PRESSURE
ON ROTARY VALVE.

TO ADJUST BEND
FLOAT ARM.

Illustration courtesy of
Tomco Industries

19.25a Float level adjustment details -
Rochester 2GV carburetors

WITH FLOAT HANGING FREELY, MEASURE
FROM GASKET SURFACE TO BOTTOM OF
METAL FLOAT OR LIP OF PLASTIC FLOAT.
(SEE DATA SPEC'S FOR MEASUREMENT.)

TO ADJUST
BEND TANG

Illustration courtesy of
Tomco Industries

19.25b Float drop adjustment details -
Rochester 2GV carburetors

22 Remove the throttle body-to-bowl attaching screws. Remove the body and gasket.
23 Further dismantling is not recommended. If it is essential to remove the idle mixture needles, pry out the plastic limiter caps, then count the number of turns to bottom the needles and fit replacements in exactly the same position. New limiter caps should be fitted after running adjustments have been made.
24 Clean all metal parts in a suitable cold solvent. Do not immerse rubber parts, plastic parts, the vacuum break assembly, or the idle stop solenoid, or permanent damage will result. Do not probe the jets, but blow them through with clean, dry compressed air. Examine all fixed and moving parts for cracks, distortion, wear and other damage; replace as necessary. Discard all gaskets and the fuel inlet filter.
25 Assembly is essentially the reverse of the removal procedure, but the following points should be noted:
a) *If new idle mixture screws are used, and the original setting was not noted, install the screws finger-tight to seat them, then back-off 4 full turns (3 full turns for 1969/1970 models).*
b) *When installing the choke valve on the shaft, the letters RP face upwards. Ensure that there is 0.020 inch clearance between the choke kick lever on the air horn before tightening the choke valve screws.*
c) *Brass float: With the air horn inverted and the air horn gasket installed, measure the distance from the gasket to the edge of the float seam at the outer edge of the float pontoon. Adjust the float level to the specified dimension by bending the float arm (see illustration). With the air horn assembly upright and float freely suspended, measure from*

the gasket to the bottom of the float pontoon. Adjust the float drop to the specified dimension by bending the tang adjacent to the float needle (see illustration).
d) *Plastic float: Refer to the procedure for the brass float, but note that for float level and float drop, the dimension is taken from the lip at the toe of the float in both instances.*
e) *Install and tighten the air horn screws evenly in the order shown (see illustration).*
f) *After reassembly, carry out an the settings and adjustments listed previously in this Chapter.*

20 Rochester 4MV carburetor - fast idle adjustment

1 With the transmission in Neutral, position the fast idle lever on the high step of the fast idle cam.
2 With the engine warm and the choke wide open, adjust the fast idle screw to obtain the specified engine speed.

21 Rochester 4MV carburetor - choke rod adjustment

1 Place the cam follower on the second step of the fast idle cam and against the high step.
2 Rotate the choke valve towards the closed position by turning the external lever counterclockwise.
3 Check that the dimension between the lower edge of the choke valve and the air horn wall (at the lever end) is as specified. Bend the choke rod if adjustment is required.

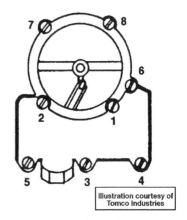

Illustration courtesy of
Tomco Industries

19.25c Air horn tightening sequence

22 Rochester 4MV carburetor - choke vacuum break adjustment

Refer to illustration 22.4

Pre-1972 models

1 Refer to the procedure given in Section 14 for the 2GV carburetor for pre-1972 models.

1972 models onwards

2 Using an external source of suction, seat the choke vacuum break diaphragm.
3 Open the throttle slightly so that the cam follower clears the fast idle cam steps, then rotate the vacuum break lever towards the direction of closed choke. Ensure that the vacuum break rod is in the outer end of the slot in the diaphragm plunger. A rubber band can be used to hold the vacuum break lever in position
4 Measure the distance from the lower edge of the choke valve to the air horn wall.

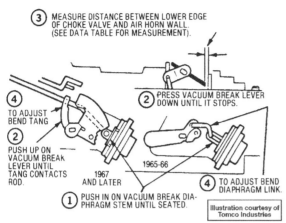

③ MEASURE DISTANCE BETWEEN LOWER EDGE OF CHOKE VALVE AND AIR HORN WALL. (SEE DATA TABLE FOR MEASUREMENT).

④ TO ADJUST BEND TANG

② PRESS VACUUM BREAK LEVER DOWN UNTIL IT STOPS.

② PUSH UP ON VACUUM BREAK LEVER UNTIL TANG CONTACTS ROD.

1967 AND LATER

1965-66

④ TO ADJUST BEND DIAPHRAGM LINK.

① PUSH IN ON VACUUM BREAK DIAPHRAGM STEM UNTIL SEATED.

22.4 Choke vacuum break adjustment details - Rochester 4MV carburetors

Illustration courtesy of Tomco Industries

Bend the vacuum break link if adjustment is required **(see illustration)**.

23 Rochester 4MV carburetor - choke coil rod adjustment

Pre-1972 models

1   Hold the choke valve closed then pull down on the coil rod to the end of its travel. The rod should contact the bracket surface
2   Bend the choke coil rod, if necessary, so that the top of the rod aligns with the bottom of the holes.
3   Connect the coil rod to the choke lever and install the retaining clip. Check that the choke operates freely over its full range of travel.

1972 models onwards

4   Rotate the choke coil lever counter-clockwise to fully close the choke.
5   With the coil rod disconnected and the cover removed, push down on the rod until it contacts the bracket surface.
6   The coil rod must fit in the choke lever notch; bend the rod to adjust if necessary.
7   Install the choke coil cover.
8   Install the coil rod in the choke coil lever slot and install the retaining clip.
9   Check that the choke operates freely over its full range of travel.

24 Rochester 4MV carburetor - air valve dashpot adjustment

Pre-1972 models

1   Seat the vacuum break diaphragm, and check that the specified clearance exists between the dashpot rod and the end of the slot in the air valve lever, when the air valve is fully closed **(see illustration)**.
2   If adjustment is necessary, bend the rod at the air valve end.

1972 models onwards

3   Seat the choke vacuum break diaphragm using an outside source of suction,
then measure the dimension between the end of the slot in the vacuum break plunger lever and the air valve when the air valve is fully closed.
4   If adjustment is necessary, bend the rod at the air valve end.

25 Rochester 4MV carburetor - servicing

Refer to illustrations 25.1, 25.33a and 25.33b
Warning: *Gasoline is extremely flammable, so take extra precautions when you work on any part of the fuel system. See the* **Warning** *in Section 1.*

1   When a carburetor develops faults after a considerable mileage, it is usually more economical to replace the complete unit rather than to completely disassemble it and replace individual components. However if it is decided to strip and rebuild the unit, first obtain a repair kit which will contain all the necessary gaskets and other renewable items, and proceed in the following sequence **(see illustration on following page)**.
2   Bend back the lockwasher tabs, then remove the idle stop solenoid.
3   Remove the larger idle stop solenoid bracket screw from the float bowl.
4   Remove the clip from the upper end of the choke rod, disconnect the rod, disconnect the rod from the upper choke shaft lever and remove the rod from the lower lever in the bowl.
5   Drive the pump lever pivot inwards to remove the roil pin then remove the pump lever from the air horn and pump rod.
6   Remove 2 long screws, 5 short screws and 2 countersunk-head screws retaining the air horn to the float bowl.
7   Remove the vacuum break hose, and the diaphragm unit from the bracket.
8   Disconnect the choke assist spring.
9   Remove the metering rod hanger and secondary rods after removing the small screw at the top of the hanger.
10   Lift off the air horn, but leave the gasket in position. Do not attempt to remove the air bleed tubes or accelerating well tubes.

11   If the choke valve is to be replaced, remove the valve attaching screws, then measure the valve and shaft.
12   The air valves and air valve shaft are calibrated and should not be removed. A shaft spring repair kit is available, and contains all the necessary instructions, if these parts require replacement.
13   Remove the pump plunger from the well.
14   Carefully remove the air horn gasket.
15   Remove the pump return spring from the pump well.
16   Remove the plastic filler over the float valve.
17   Press the power piston down and release it to remove it. Remove the spring from the well. Note the power piston plastic retainer which is used for ease of assembly.
18   Remove the metering rods from the power piston by disconnecting the spring from the top of each rod, then rotating the rod to remove it from the hanger.
19   Remove the float assembly by pulling up on the retaining pin until it can be removed, then sliding the float towards the front of the bowl to carefully disengage the needle pull clip.
20   Remove the pull clip and the fuel inlet needle, then unscrew the needle seat and remove the gasket.
21   Unscrew the primary metering jets; do not attempt to remove the secondary metering jets.
22   Remove the discharge ball retainer and the check ball.
23   Remove the baffle from the secondary side of the bowl.
24   Remove the choke assembly after removing the retaining screw on the side of the bowl Remove the secondary lockout lever from the cast boss on the bowl.
25   Remove the fast idle cam and the choke assembly.
26   Remove the intermediate choke rod and actuating lever from the float bowl.
27   Remove the fuel inlet filter nut, gasket, filter and spring.
28   Remove the throttle body to bowl screws. Remove the throttle body.
29   Remove the throttle body to bowl insulator gaskets.
30   Remove the pump rod from the throttle lever by rotating the rod out of the primary lever.
31   Further dismantling is not recommended. If it is essential to remove the idle mixture needles, pry out the plastic limiter caps then count the number of turns to bottom the needles and fit replacements in exactly the same position. New limiter caps should be fitted after running adjustments have been made.
32   Clean all metal parts in a suitable solvent. Do not immerse rubber parts, plastic parts, the vacuum break assembly or the idle stop solenoid, or permanent damage will result, Do not probe the jets, but blow them through with clean dry compressed air. Examine all fixed and moving parts for cracks,

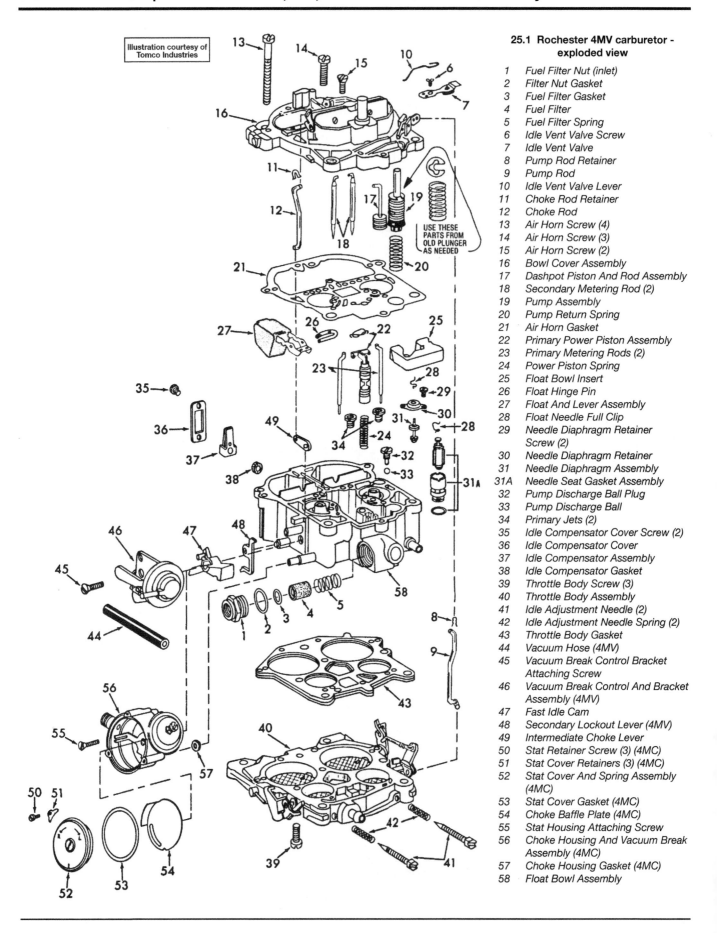

Illustration courtesy of
Tomco Industries

**25.1 Rochester 4MV carburetor -
exploded view**

1   Fuel Filter Nut (inlet)
2   Filter Nut Gasket
3   Fuel Filter Gasket
4   Fuel Filter
5   Fuel Filter Spring
6   Idle Vent Valve Screw
7   Idle Vent Valve
8   Pump Rod Retainer
9   Pump Rod
10   Idle Vent Valve Lever
11   Choke Rod Retainer
12   Choke Rod
13   Air Horn Screw (4)
14   Air Horn Screw (3)
15   Air Horn Screw (2)
16   Bowl Cover Assembly
17   Dashpot Piston And Rod Assembly
18   Secondary Metering Rod (2)
19   Pump Assembly
20   Pump Return Spring
21   Air Horn Gasket
22   Primary Power Piston Assembly
23   Primary Metering Rods (2)
24   Power Piston Spring
25   Float Bowl Insert
26   Float Hinge Pin
27   Float And Lever Assembly
28   Float Needle Full Clip
29   Needle Diaphragm Retainer
     Screw (2)
30   Needle Diaphragm Retainer
31   Needle Diaphragm Assembly
31A Needle Seat Gasket Assembly
32   Pump Discharge Ball Plug
33   Pump Discharge Ball
34   Primary Jets (2)
35   Idle Compensator Cover Screw (2)
36   Idle Compensator Cover
37   Idle Compensator Assembly
38   Idle Compensator Gasket
39   Throttle Body Screw (3)
40   Throttle Body Assembly
41   Idle Adjustment Needle (2)
42   Idle Adjustment Needle Spring (2)
43   Throttle Body Gasket
44   Vacuum Hose (4MV)
45   Vacuum Break Control Bracket
     Attaching Screw
46   Vacuum Break Control And Bracket
     Assembly (4MV)
47   Fast Idle Cam
48   Secondary Lockout Lever (4MV)
49   Intermediate Choke Lever
50   Stat Retainer Screw (3) (4MC)
51   Stat Cover Retainers (3) (4MC)
52   Stat Cover And Spring Assembly
     (4MC)
53   Stat Cover Gasket (4MC)
54   Choke Baffle Plate (4MC)
55   Stat Housing Attaching Screw
56   Choke Housing And Vacuum Break
     Assembly (4MC)
57   Choke Housing Gasket (4MC)
58   Float Bowl Assembly

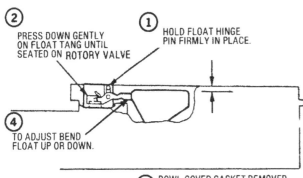

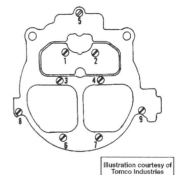

25.33a Float level adjustment details - Rochester 4MV carburetors

25.33b Air horn tightening sequence

26.1 The same procedure is used for setting the float level in both the front and rear float bowls of the Holley 4150 carburetor; first, remove the sight plug . . .

26.4 . . . then loosen the lock screw slightly to allow the adjustment nut to turn so you can raise or lower the float. Now turn the adjustment nut so the fuel level just comes up to and possibly trickles over the threads of the sight hole

27.1 Location of the secondary throttle valve stop screw (arrow) - Holley 4150 carburetor

distortion, wear and other damage; replace as necessary. Discard all gaskets and the fuel inlet filter.

33   Assembly is essentially the reverse of the removal procedure, but the following points should be noted:

a) If new idle mixture screws are used, and the original setting was not noted, install the screws finger-tight to seat them, then back off 4 full turns.

b) Having installed the float, measure from the top of the float bowl gasket surface (gasket not fitted) to the top of the float at a point 3/16 inch from the toe. Bend the f/oat up, or down, to obtain the specified dimension (see illustration).

c) Tighten the air horn retaining screws in the correct sequence (see illustration).

d) When connecting the pump lever to the upper pump rod, install the rod in the inner hole.

e) After reassembly, carry out all the settings and adjustments listed previously in this Chapter.

26 Holley 4150 carburetor - float adjustment

Refer to illustrations 26.1 and 16.4
Warning: *Gasoline is extremely flammable, so take extra precautions when you work on any part of the fuel system. See the* **Warning** *in Section 1.*

1   Remove the air cleaner then unscrew the fuel level sight plugs (see illustration). Place a rag under the exposed hole to catch any fuel that may spill out during the adjustment procedure.

2   Start the engine and allow it to idle.

3   With the vehicle on a level surface the fuel level should be within +/- 1/32 inch of the threads at the bottom of the sight plug port.

4   If adjustment is required, loosen the inlet needle lock screw and turn the adjusting nut clockwise to lower, or counter clockwise to raise, the fuel level (see illustration). Tighten the lock screw afterwards. (1/6 turn of the nut gives approximately 1/16 inch fuel level change).

5   Allow approximately one minute for the fuel level to stabilize, then recheck.

6   To ensure a proper secondary float level setting, accelerate the primary throttles slightly, and hand operate the secondary throttles.

7   Install the sight plugs and air cleaner on completion.

27 Holley 4150 carburetor - secondary throttle valve stop screw adjustment

Refer to illustration 27.1

1   Back off the adjustment screw until the throttle plates are fully closed (see illustration).

2   Rotate the adjustment screw until it just touches the throttle lever then rotate it an additional half-turn.

28 Holley 4150 carburetor - fast idle cam adjustment

1   Open the throttle slightly then close the choke plate, positioning the fast idle lever against the top step of the fast idle cam.

29.2 Measure the gap between the spring adjusting nut and the pump lever arm (arrow) - Holley 4150 carburetor

2   Adjust the fast idle to give the specified throttle opening on the idle transfer side of the carburetor. Adjust, if necessary, by bending the fast idle lever.

29   Holley 4150 carburetor - accelerator pump adjustment

Refer to illustration 29.2

1   Hold the throttle lever wide open (using a rubber band to keep it in place), then fully compress the pump lever (lever pressed down).
2   Measure the gap between the spring adjusting nut and the pump lever arm **(see illustration)**. Adjust, if necessary, to obtain a gap of 0.015 in. by turning the nut or screw while preventing the opposite end from turning.
3   After adjustment, close the throttle lever then partly open it again. Any movement of the throttle lever should be noticed at the operating lever spring end, indicating correct pump tip-in.

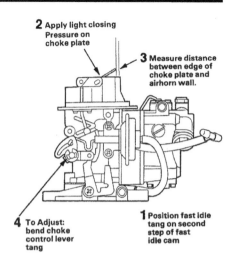

30.2 Choke unloader adjustment details - Holley 4150 carburetor

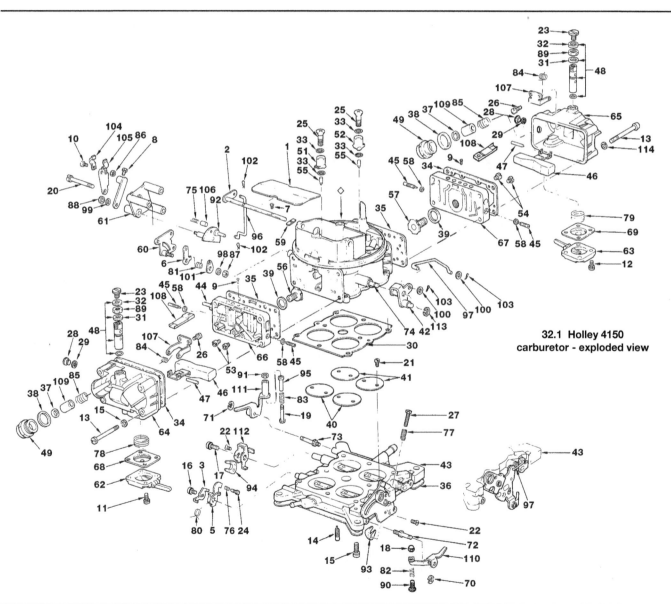

32.1 Holley 4150 carburetor - exploded view

30 Holley 4150 carburetor - choke unloader adjustment

Refer to illustration 30.2

1   Hold the throttle lever wide open (using a rubber band to keep it in place).
2   Hold the choke valve towards the closed position against the unloader tang of the throttle shaft. Measure the gap between the choke valve lower edge and main body (see illustration). Adjust, if necessary, to obtain the specified gap by bending the choke rod at the offset bend.

31 Holley 4150 carburetor - vacuum break adjustment

1   Hold the choke valve closed (using a rubber band to keep it in place).
2   Hold the vacuum break against the stop then measure the gap between the choke valve lower edge and the main body. Adjust, if necessary, to obtain the specified gap by bending the vacuum break link.

32 Holley 4150 carburetor - servicing

Refer to illustration 32.1

Warning: *Gasoline is extremely flammable, so take extra precautions when you work on any part of the fuel system. See the* **Warning** *in Section 1.*

1   When a carburetor fault develops after a considerable mileage, it is usually more economical to replace the complete unit rather than to replace individual components **(see illustration)**. However, if it is decided to strip and rebuild the unit, first obtain a repair kit which will contain all the necessary gaskets and other renewable items, and proceed in the following manner:
2   Initially loosen the fuel inlet fitting, fuel bowl sight plugs, and the needle and seat lock screws.
3   Remove the primary fuel bowl (4 screws), the metering body, splash shield and gasket.
4   Remove the secondary fuel bowl (4 screws), the metering body and gasket.
5   Disconnect the secondary throttle operating rod at the throttle lever, followed by the throttle operating assembly and gasket.
6   Disconnect the hose at the vacuum break.
7   Remove the throttle body to main body screws, the throttle body and the gasket.
8   Loosen the inlet needle and seat lock screw, then turn the adjusting nut counter-clockwise to remove the needle and seat assembly.
9   Remove the hinge pin retainer and slide out the float. Remove the spring if necessary.
10   Remove the sight plug and gasket, and the inlet fitting, fuel filter, spring and gaskets.
11   From the primary bowl remove the air vent valve assembly (where applicable) and the pump diaphragm screws, pump housing, diaphragm and spring. Check the pump inlet ball for damage and correct operation.
12   Using a wide-bladed screwdriver, remove the main metering jets. Remove the power valves using a 1-inch 12 point socket. From the primary side only, remove the idle mixture screws and seals.
13   To disassemble the secondary throttle operating assembly, remove the diaphragm cover, spring and diaphragm.
14   From the main body, remove the vacuum break retaining screws. Remove the assembly, disconnecting the link at the choke lever.
15   Remove the choke lever retaining clip followed by the lever itself and the fast idle cam.
16   Remove the pump discharge nozzle screw, nozzle and gasket, then invert the body to remove the discharge check valve.

Index Number	Part Name	Index Number	Part Name
1	Choke Plate	56	Power Valve Assy. Pri.
2	Choke Shaft & Lever Assy.	57	Power Valve Assy. or Plug Sec.
3	Fast Idle Pick-up Lever	58	Idle Needle Seal Pri. & Sec.
4	Choke Lever & Swivel Assy.	59	Choke Rod Seal
5	Fast Idle Cam Lever	60	Back-up Plate & Stud Assy.
6	Choke Rod Lever & Bushing Assy.	61	Fast Idle Cam Plate
7	Choke Plate Screw	62	Pump Cover Assy. Pri.
8	Choke Swivel Screw	63	Pump Cover Assy. Sec.
9	Drive Screw	64	Fuel Bowl & Plugs Assy. Pri.
10	Clamp Screw	65	Fuel Bowl & Plugs Assy. Sec.
11	Fuel Pump Cover Screw Pri.	66	Metering Body & Plugs Assy. Pri.
12	Fuel Pump Cover Screw Sec.	67	Metering Body & Plugs Assy. Sec.
13	Fuel Bowl Screw	68	Pump Diaphragm Assy. Pri.
14	Secondary Idle Adjusting Screw	69	Pump Diaphragm Assy. Sec.
15	Throttle Body Screw & L.W.	70	Pump Operating Lever Retainer Pri.
16	Fast Idle Cam Lever Screw & L.W.	71	Pump Operating Lever Retainer Sec.
17	Secondary Pump Cam Lever Screw & L.W.	72	Pump Lever Stud Pri.
18	Pump Lever Adjusting Screw Pri.	73	Pump Lever Stud Sec.
19	Pump Lever Adjusting Screw Sec.	74	Cam Follower Lever Stud
20	Fast Idle Cam Plate Screw & L.W.	75	Plunger Spring
21	Throttle Plate Screw Pri. & Sec.	76	Fast Idle Cam Lever Screw Spring
22	Pump Cam Lock Screw Pri. & Sec.	77	Throttle Stop Screw Spring
23	Fuel Valve Seat Lock Screw	78	Diaphragm Return Spring Pri.
24	Fast Idle Cam Lever Adj. Screw	79	Diaphragm Return Spring Sec.
25	Pump Discharge Nozzle Screw Pri. & Sec.	80	Fast Idle Cam Lever Spring
26	Float Shaft Bracket Screw & L.W. Pri. & Sec.	81	Choke Spring
27	Throttle Stop Screw	82	Pump Lever Adj. Screw Spring Pri.
28	Fuel Level Check Plug Pri. & Sec.	83	Pump Lever Adj. Screw Spring Sec.
29	Fuel Level Check Plug Gasket Pri. & Sec.	84	Float Spring Pri. & Sec.
30	Throttle Body Gasket	85	Fuel Inlet Filter Spring
31	Fuel Valve Seat Adjusting Nut Gasket	86	Bracket Clamp Screw Nut
32	Fuel Valve Seat Lock Screw Gasket	87	Choke Spring Nut
33	Pump Discharge Nozzle Gasket	88	Choke Lever Nut
34	Fuel Bowl Gasket Pri. & Sec.	89	Fuel Valve Seat Adjusting Nut Pri. & Sec.
35	Metering Body Gasket Pri. & Sec.	90	Pump Lever Adjusting Screw Nut Pri.
36	Flange Gasket	91	Pump Lever Adjusting Screw Nut Sec.
37	Fuel Inlet Filter Gasket	92	Fast Idle Cam & Shaft Assy.
38	Fuel Inlet Fitting Gasket	93	Pump Cam Pri.
39	Power Valve Gasket	94	Pump Cam Sec.
40	Throttle Plate Primary	95	Pump Operating Lever Screw Sleeve
41	Throttle Plate Secondary	96	Choke Rod
42	Cam Follower Lever Assy.	97	Secondary Connecting Rod
43	Throttle Body & Shaft Assy.	98	Choke Spring Nut L.W.
44	Spark Tube	99	Choke Control Lever L.W.
45	Idle Adjusting Needle Pri. & Sec.	100	Connecting Rod Washer
46	Float & Hinge Assy. Pri. & Sec.	101	Choke Spring Washer
47	Float Shaft Pri. & Sec.	102	Choke Rod Retainer
48	Fuel Inlet Needle & Seat Assy. Primary & Sec.	103	Cotter Pin
49	Fuel Inlet Fitting Primary	104	Choke Control Wire Bracket Clamp
50	Fuel Inlet Fitting Secondary	105	Choke Control Wire Bracket
51	Pump Discharge Nozzle Pri.	106	Fast Idle Cam Plunger
52	Pump Discharge Nozzle Sec.	107	Float Shaft Retaining Bracket Pri. & Sec.
53	Main Jet Pri.	108	Fuel Bowl Vent Baffle Pri. & Sec.
54	Main Jet Sec.	109	Fuel Inlet Filter Pri. & Sec.
55	Pump Discharge Needle Valve Pri. & Sec.	110	Pump Operating Lever Pri.
		111	Pump Operating Lever Sec.
		112	Pump Cam Lever Sec.
		113	E. Ring Retainer
		114	Fuel Bowl Screw Gasket

Parts list for Holley 4150 carburetor

17   If further dismantling is required, file off the staked ends of the shaft screws and remove them. Remove the choke rod (upwards, through the seal), followed by the seal. Remove the valve from the shaft slot and slide the shaft from the main body.

18   To disassemble the throttle body, if required, remove the pump operating lever assembly and disassemble the spring, bolt and nut.

19   Remove the idle speed screw and spring.

20   Remove the secondary throttle shaft diaphragm lever and the primary throttle shaft fast idle cam lever.

21   Remove the key and disconnect the secondary lockout throttle connecting link from the shaft levers.

22   File off the staked ends of the throttle plate attaching screws, then remove the screws and plates. Remove any burrs from the shafts and withdraw them out of the flange.

23   Remove the throttle lever accelerator pump cam and the vacuum break hose.

24   Clean all metal parts in a suitable cold solvent (this includes the choke rod seal - see paragraph 17). Do not immerse rubber parts, plastic parts (eg: secondary throttle shaft bushings and accelerator pump cam), vacuum break unit and other non-metallic parts. Do not probe the jets, but blow through them with clean, dry compressed air. Examine all fixed and moving parts for cracks, distortion, wear and other damage; replace as necessary. Discard all gaskets. Check the secondary throttle operating diaphragm by moving the diaphragm rod to the up position, plugging the vacuum passage opening and checking that the diaphragm holds upwards until the passage is unplugged.

25   Assembly is essentially the reverse of the removal procedure, but the following points should be noted:

a) *Throttle shaft plastic bushings should be rolled between the finger and thumb to help shape them.*

b) *When installing throttle valves, the identification numbers should be downwards (to the manifold side).*

c) *Install the idle speed screw to just contact the throttle lever, then turn 1-1/2 turns further.*

d) *Ensure that the choke valve can fall freely under its own weight*

e) *When installing the idle mixture screws, use new seals. Preliminary adjustment is made by screwing them in lightly to just seat, then screwing out one turn.*

f) *Adjust the floats by inverting the fuel bowls and turning the adjustable needle seat until the top of the float is the specified distance from the top of the fuel bowl.*

g) *On completion of assembly, install the carburetor and adjust the float, the secondary throttle stop valve, fast idle cam, accelerator pump, choke unloader and vacuum break, as necessary.*

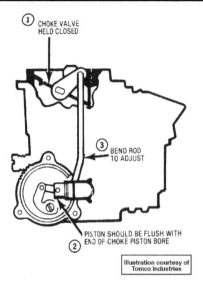

① CHOKE VALVE HELD CLOSED
③ BEND ROD TO ADJUST
PISTON SHOULD BE FLUSH WITH END OF CHOKE PISTON BORE ②

Illustration courtesy of Tomco Industries

36.1 Intermediate choke rod adjustment details - Rochester 2GC carburetor

33 Rochester 2GC carburetor - accelerator pump rod adjustment

1   Refer to the procedure given in Section 18 for the Rochester 2GV carburetor.

34 Rochester 2GC carburetor - fast idle cam adjustment

1   Turn the idle speed screw in until it just contacts the low step of the fast idle cam, then screw it in one more turn.

2   Place the idle speed screw on the second step of the cam, against the high step.

3   Check the dimension between the upper edge of the choke valve and the air horn wall.

4   If adjustment is necessary to obtain the specified dimension, bend the choke lever tang.

35 Rochester 2GC carburetor - choke unloader adjustment

1   Refer to the procedure given in Section 16 for the Rochester 2GV carburetor.

36 Rochester 2GC carburetor - intermediate choke rod adjustment

Refer to illustration 36.1

1   Remove three screws and retainers, and remove the thermostatic coil cover, gasket and inside baffle plate assembly.

2   Place the idle speed screw on the highest step of the fast idle cam.

3   Close the choke valve by pushing up on the intermediate choke lever.

ROTATE STAT COVER AGAINST SPRING TENSION. SET MARK ON COVER TO SPECIFIED POINT ON CHOKE HOUSING.

RICH LEAN

ALLOWABLE VARIATIONS - 2 NOTCHES EITHER WAY FROM INITIAL SETTING

Illustration courtesy of Tomco Industries

37.3 Automatic choke adjustment details - Rochester 2GC carburetor

4   The edge of the piston inside the choke housing should be flush with the end of the piston bore **(see illustration)**.

5   If necessary, bend the choke rod to adjust.

37 Rochester 2GC carburetor - automatic choke coil adjustment

Refer to illustration 37.3

1   Place the idle speed screw on the highest step of the fast idle cam.

2   Loosen the choke coil cover retaining screws.

3   Rotate the cover against the coil tension until the choke begins to close **(see illustration)**. Continue rotating until the index mark aligns with the specified point on the choke housing.

4   Tighten the choke cover retaining screws.

38 Rochester 2GC carburetor - vacuum break adjustment

1   Using a separate source of suction, seat the vacuum break diaphragm.

2   Cover the vacuum break bleed hole with a piece of masking tape.

3   Place the idle speed screw on the high step of the fast idle cam.

4   Hold the choke coil lever inside the choke housing towards the closed choke position.

5   Check the dimension between the upper edge of the choke valve and the air horn wall. If adjustment is required to obtain the specified dimension, bend the vacuum break rod.

6   Remove the masking tape on the vacuum unit bleed hole and reconnect the vacuum hose.

39 Rochester 2GC carburetor - servicing

Refer to illustration 39.1
Warning: *Gasoline is extremely flammable,*

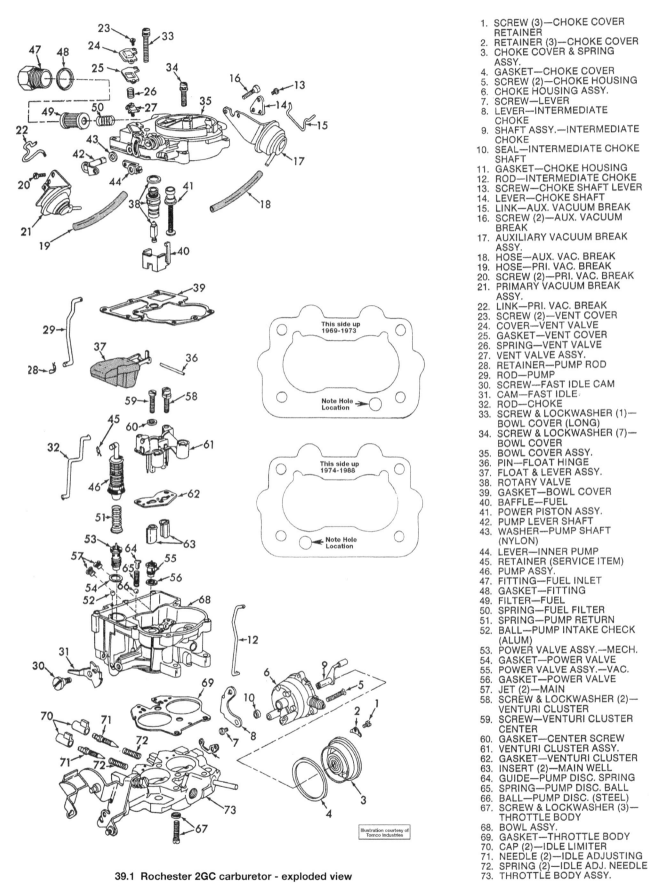

1. SCREW (3)—CHOKE COVER RETAINER
2. RETAINER (3)—CHOKE COVER
3. CHOKE COVER & SPRING ASSY.
4. GASKET—CHOKE COVER
5. SCREW (2)—CHOKE HOUSING
6. CHOKE HOUSING ASSY.
7. SCREW—LEVER
8. LEVER—INTERMEDIATE CHOKE
9. SHAFT ASSY.—INTERMEDIATE CHOKE
10. SEAL—INTERMEDIATE CHOKE SHAFT
11. GASKET—CHOKE HOUSING
12. ROD—INTERMEDIATE CHOKE
13. SCREW—CHOKE SHAFT LEVER
14. LEVER—CHOKE SHAFT
15. LINK—AUX. VACUUM BREAK
16. SCREW (2)—AUX. VACUUM BREAK
17. AUXILIARY VACUUM BREAK ASSY.
18. HOSE—AUX. VAC. BREAK
19. HOSE—PRI. VAC. BREAK
20. SCREW (2)—PRI. VAC. BREAK
21. PRIMARY VACUUM BREAK ASSY.
22. LINK—PRI. VAC. BREAK
23. SCREW (2)—VENT COVER
24. COVER—VENT VALVE
25. GASKET—VENT COVER
26. SPRING—VENT VALVE
27. VENT VALVE ASSY.
28. RETAINER—PUMP ROD
29. ROD—PUMP
30. SCREW—FAST IDLE CAM
31. CAM—FAST IDLE
32. ROD—CHOKE
33. SCREW & LOCKWASHER (1)—BOWL COVER (LONG)
34. SCREW & LOCKWASHER (7)—BOWL COVER
35. BOWL COVER ASSY.
36. PIN—FLOAT HINGE
37. FLOAT & LEVER ASSY.
38. ROTARY VALVE
39. GASKET—BOWL COVER
40. BAFFLE—FUEL
41. POWER PISTON ASSY.
42. PUMP LEVER SHAFT
43. WASHER—PUMP SHAFT (NYLON)
44. LEVER—INNER PUMP
45. RETAINER (SERVICE ITEM)
46. PUMP ASSY.
47. FITTING—FUEL INLET
48. GASKET—FITTING
49. FILTER—FUEL
50. SPRING—FUEL FILTER
51. SPRING—PUMP RETURN
52. BALL—PUMP INTAKE CHECK (ALUM)
53. POWER VALVE ASSY.—MECH.
54. GASKET—POWER VALVE
55. POWER VALVE ASSY.—VAC.
56. GASKET—POWER VALVE
57. JET (2)—MAIN
58. SCREW & LOCKWASHER (2)—VENTURI CLUSTER
59. SCREW—VENTURI CLUSTER CENTER
60. GASKET—CENTER SCREW
61. VENTURI CLUSTER ASSY.
62. GASKET—VENTURI CLUSTER
63. INSERT (2)—MAIN WELL
64. GUIDE—PUMP DISC. SPRING
65. SPRING—PUMP DISC. BALL
66. BALL—PUMP DISC. (STEEL)
67. SCREW & LOCKWASHER (3)—THROTTLE BODY
68. BOWL ASSY.
69. GASKET—THROTTLE BODY
70. CAP (2)—IDLE LIMITER
71. NEEDLE (2)—IDLE ADJUSTING
72. SPRING (2)—IDLE ADJ. NEEDLE
73. THROTTLE BODY ASSY.

This side up 1969-1973

Note Hole Location

This side up 1974-1988

Note Hole Location

Illustration courtesy of Tomco Industries

39.1 Rochester 2GC carburetor - exploded view

so take extra precautions when you work on any part of the fuel system. See the **Warning** in Section 1.

1   When a carburetor fault develops after a considerable mileage, it is usually more economical to replace the complete unit rather than to completely disassemble it and replace individual components **(see illustration)**. However, if it is decided to strip and rebuild the unit, first obtain a repair kit which will contain all the necessary gaskets and other renewable items and proceed in the following manner.

2   Remove the fuel inlet filter nut, gasket, filter and spring.

3   Disconnect the lower end of the pump rod from the throttle lever.

4   Remove the upper end of the pump rod from the pump lever.

5   Remove the vacuum break diaphragm hose.

6   Remove the vacuum break diaphragm assembly (2 screws) and disconnect it from the lever on the end of the choke shaft.

7   Remove the vacuum break lever from the end of the choke shaft (1 screw), then remove the intermediate choke rod from the vacuum break lever on the coil housing.

8   Remove the fast idle cam retaining screw, then remove the cam from the end of the choke rod. The upper end of the rod cannot be removed until the air horn has been removed from the float bowl.

9   Remove the 8 air horn attaching screws and lockwashers, then lift off the air horn.

10   Remove the float hinge pin and lift off the float. The float needle and pull clip (where applicable) can now be removed from the float arm.

11   Unscrew the float needle seat and remove the gasket.

12   Depress the power piston and release it to allow it to snap free.

13   Remove the pump plunger assembly and inner pump lever from the shaft by loosening the set screws on the inner lever.

14   If the pump assembly is to be overhauled, break off the swaged or flattened end of the pump plunger stem; the service pump uses a grooved pump plunger stem and retaining clip. After removing the inner pump lever and pump assembly, remove the outer lever and shaft assembly from the air horn. Remove the plastic washer from the pump plunger shaft.

15   Remove the gasket from the air horn.

16   Remove the fuel inlet baffle (next to the needle seat).

17   Taking care not to bend the choke shaft, remove the choke valve. The ends of the retaining screws may need to be filed off to permit removal.

18   Remove the choke valve shaft. Remove the fast idle cam rod and lever from the shaft.

19   Remove the pump plunger return spring from the float bowl pump well, then invert the bowl and remove the check ball.

20   Remove the main metering jets, power valve and gasket from inside the float bowl.

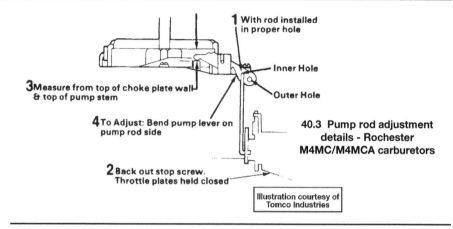

40.3 Pump rod adjustment details - Rochester M4MC/M4MCA carburetors

Illustration courtesy of Tomco Industries

21   Remove the three screws which retain the venturi cluster; remove the cluster and gasket.

22   Use needle-nosed pliers to remove the pump discharge spring retainer, then remove the spring and check ball from the discharge passage.

23   Remove the three large throttle body to bowl attaching screws and lockwashers. Remove the throttle body and gasket.

24   Remove the thermostatic choke coil cover (3 screws arid retainers) and gasket from the choke housing. Do not remove the cap baffle from beneath the coil cover.

25   Remove the choke housing baffle plate.

26   From inside the choke housing, remove the 2 attaching screws; remove the housing and gasket.

27   Remove the screw from the end of the intermediate choke shaft, then remove the choke lever from the shaft. Remove the inner choke coil lever and shaft assembly from the choke housing, followed by the rubber dust seal.

28   Further dismantling is not recommended, particularly with regard to the throttle valves or shaft since it may be impossible to reassemble the valves correctly in relation to the idle discharge orifices. If it is essential to remove the idle mixture needles, pry out the plastic limiter caps then count the number of turns to bottom the needles and fit replacements in exactly the same position. New limiter caps should be fitted after running adjustments have been made.

29   Clean all metal parts in carburetor cleaner. Do not immerse rubber parts, plastic parts. diaphragm assemblies or pump plungers, as permanent damage will result. Do not probe the jets, but blow through with clean, dry compressed air. Examine all fixed and moving parts for cracks, distortion, wear and other damage; replace as necessary, Discard all gaskets and the fuel inlet filter.

30   Assembly is essentially the reverse of the removal procedure, but the following points should be noted:

a) If new idle mixture screws were used, and the original setting was not noted, install the screws finger-tight to seat them, then back off 4 full turns.

b) When installing the rubber dust seal in the choke housing cavity, the seal lip faces towards the carburetor after the housing is installed.

c) Before installing the choke cover coil and baffle plate assembly carry out the Intermediate Choke Rod adjustment (Section 36).

d) When installing the choke coil and cover assembly, the end of the coil must be below the plastic tang on the inner choke housing lever. At this stage carry out the Automatic Choke Coil adjustment (Section 37).

e) When installing the venturi cluster, ensure that a gasket is fitted on the center screw.

f) Install the choke valve with the letters 'RP' or the part number, facing upwards.

g) Carry out float level and float drop checks as specified for the 2GV carburetor in Section 19.

h) Install and tighten the air horn screws as shown for 2GV carburetors in **(see illustration 19.25c)**.

j) After reassembly, carry out the relevant settings and adjustments listed previously in this Chapter.

40   Rochester M4MC/M4MCA carburetor - pump rod adjustment

Refer to illustration 40.3

1   With the fast idle cam follower off the steps of the fast idle cam, back out the idle speed screw until the throttle valves are completely closed in the bore. Make sure that the secondary actuating rod is not restricting movement; bend the secondary closing tang if necessary then readjust it after pump adjustment.

2   Place the pump rod in the specified hole in the lever.

3   Measure from the top of the choke valve wall (next to the vent stack) to the top of the pump stem **(see illustration)**.

4   If necessary, adjust to obtain the specified dimension by bending the lever while supporting it with a screwdriver.

5   Adjust the idle speed (Section 7).

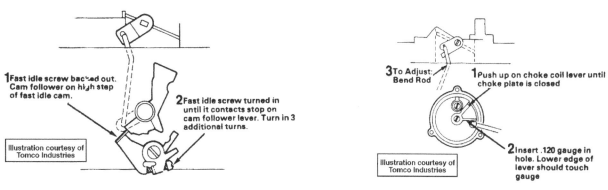

41.1 Fast idle setting details -
Rochester M4MC/M4MCA carburetors

42.4 Choke coil lever adjustment details -
Rochester M4MC/M4MCA carburetors

41   Rochester M4MC/M4MCA carburetor - fast idle adjustment

Refer to illustration 41.1

1   Place the cam follower lever on the highest step of the fast idle cam **(see illustration)**.
2   Turn the fast idle screw out until the primary throttle valves are closed.
3   Turn in the fast idle screw to contact the lever then screw in a further 3 full turns.
4   Refer to the tune-up label and readjust if necessary to obtain the correct idle speed.

42   Rochester M4MC/M4MCA carburetor - choke coil lever adjustment

Refer to illustration 42.4

1   Loosen - the 3 retaining screws and remove the cover and coil assembly from the choke housing.
2   Push up on the thermostatic coil tang (counterclockwise) until the choke valve is closed.
3   Check that the choke rod is at the bottom of the slot in the choke lever.
4   Insert a plug gauge (an unmarked drill shank is suitable) of the specified size in the hole in the choke housing **(see illustration)**.
5   The lower edge of the choke coil lever should just contact the side of the plug gauge.
6   If necessary, bend the choke rod at the point shown to adjust.

43   Rochester M4MC/M4MCA carburetor - fast idle cam (choke rod) adjustment

1977 and earlier models

Refer to illustration 43.2

1   Turn the fast idle screw in until it contacts the fast idle cam follower lever, then turn in 3 full turns more.
2   Place the lever on the second step of

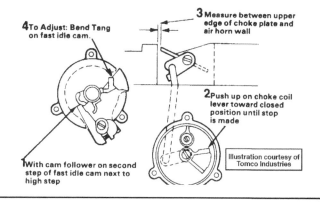

43.2 Choke rod adjustment details - Rochester M4MC/M4MCA carburetors

the fast idle cam against the rise of the high step **(see illustration)**.
3   Push upwards on the choke coil lever inside the housing to close the choke valve.
4   Measure between the upper edge of the choke valve and the air horn wall.
5   If necessary, bend the tang on the fast idle cam to adjust, but ensure that the tang lies against the cam after bending.
6   Recheck the fast idle adjustment (Section 41).

1978 and later models

Refer to illustration 43.8
Note: *This procedure requires a special choke angle gauge, which can be obtained at some auto parts stores.*
7   Be sure the choke coil lever and fast idle adjustments are correct.

8   With the choke plate completely closed, place an appropriate choke valve angle measuring gauge onto the choke plate and rotate the degree wheel so the zero is opposite the pointer **(see illustration)**. **Note:** *If the carburetor is off of the engine, be sure it is securely supported so it will not move once the gauge is in place.*
9   With the choke valve completely closed, place the magnet squarely on top of the choke valve.
10   Rotate the leveling bubble until it is centered.
11   Rotate the scale so that the number of degrees specified for adjustment (see Specifications) is opposite the pointer.
12   Place the cam follower on the second step of the cam so it is against the rise of the high step.

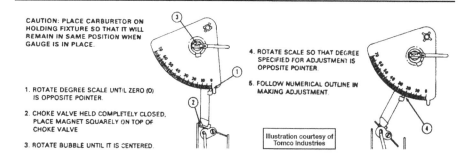

43.8 Typical choke plate angle gauge details

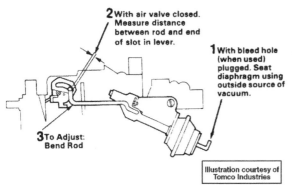

44.1 Air valve rod adjustment details - Rochester M4MC/M4MCA carburetors

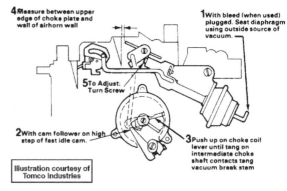

45.3 Front vacuum break adjustment details - Rochester M4MC/M4MCA carburetors

13   Push upward on the choke coil lever or the vacuum break lever tang to close the choke, and secure it in this position with a rubber band.

14   If adjustment is needed, bend the tang on the fast idle cam until the leveling bubble is centered.

15   Remove the gauge.

44 Rochester M4MC/M4MCA carburetor - air valve dashpot adjustment

Refer to illustration 44.1

1   Using a hand-held vacuum pump, seat the front vacuum break diaphragm **(see illustration)**.

2   Ensure that the air valves are completely closed then measure between the air valve dashpot and the end of the slot in the air valve lever.

3   Bend the air valve dashpot rod at the point shown, if adjustment is necessary.

45 Rochester M4MC/M4MCA carburetor - front vacuum break adjustment

1976 and earlier models

Refer to illustration 45.3

1   Loosen the 3 retaining screws and remove the choke coil cover and coil assembly from the choke housing.

2   Place the cam follower lever on the highest step of the fast idle cam.

3   Using a hand-held vacuum pump, seat the diaphragm unit **(see illustration)**.

4   Push up on the inside choke coil lever until the tang on the vacuum break lever contacts the tang on the plunger.

5   Measure between the upper edge of the choke valve and the inside of the air horn wall.

6   Turn the adjustment screw on the vacuum break plunger to obtain the specified dimension.

7   Install the vacuum hose on completion.

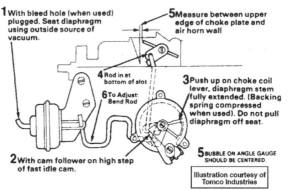

46.1 Rear vacuum break adjustment details - Rochester M4MC/M4MCA carburetors

1977 and later models

Note: *This procedure requires a special choke angle gauge, which can be obtained at some auto parts stores.*

8   Insert an appropriate choke valve angle gauge into position on the choke plate **(see illustration 43.8)**. **Note:** *If the carburetor is off of the engine, be sure it is securely supported so it will not move during the adjustment.*

9   Rotate the degree scale until the zero is opposite the pointer.

10   With the choke valve completely closed, place the magnet squarely on top of the choke plate.

11   Rotate the bubble until it is centered.

12   Rotate the scale so that the number of degrees specified for adjustment (see Specifications) is opposite the pointer.

13   Attach a vacuum pump to the vacuum diaphragm unit and apply enough vacuum to seat it. **Note:** *On models that use an air bleed cover the end cover with masking tape to cut off the bleed. On delay models with an air bleed, the rubber cover over the filter element must be removed before applying the tape.*

14   Push upward on the choke coil lever or vacuum break lever tang, so the choke valve is in the closed position and secure it in that position with a rubber band.

15   If necessary, turn the adjusting screw until the leveling bubble is centered.

16   Remove the gauge. On air bleed models, also remove the masking tape.

46 Rochester M4MC/M4MCA carburetor - rear vacuum break adjustment

Refer to illustration 46.1

1   Initially follow the procedure of paragraphs 1 thru 3 in the previous Section, but additionally plug the bleed hose in the end cover of the vacuum break unit using adhesive tape **(see illustration)**.

2   Push up on the choke coil lever inside the choke housing towards the closed choke position.

3   With the choke rod in the bottom of the slot in the choke lever, measure between the upper edge of the choke valve and the air horn wall.

4   If necessary, bend the vacuum break rod to obtain the specified dimension.

5   On completion, remove the adhesive tape and install the vacuum hose.

47 Rochester M4MC/M4MCA carburetor - automatic choke coil adjustment

1   Install the choke coil and cover assembly with a gasket between the cover and housing. The tang in the coil must be installed in the slot inside the choke coil lever pick-up arm.

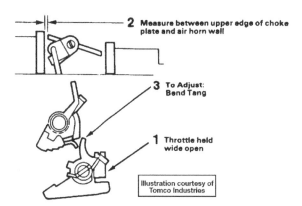

48.3 Choke unloader adjustment details -
Rochester M4MC/M4MCA carburetors

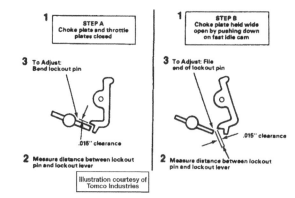

49.1 Secondary throttle lockout adjustment details -
Rochester M4MC/M4MCA carburetors

2   Place the fast idle cam follower on the highest step of the fast idle cam then rotate the cover counterclockwise until the choke just closes.
3   Align the index mark on the cover with the specified point on the choke housing then tighten the retaining screws.

48   Rochester M4MC/M4MCA carburetor - unloader adjustment

1976 and earlier models

Refer to illustration 48.3
1   Adjust the choke coil, as described in the previous Section.
2   Hold the throttle valves wide open and the choke plates fully closed. A rubber band can be used on the tang of the intermediate choke lever if the engine is warm.
3   Measure between the upper edge of the choke valve and the air horn wall **(see illustration)**.
4   If adjustment is necessary, bend the tang on the fast idle lever to obtain the specified dimension. Ensure that the tang on the fast idle cam lever contacts the center point of the fast idle cam after adjustment.

1978 and later models

Note: *This procedure requires a special choke angle gauge, which can be obtained at some auto parts stores.*
5   Insert an appropriate choke valve angle gauge into position on the choke plate **(see illustration 43.8)**. Note: *If the carburetor is off of the engine, be sure it is securely supported so it will not move during the adjustment.*
6   Rotate the degree scale until the zero is opposite the pointer.
7   With the choke valve completely closed, place the magnet squarely on top of the choke valve.
8   Rotate the bubble until it is centered.
9   Rotate the scale so that the number of degrees specified for adjustment (see Specifications) is opposite the pointer.
10   Install the choke thermostatic cover and

coil assembly in the housing, and align the index mark with the specified point on the housing.
11   Hold the primary throttle valves wide open.
12   Close the choke valve by pushing up on the tang in the vacuum break lever, and secure it in that position with a rubber band.
13   If adjustment is necessary, bend the tang on the fast idle lever until the leveling bubble is centered.
14   Remove the gauge.

49   Rochester M4MC/M4MCA carburetor - secondary throttle valve lock-out adjustment

Refer to illustration 49.1

Lock-out lever clearance

1   Hold the choke valves and secondary lockout valves closed then measure the clearance between the lock-out pin and lock-out lever **(see illustration)**.
2   If adjustment is necessary, bend the lock-out pin to obtain the specified clearance.

Opening clearance

3   Push down on the tail of the fast idle cam to hold the choke wide open.
4   Hold the secondary throttle valves partly

open then measure between the end of the lock-out pin and the toe of the lock-out lever **(see illustration 49.1)**.
5   If adjustment is necessary, file the end of the lock-out pin but ensure that no burrs remain afterwards.

50   Rochester M4MC/M4MCA carburetor - secondary closing adjustment

Refer to illustration 50.3
1   Adjust the engine idle speed, as described previously in this Chapter.
2   Hold the choke valve wide open with the cam follower lever off the steps of the fast idle cam.
3   Measure the clearance between the slot in the secondary throttle valve pick-up lever and the secondary actuating rod **(see illustration)**.
4   If adjustment is necessary, bend the secondary closing tang on the primary throttle lever to obtain the specified clearance.

51   Rochester M4MC/M4MCA carburetor - secondary opening adjustment

Refer to illustration 51.1
1   Lightly open the primary throttle lever

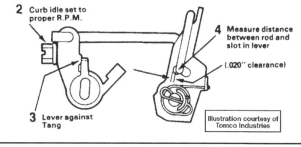

50.3 Secondary closing adjustment details - Rochester M4MC/M4MCA carburetors

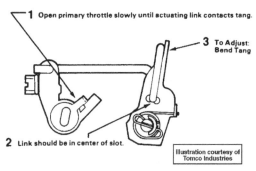

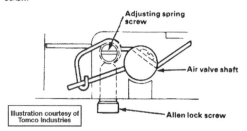

51.1 Secondary throttle opening adjustment details - Rochester M4MC/M4MCA carburetors

52.2 Air valve spring adjustment details - Rochester M4MC/M4MCA carburetors

until the link just contacts the tang on the secondary lever **(see illustration)**.

2   Bend the tang on the secondary lever, if necessary, to position the link in the center of the secondary lever slot.

52 Rochester M4MC/M4MCA carburetor - air valve spring wind-up adjustment

Refer to illustration 52.2

1   Remove the front vacuum break diaphragm unit and the air valve dashpot rod.

2   Loosen the lock screw then turn the tension adjusting screw counterclockwise until the air valve is partly open **(see illustration)**.

3   Hold the air valve closed then turn the tension adjusting screw clockwise the specified number of turns after the spring contacts the pin.

4   Tighten the lock screw and install the air valve dashpot rod, and the front vacuum break diaphragm unit and bracket.

53 Rochester M4MC/M4MCA carburetor - servicing

Refer to illustrations 53.6, 53.7, 53.8, 53.12 and 53.34

Warning: *Gasoline is extremely flammable, so take extra precautions when you work on any part of the fuel system. See the* **Warning** *in Section 1.*

Note: *This procedure also applies to the Rochester M2MC carburetor. It is essentially half of an M4MC carburetor; simply ignore the differences in the text.*

1   When a carburetor fault develops after a considerable mileage, it is usually more economical to replace the complete unit rather than to completely disassemble it and replace individual components **(see illustration 25.1)**. However, if it is decided to strip and rebuild the unit, first obtain a repair kit which will contain all the necessary gaskets and other renewable items, and proceed in the following manner.

2   If the carburetor has an idle stop solenoid, remove the bracket retaining screws and lift away the solenoid and bracket assembly.

3   Remove the upper choke lever from the end of the choke shaft (1 screw) then rotate the lever to remove it, to disengage it from the choke rod.

4   Remove the choke rod from the lower lever inside the float bowl by holding the lever outwards with a small screwdriver and twisting the rod counter clockwise.

5   Remove the vacuum hose from the front vacuum break unit.

6   Remove the small screw at the top of the metering rod hanger, and remove the secondary metering rods and hanger **(see illustration)**.

7   Using a suitable drift, drive the small pump lever pivot roll pin inwards to permit removal of the lever **(see illustration)**.

8   Remove 2 long screws, 5 short screws and 2 countersunk head screws to detach the air horn from the float bowl **(see illustration)**. Remove the secondary air baffle deflector (where applicable) from beneath the 2 center air horn screws.

9   Remove the float bowl but leave the gasket in position at this stage. Do not attempt to remove the small tubes protruding from the air horn.

10   Remove the front vacuum break bracket screws and lift off the unit. Detach the air valve dashpot rod from the diaphragm assembly and the air valve lever.

11   If considered necessary, remove the staked choke valve attaching screws then remove the choke valve and shaft from the air horn. Do not remove the air valve and the air valve shaft. The air valve closing spring or center plastic cam can be replaced by following the instructions in the appropriate repair kit.

12   Remove the air horn gasket from the float bowl taking care not to distort the springs holding the main metering rods **(see illustration)**.

13   Remove the pump plunger and pump return spring from the pump well.

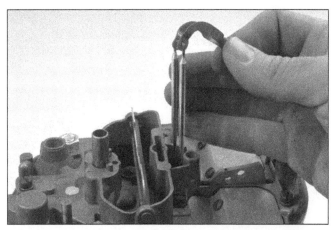

53.6 Remove the screw and lift out the secondary metering rods and hanger

53.7 Using a very small punch and a hammer, drive the roll pin through the lever just far enough to lift the lever from the air horn, then disconnect the lever from the rod

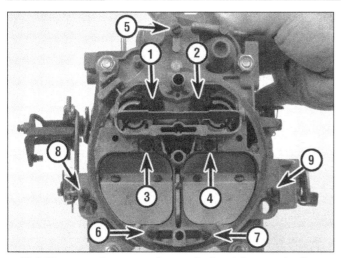

53.8 These are the screws that retain the air horn to the float bowl (the numbers indicate the recommended tightening sequence)

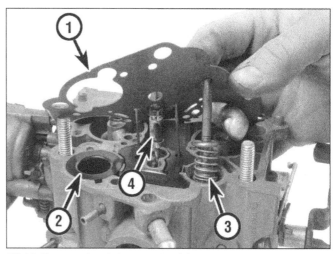

53.12 Remove the air horn gasket (1), the plastic cup, if equipped (2), and the accelerator pump assembly (3). The power piston (4) and the metering rods can be removed with the gasket

14 Depress the power piston stem and allow it to snap free, withdrawing the metering rods with it. Remove the power piston spring from the well.

15 Taking care to prevent distortion, remove the metering rods from the power piston by disconnecting the tension springs then rotating the rods.

16 Remove the plastic filler block over the float valve then remove the float assembly and needle by pulling up on the pin. Remove the needle seat and gasket.

17 Remove the 2 cover screws and carefully lift out the metering rod and filler spool (metering rod and aneroid on M4MCA carburetors) from the float bowl. **Note:** *The adjustable part throttle (A.P.T.) metering rod assembly is extremely fragile and must not be interfered with.* If replacement is necessary, refer to paragraphs 35 thru 44.

18 Remove the primary main metering jets. Do not attempt to remove the A.P.T. metering jet or secondary metering orifice plates.

19 Remove the pump discharge check ball retainer and the ball.

20 Remove the rear vacuum break hose and the bracket retaining screws. Remove the vacuum break rod from the slot in the plunger head.

21 Press down on the fast idle cam and remove the vacuum break rod. Move the end of the rod away from the float bowl, then disengage the rod from the hole in the intermediate lever.

22 Remove the choke cover attaching screws and retainers. Pull off the cover and remove the gasket.

23 Remove the choke housing assembly from the float bowl by removing the retaining screw and washer.

24 Remove the secondary throttle valve lock-out lever from the float bowl.

25 Remove the lower choke lever by inverting the float bowl.

26 Remove the plastic tube seal from the choke housing.

27 If it is necessary to remove the intermediate choke shaft from the choke housing, remove the coil lever retaining screw and withdrew the lever. Slide out the shaft and (if necessary), remove the fast idle cam.

28 Remove the fuel inlet filter nut, gasket and filter from the float bowl.

29 If necessary, remove the pump well fill slot baffle and the secondary air baffle.

30 Remove the throttle body attaching screws and lift off the float bowl. Remove the insulator gasket.

31 Remove the pump rod from the lever on the throttle body.

32 If it is essential to remove the idle mixture needles, pry out the plastic limiter caps then count the number of turns to bottom the needles and fit replacements in exactly the same position. New limiter caps should be fitted after running adjustments have been made.

33 Clean all metal parts in a suitable cold solvent. Do not immerse rubber parts, plastic parts, pump plungers, filler spools or aneroids, or vacuum breaks. If the choke housing is to be immersed, remove the cup seal from inside the choke housing shaft hole. If the bowl is to be immersed remove the cup seal from the plastic insert; do not attempt to remove the plastic insert. Do not probe the jets, but blow through with clean, dry compressed air. Examine all fixed and moving parts for cracks, distortion, wear and

other damage; replace as necessary. Discard all gaskets and the fuel inlet filter.

34 Assembly is essentially the reverse of the removal procedure, but the following points should be noted:

a) *If new idle mixture screws were used, and the original setting was not noted, install the screws finger-tight to seat them, then back off 4 full turns.*

b) *The lip on the plastic insert cup seal (on the side of the float bowl) faces outwards.*

c) *The lip on the inside choke housing shaft hole cup seal faces inwards towards the housing.*

d) *When installing the assembled choke body, install the choke rod lever into the cavity in the float bowl. Install the plastic tube seal into the housing cavity before installing the housing. Ensure that the intermediate choke shaft engages into the lower choke lever. The choke coil is installed at the last stage of assembly.*

e) *Where applicable, the notches on the secondary float bowl air baffle are towards the top, and the top edge of the baffle must be flush with the bowl casting.*

f) *To adjust the float, hold the retainer firmly in place and push down lightly against the needle. Measure from the top of the float bowl casting (air horn gasket removed) to a point on the top of*

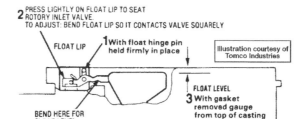

53.34 Float adjustment details - Rochester M4MC/M4MCA carburetors

the float 1/16 inch back from the toe (1976 and earlier models) or 3/16 inch back from the toe (1977 and later models). Bend the float arm to obtain the specified dimension by pushing on the pontoon (see illustration).

g) *Tighten the air horn screws in the order shown in illustration 53.8.*

h) *On completion of assembly, adjust the front and rear vacuum breaks, fast idle cam (choke rod), choke coil lever and automatic choke coil.*

A.P.T. metering rod replacement

35   Replacement of the metering rod must only be carried out if the assembly is damaged or the aneroid has failed.

36   Lightly scribe the cover to record the position of the adjusting screw slot.

37   Remove the cover screws then carefully lift out the rod and cover assembly.

38   Hold the assembly upright then turn the adjusting screw counterclockwise, counting the number of turns until the metering-rod bottoms in the cover.

39   Remove the E-clip from the threaded end of the rod then turn the rod clockwise until it disengages from the cover.

40   Install the tension spring on the replacement metering rod assembly and screw the rod and spring assembly into the cover until the assembly bottoms.

41   Turn the adjusting screw clockwise the number of turns noted at paragraph 38.

42   Install the E-clip. **Note:** *It will not matter if the scribed line (paragraph 36) does not align exactly provided that the assembly sequence has been followed.*

43   Carefully install the cover and metering rod assembly onto the float bowl, aligning the tab on the cover assembly with the float bowl slot closest to the fuel inlet nut.

44   Install the cover attaching screws and nut.

54   Fuel tank - removal and installation

Refer to illustration 54.6

Warning: *Gasoline is extremely flammable, so take extra precautions when you work on any part of the fuel system. See the **Warning** in Section 1.*

Note: *For the sake of safety and convenience, the fuel tank should be empty before removal if at all possible. If circumstances permit, run the tank dry, or almost dry, of fuel by normal vehicle usage. If it is necessary to siphon fuel fromm the tank, use a siphoning kit, available at most auto parts stores.*

1   Disconnect the battery ground cable.

2   Disconnect the meter wire at the rear harness connector. Where applicable, push out the grommet in the trunk compartment floor pan and feed the wire through.

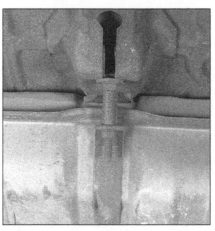

54.6 The fuel tank its retained by two straps like this. Remove the nuts and allow the straps to pivot down

3   Raise the vehicle and support it securely on jackstands.

4   Disconnect the fuel hose at the gauge unit fitting.

5   Disconnect the gauge unit ground wire.

6   Remove the tank retaining strap bolts and lower the tank carefully **(see illustration)**.

7   Installation is the reverse of the removal procedure.

55   Evaporation control system (ECS) - description and maintenance

1   The system is designed to minimize the escape of fuel vapor from the fuel tank and carburetor bowl to atmosphere.

2   The vapor is retained in a canister filled with carbon. The canister is purged by manifold vacuum (when the engine is operating) and the vapor is burned in the combustion chambers. The canister is located at the front left side of the engine compartment.

3   Check the connecting hoses of the system regularly both with respect to condition and security of connections.

4   Every 24 months or 24, 000 miles the carbon canister filter must be replaced, To do this, raise the vehicle, note the installed position of the hoses, disconnect them, loosen the canister retaining clamps and remove the canister.

5   Remove the old filter and fit the new one to the lower end of the canister **(see illustration)**.

6   The canister incorporates a purge valve which can be renewed if necessary by first disconnecting the hoses from the valve and then prying off the cap.

7   Remove the cap, diaphragm, spring retainer and spring.

8   The separator is mounted as an independent unit, and can be removed by disconnecting the lines and removing the retaining screw.

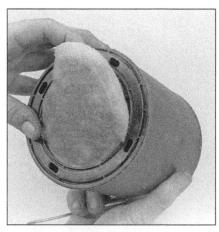

55.5 To replace the canister filter, peel it out and install a new one

56   Inlet and exhaust manifolds - removal and installation

Refer to Chapter 1 for details of these operations.

57   Exhaust system - removal and installation

Muffler and tailpipe

1   Raise the vehicle and support it securely on jackstands.

2   Single exhaust: remove the muffler and tailpipe by releasing the clamp.

3   Dual exhaust: remove the appropriate muffler and tailpipe by either cutting it off or releasing the clamp.

4   Install replacements by aligning the exhaust pipes with the slot in the muffler.

5   Attach the tailpipes to the hangers, then tighten all clamps and attachments. Make sure that at least 3/4 inch clearance exists between the floorpan and any part of the exhaust.

6   Lower the vehicle to the ground.

58   Catalytic converter - servicing

Warning: *The catalytic converter can remain hot for several hours after vehicle operation. Allow it to cool completely before inspection or servicing.*

The catalytic converter fitted to the exhaust system of 1975 and later models can be removed by unbolting at the front and rear after lifting the vehicle for access, When replacing, the use of new nuts, bolts and washers is recommended.

59   Transmission controlled spark (T.C.S.) system - description and maintenance

1   This system is installed on many models in the Nova range from 1970 onwards, but

varies in type according to the year of manufacture.

2   Vacuum advance is controlled by a solenoid-operated switch which is actuated by a transmission switch. When the switch is in operation, the vacuum normally applied to the distributor is vented to atmosphere and its advance characteristic is obtained solely by the mechanical counter weights.

3   A thermostatically controlled coolant temperature switch is wired into the solenoid circuit to prevent vacuum cut off to the distributor at engine temperatures below 93°F.

4   In order to compensate for the retarded spark, and possibly engine run-on (dieseling) when the engine is switched off, an idle stop solenoid is provided on many carburetors to permit the throttle valve to close further than the normally (slightly open) idling position of the valve plate.

5   1971 models employ a solenoid vacuum switch to prevent run-on (dieseling) when the engine is switched off. This solenoid alters the carburetor valve plate closed position in a similar manner to the idle stop solenoid referred to in paragraph 4. Air-conditioned models with automatic transmission have a semi-conductor device which operates the air conditioning compressor for approximately 3 seconds to provide an additional engine braking load, The T.C.S. system used for 1971 models is also known as the Combination Emission Control (C.E.C.) system.

6   After a period of approximately 20 seconds (1972 models onwards) or 15 seconds (1971 models), a time relay operates to open the advance unit to atmosphere. This is only effective in low gear with an engine temperature above 93°F. Where the vehicle is not started within this time, the ignition must be switched off to allow the heat operated time relay to cool, or vacuum advance will be denied.

7   Maintenance consists of occasionally checking the security of the electrical connections and the vacuum pipes. In the event of malfunction of the system, refer to Section 63.

---

60   Air Injection Reactor (AIR) system - description and servicing

---

1   The system comprises a belt-driven air injection pump and the necessary valves and connecting hoses to provide a controlled supply of air to the area of the exhaust valves (and upstream of the catalytic converter on 1975 models), where it dilutes the exhaust gases and helps to burn the unburned portion of the exhaust gases in the exhaust system.

2   The diverter valve is actuated by a sharp increase in manifold vacuum and under these conditions, the valve shuts off the injected air to prevent backfiring during this period of richer mixture.

3   On engine overrun, the air generated by the air pump is expelled to atmosphere

through the diverter valve and muffler. At high engine speeds, excess air pressure is released through the pressure relief valve of the diverter valve.

Drivebelt - replacement and adjustment

4   Loosen the air injection pump mounting bolt and adjustment bracket then move the pump to permit the belt to be removed.

5   Install the new drivebelt and adjust the tension so that there is approximately 1/2 inch free-play in the longest run of the belt, then tighten the adjustment bracket and mounting bracket bolts.

Pump filter - replacement

Refer to illustration 60.8

6   Compress the pump drivebelt to prevent the pulley turning then loosen the pulley bolts.

7   Remove the drivebelt and the pulley.

8   Where applicable, pry loose the filter fan outer disc. Pull off the filter with pliers, taking care that no debris enters the air intake hole **(see illustration)**.

9   Carefully and evenly draw the new filter onto the pump using the pulley and pulley bolts. The filter must slip squarely into the housing; a slight interference is normal and it may initially squeal when running until the sealing lip has worn in.

10   The remainder of the installation is the reverse of the removal procedure.

Air manifold, hose and tube

11   Periodically inspect all the connections, pipes and hoses for deterioration and for loose connections.

12   Leakage on the pressure side may be detected by brushing or pouring a soapy water solution around the joints and connections with the engine running.

13   Replacement of hoses and tubes is straightforward. Anti-seize compound should be used on air manifold and exhaust manifold screw threads.

Check valve

14   To test the operation of the check valve, it should be possible to blow through it (by mouth) towards the manifold, but not possible to blow through it the other way. **Warning:** *Don't attempt to suck through the valve; you might inhale carbon debris or get it in your mouth.*

15   The check valve can be unscrewed from the manifold when replacement is required.

Diverter valve

16   Check the condition of all the lines and connections.

17   To check the operation, disconnect the signal line at the valve and check that suction can be felt when the engine is running. Under normal operating conditions and with the engine idling, no air should be escaping through the muffler; when the throttle is manually opened, then quickly closed, an air blast

60.8 The filter can be pulled out of the pump with a pair of needle-nose pliers

lasting at least one second should be discharged through the muffler. A defective valve must be replaced.

Air injection tubes

18   No periodic inspection is required, but when exhaust manifolds are removed any carbon build-up in the tubes should be removed with a wire brush. Warped or burned tubes must be replaced.

19   To remove the tubes, clamp the manifold in a vise, clean away as much carbon as possible and work the tubes out of the manifold using penetrating oil.

Air injection pump

20   To check the pump operation, accelerate the engine to approximately 1500 rpm and check that airflow from the hose(s) increases.

21   The pump can be replaced by disconnecting the hoses, removing the pulley, then removing the pump mounting bolts.

22   Installation is the reverse of removal, following which the belt tension must be adjusted (see Step 5).

---

61   Exhaust Gas Recirculation (EGR) system - description and servicing

---

1   This system is designed to reduce the emissions of oxides of nitrogen from the exhaust pipe. Formation of this pollutant occurs during the period of highest temperature during the combustion cycle.

2   To accomplish a reduction in these toxic products, a small quantity of inert gas is introduced into the combustion process,

3   A 'U' connection between the exhaust and inlet manifolds is used for the purpose and the continuous supply of exhaust gas is tapped as the source of inert gas.

4   A vacuum controlled EGR combined shut off and metering valve controls the flow of exhaust (inert) gas according to engine operating conditions. The system is ineffective at coolant temperatures below 100° F.

5   Every 12000 miles or 12 months, the EGR valve should be removed and its base cleaned with a wire brush. The valve seat should be cleaned using an abrasive type spark plug cleaning machine, with the valve in both open and closed positions. Finally apply compressed air to remove all traces of abrasive material.

6   Other maintenance consists of checking the security and condition of the hoses in the system.

62   Early Fuel Evaporation (EFE) system - description

1   This system, introduced for 1975 models, provides improved driveability while reducing exhaust emissions.

2   The system consists of an EFE valve at the exhaust manifold flange, an actuator and a Thermal Vacuum Switch (TVS).

3   The TVS, mounted in the coolant outlet housing, controls the vacuum to the actuator which in turn operates the EFE valve.

4   At coolant temperatures below 180°F, manifold vacuum is applied to the actuator, which closes the EFE valve. This routes hot air to the carburetor.

5   At coolant temperatures below 180°F, vacuum to the actuator is denied. This allows the actuator to return to the "at rest" position and the EFE valve is opened.

63   Troubleshooting - carburetion, fuel, exhaust and emission control systems

Symptom

Excessive fuel consumption * ....

Reason

Air filter clogged
Leaks in fuel tank, carburetor or fuel lines
Fuel level in float chamber too high
Mixture too rich
Incorrect valve clearances
Dragging brakes
Tires under inflated

* May also be caused by a faulty condenser or counterweights in the ignition distributor.

Insufficient fuel delivery or weak mixture ....

Clogged fuel line filter
Stuck carburetor inlet needle valve
Faulty fuel pump
Leaking fuel pipe unions
Leaking inlet manifold gasket
Leaking carburetor mounting flange gasket
Incorrect carburetor adjustment

PCV system

Escaping fumes from engine..

Clogged PCV valve
Split or collapsed hoses

Fuel evaporative emission control system (E.C.S.)

Fuel odor and/or rough running engine...

Choked carbon canister
Stuck filler cap valve
Collapsed or split hoses

Transmission controlled spark (T.C.S.) system

Idle speed too low or too high or dieseling ....
Poor high gear performance
Excessive fuel consumption
Backfire on deceleration
Difficult cold start

Faulty idle stop solenoid
Blown fuse
Loose connections or broken leads
Faulty coolant temperature switch
Faulty transmission switch
Faulty vacuum advance solenoid

Air injection reactor (AIR) system

Fume emission from exhaust pipe ....

Loose air pump drive belt
Split or broken hoses
Clogged pump air filter
Defective air pump
Leaking pressure relief valve

Exhaust gas recirculation (EGR) system

Rough idling ....

Faulty or dirty EGR valve Broken valve diaphragm spring
Disconnected or split vacuum hose
Split valve diaphragm
Leaking valve gasket

Chapter 4   Ignition system

Contents

Specifications

Year	Engine and transmission	Distributor part no.	Centrifugal advance	Vacuum advance	Ignition timing at idle speed *	Spark plug type and gap
1969	307 cu in. 200 HP, Manual and Automatic	1111481	0 @ 1000 rpm 10° @ 1600 rpm 28° @ 4300 rpm	0° @ 6 in. Hg 15° @ 12 in. Hg	2° BTDC @ 600 rpm Auto. 700 rpm Manual	AC R45 0.035 in.
1969	350 cu in. RPO LM-1 255 HP Automatic	1111955	0 @ 900 rpm 17° @ 1900 rpm 22° @ 2200 rpm 32° @ 4400 rpm	0 @ 7 in. Hg 13 @ 17 in. Hg	4° BTDC @ 600 rpm 0.035 in.	AC R44
1969	350 cu in. RPO LM-1 255 HP Manual	1111956	0 @ 900 rpm 10° @ 1200 rpm 21° @ 2000 rpm 32° @ 4400 rpm	0 @ 7 in. Hg 24 @ 13 in. Hg	TDC @ 700 rpm	AC R44 0.035 in.
1969	350 cu in. RPO L-65 250 HP Manual	1111486	0 @ 800 rpm 3° @ 1000 rpm 15° @ 1800 rpm 36° @ 4100 rpm	0 @ 7 in. Hg 13 @ 17 in. Hg	TDC @ 700 rpm	AC R44 0.035 in.
1969	350 cu in. RPO L-65 250 HP Automatic	1111487	0 @ 900 rpm 2° @ 1100 rpm 8° @ 1400 rpm 32° @ 4400 rpm	0 @ 7 in. Hg 13 @ 17 in. Hg	4° BTDC @ 600 rpm	AC R44 0.035 in.
1969	350 cu in. RPO L-48 300 HP Manual	1111488	0 @ 950 rpm 14° @ 1400 rpm 20° @ 1800 rpm 30° @ 4700 rpm	0 @ 10 in. Hg 10° @ 17 in. Hg	TDC @ 700 rpm	AC R44 0.035 in.
1969	350 cu in. RPO L-48 300 HP Automatic	1111489	0 @ 900 rpm 9° @ 1300 rpm 15° @ 1700 rpm 26° @ 4700 rpm	0 @ 10 in. Hg 10° @ 17 in. Hg	4° BTDC @ 600 rpm	AC R44 0.035 in.

Year	Engine and transmission	Distributor part no.	Centrifugal advance	Vacuum advance	Ignition timing at idle speed *	Spark plug type and gap
1969	396 cu in. RPO L-34 350 HP Manual	1111498	0 @ 900 rpm 13° @ 1275 rpm 21 ° @ 2000 rpm 36° @ 5000 rpm	0 @ 8 in. Hg 15° @ 15.5 in. Hg	TDC @ 800 rpm	AC R44N 0.035 in.
1969	396 cu in. RPO L-34 350 HP Automatic	1111499	0 @ 900 rpm 9° @ 1250 rpm	0 @ 6 in. Hg 15° @ 12 in. Hg	4° BTDC @ 600 rpm (600 HP)	AC R43N 0.035 in.
1969	396 cu in. RPO L-78 375 HP Manual		17° @ 2000 rpm 32° @ 6000 rpm		4° BTDC @ 750 rpm (375 HP)	
1970	307 cu in. 200 HP Manual	1111995	0 @ 1000 rpm 3° @ 1200 rpm 10° @ 1600 rpm 28° @ 4300 rpm	0 @ 6 in. Hg 15° @ 12 in. Hg	2° BTDC	AC R45 0.035 in.
1970	307 cu in. 200 HP Automatic	1112005	0 @ 800 rpm 2° @ 1200 rpm 12° @ 2200 rpm 24° @ 4300 rpm	0 @ 8 in. Hg 20° @ 17 in. Hg	8° BTDC	AC R45 0.035 in.
1970	350 cu in. RPO L-65 250 HP Manual	1112001	0 @ 800 rpm 3° @ 1000 rpm 15° @ 1800 rpm 36° @ 4100 rpm	0 @ 7 in. Hg 24° @ 17 in. Hg	TDC	AC R44 0.035 in.
1970	350 cu in. RPO L-65 250 HP Automatic	1112002	0 @ 900 rpm 2° @ 1100 rpm 8° @ 1400 rpm 32° @ 4400 rpm	0 @ 7 in. Hg 24° @ 17 in. Hg	4° BTDC	AC R44 0.035 in.
1970	350 cu in. RPO L-48 300 HP Manual	1111996	0 @ 950 rpm 14° @ 1400 rpm 20° @ 1800 rpm 30° @ 4700 rpm	0 @ 10 in. Hg 15° @ 17 in. Hg	TDC	AC R44 0.035 in.
1970	350 cu in. RPO L-48 300 HP Automatic	1111997	0 @ 900 rpm 9° @ 1300 rpm 15° @ 1700 rpm 26° @ 4700 rpm	0 @ 10 in. Hg 20° @ 17 in. Hg	4° BTDC	AC R44 0.035 in.
1970	400 cu in. RPO L-34 350 HP Manual	1111999	0 @ 900 rpm 13° @ 1275 rpm 21 ° @ 2000 rpm 36° @ 5000 rpm	0 @ 8 in. Hg 15° @ 15.5 in. Hg	TDC	AC R44T 0.035 in.
1970	400 cu in. RPO L-34 350 HP Automatic	1112000	0 @ 900 rpm 9° @ 1250 rpm	0 @ 6 in. Hg 15° @ 12 in. Hg	4° BTDC	AC R44T (350 HP)
1970	400 cu in. RPO L-78 375 HP Manual and Automatic		17° @ 2000 rpm 32° @ 5000 rpm			AC R43T (375 HP) 0.035 in.
1971, 1972	307 cu in. 200 HP Automatic	1112039	0 @ 680 rpm

2° @ 1320 rpm 20° @ 4200 rpm | 0 @ 8 in. Hg

20° @ 17 in. Hg | 8° BTDC @

550 rpm | AC R45TS (1971)

AC R44T (1972) 0.035 in. |
| 1971 | 307 cu in. 200 HP Manual | 1112005 | 0 @ 800 rpm 2° @ 1200 rpm | 0 @ 8 in. Hg 20° @ 17 in. Hg | 4° BTDC @ 600 rpm (200 HP) | AC R45TS 0.035 in. |
| 1971 | 350 cu in. 250 HP and RPO L-65 Automatic | | 12° @ 2200 rpm

24° @ 4300 rpm | | 6° BTDC @ 550 rpm (250 HP and RPO L-65) Automatic | |
1971	350 cu in. 250 HP and RPO L-65 Manual	1112042	0 @ 880 rpm 2° @ 1120 rpm 10° @ 1600 rpm 15° @ 2200 rpm 28° @ 4300 rpm	0 @ 8 in. Hg 20° @ 17 in. Hg	2° BTDC @ 600 rpm	AC R45TS 0.035 in.
1971	350 cu in. RPO L-48 Automatic	1112045	0 @ 865 rpm 2°@1335 rpm	0 @ 8 in. Hg 8° 15° @ 15.5 in Hg	BTDC @ 550 rpm	AC R45TS 0.035 in.
1972	350 cu in. (4 bbl) Automatic		11 ° @ 2400 rpm 18° @ 4200 rpm			AC R44T 0.035 in.

Year	Engine and transmission	Distributor part no.	Centrifugal advance	Vacuum advance	Ignition timing at idle speed *	Spark plug type and gap
1972	350 cu in. (2-bbl) Manual and Automatic	1112005	0 @ 800 rpm 2° @ 1200 rpm	0 @ 8 in. Hg 20° @ 17 in. Hg	6° BTDC (350 cu in.)	AC R44T 0.035 in.
1972	307 cu in. Manual		12° @ 2200 rpm 24° @ 4300 rpm		4° BTDC (307 cu in.)	
1972	350 cu in. (4-bbl) Manual	1112044	0 @ 840 rpm 2° @ 1160 rpm	0 @ 8 in. Hg 15° @ 15.5 in. Hg	4° BTDC	AC R44T (1972)
1971	350 cu in. RPO L-48 Manual		10° @ 1800 rpm 15° @ 2400 rpm 22° @ 4200 rpm			AC R44TS (1971) 0.035 in.
1973	307 cu in. Automatic	1112102	0 @ 1000 rpm 2° @ 1320 rpm 10° @ 2100 rpm 20° @ 4200 rpm	0 @ 6 in. Hg 15° @ 12 in. Hg	8° BTDC	AC R44T 0.035 in.
1973	307 cu in. Manual	1112227	0 @ 1000 rpm 2° @ 1200 rpm 12° @ 2200 rpm 24° @ 4300 rpm	0 @ 6 in. Hg 15° @ 12 in. Hg	4° BTDC	AC R44T 0.035 in.
1973	350 cu in. (2-bbl) Manual and Automatic	1112168	0 @ 1000 rpm 2° @ 1300 rpm 10° @ 2600 rpm 20° @ 4200 rpm	0 @ 4 in Hg 14° @ 7 in. Hg	8° BTDC	AC R44T 0.035 in.
1973	350 cu in. (4 bbl) Manual	1112093	0 @ 1100 rpm 2° @ 1320 rpm 6° @ 1800 rpm 11° @ 2400 rpm 18° @ 4200 rpm	0 @ 6 in. Hg 15° @ 14 in. Hg	8° BTDC	AC R44T 0.035 in.
1973	350 cu in. (4 bbl) Manual	1112094	0 @ 1100 rpm 2° @ 1550 rpm 6° @ 2410 rpm 12° @ 3300 rpm 14° @ 4200 rpm	0 @ 6 in. Hg 15° @ 14 in. Hg	12° BTDC	AC R44T 0.035 in.
1974	350 cu in. (2-bbl) and RPO L-65, Manual and Automatic (Federal)	1112168	0 @ 1000 rpm ** 20° @ 2400 rpm **	0 @ 2 to 4 in. Hg 14° @ 7.5 to 8.5 in. Hg	—	AC R44T 0.035 in.
1974	350 cu in. (4 bbl) and RPO L-48, Manual and Automatic (Federal)	1112093	0 @ 1100 rpm ** 11° @ 2400 rpm ** 18 @ 4200 rpm **	0 @ 5 to 7 in. Hg 15° @ 13 to 14 in. Hg	—	AC R44T 0.035 in.
1974	350 cu in.(4 bbl) and RPO LM-1, Automatic (California)	1112847	0 @ 1100 rpm ** 11° @ 2400 rpm ** 18° @ 4200 rpm **	0 @ 2 to 4 in. Hg 14° @ 7.5 to 8.5 in. Hg	—	AC R44T 0.035 in.
1974	350 cu in.(4 bbl) RPO L-48 and RPO LM-1 Manual (California)	1112849	0 @ 1000 rpm ** 10° @ 1800 rpm ** 15° @ 2400 rpm ** 22° @ 4200 rpm **	0 @ 24 in. Hg 14° @ 7.5 to 8.5 in. Hg	—	AC R44T 0.035 in.
1975	262 cu in. (4.3 liter) LV-1	1112933	0 @ 1200 rpm 9° @ 2000 rpm 22° @ 4000 rpm	0 @ 3 in. Hg 16° @ 8 in. Hg	8° BTDC	AC R44TX 0.060 in.
1975	350 cu in.(2-bbl) L-65 Nationwide 350 cu in. (4 bbl) LM-1 California	1112880	0 @ 1200 rpm 12° @ 2000 rpm 22° @ 4200 rpm	0 @ 4 in. Hg 18° @ 12 in. Hg	6° BTDC	AC R44TX 0.060 in.
1976	305 cu in LG-3 Automatic (California)	1112999	0° @ 1000 10° @ 1700 20° @ 3800	0° @ 4 in Hg 10° @ 8 in Hg	TDC	AC R45TS 0.045 in.
1976	305 cu in LG-3	1112977	0° @ 1000	0° @4 in Hg	6° BTDC (manual)	AC R45TS

Year	Engine and transmission	Distributor part no.	Centrifugal advance	Vacuum advance	Ignition timing at idle speed	Spark plug type and gap
	Manual & Automatic (Federal)		10° @ 1700 20° @ 3800	18° @ 12 in Hg	or 8° BTDC (automatic)	0.045 in.
1976	350 cu in LM-1 Manual & Automatic (California)	1112905	0° @ 1200 12° @ 2000 22° @ 4200	0° @ 6 in Hg 15(@ 12 in Hg	6° BTDC	AC R45TS 0.045 in.
1976	350 cu in Manual & Automatic (Federal)	1112888	0° @ 1100 12 @ 1600 16° @ 2400 22° @ 4600	0° @ 4 in Hg 18° @ 12 in Hg	8° BTDC	AC R45TS 0.045 in.
1977	305 cu in LG-3 Manual (Federal)	1103239	0° @ 1200 15° @ 2700 20° @ 4200	0° @ 4 in Hg 15° @ 10 in Hg	8° BTDC @ 600 rpm	AC R45TS 0.045 in.
1977	305 cu in LG-3 Automatic (Federal)	1103239	0° @ 1200 15° @ 2700 20° @ 4200	0° @ 4 in Hg 15° @ 10 in Hg	8° BTDC @ 500 rpm	AC R45TS 0.045 in.
1977	305 cu in LG-3 Automatic (California)	1103244	0° @ 1000 10° @ 1700 20° @ 3800	0° @ 4 in Hg 20° @ 10 in Hg	6° BTDC @ 500 rpm	AC R45TS 0.045 in.
1977	350 cu in LM-1 Manual (Federal)	1103246	0° @ 1200 12° @ 2000 22° @ 4200	0° @ 4 in Hg 18° @ 12 in Hg	8° BTDC @ 700 rpm	AC R45TS 0.045 in.
1977	350 cu in LM-1 Automatic (Federal)	1103246	0° @ 1200 12° @ 2000 22° @ 4200	0° @4 in Hg 18° @ 12 in Hg	8° BTDC @ 500 rpm	AC R45TS 0.045 in.
1977	350 cu in LM-1 Automatic (high altitude)	1103246	0° @ 1200 12° @ 2000 22° @ 4200	0° @ 4 in Hg 18° @ 12 in Hg	8° BTDC @ 600 rpm	AC R45TS 0.045 in.
1977	350 cu in LM-1 Automatic (California)	1103248	0° @ 1200 12° @ 2000 22° @ 4200	0° @ 4 in Hg 10° @ 8 in Hg	8° BTDC @ 500 rpm	AC R45TS 0.045 in.
1978	305 cu in LG-3 Manual (Federal)	1103281	0° @ 1000 10° @ 1700 20° @ 3800	0° @ 4 in Hg 18° @ 12 in Hg	4° BTDC @ 600 rpm	AC R45TS 0.045 in.
1978	305 cu in LG-3 Automatic (Federal)	1103282	0° @ 1000 10° @ 1700 20° @ 3800	0° @ 4 in Hg 20° @ 10 in Hg	4° BTDC @ 500 rpm	AC R45TS
1978	305 cu in LG-3 Automatic (California)	1103282	0° @ 1000 10° @ 1700 20° @ 3800	0° @4 in H9 20° @ 10 in Hg	6° BTDC @ 500 rpm	AC R45TS 0.045 in.
1978	350 cu in LM-1 Automatic (high altitude)	1103353	0° @ 1100 12° @ 1600 16° @ 2400	0° @ 4 in Hg 20° @ 10 in Hg	8° BTDC @ 500 rpm	AC R45TS 0.045 in.
1978	350 cu in LM-1 Automatic (California)	1103285	0° @ 1200 12(@ 1200 22° @ 4200	0° @ 4 in Hg 10° @ 8 in Hg	8° BTDC @ 500 rpm	AC R45TS 0.045 in.
1979	305 cu in LG-3 Manual (Federal)	1103281	0° @ 1000 10° @ 1700 20° @ 3800	0° @ 4 in Hg 18° @ 12 in Hg	4° BTDC @ 600 rpm	AC R45TS 0.045 in.
1979	305 cu in LG-3 Automatic (Federal)	1103379	0° @ 1000 10° @ 1700 20° @ 3800	0° @ 3 in Hg 20° @ 8.5 in Hg	4° BTDC @ 600 rpm	AC R45TS 0.045 in.
1979	305 cu in LG-3 Automatic (California)	1103285	0° @ 1200 12° @ 2000 22° @ 4200	0° @ 4 in Hg 0° @ 8 in Hg	4° BTDC @ 600 rpm	AC R45TS 0.045 in.
1979	350 cu in LM-1 Automatic (California)	1103285	0° @ 1200 12° @ 2000 22° @ 4200	0° @ 4 in Hg 10° @ 8 in Hg	8° BTDC @ 600 rpm	AC R45TS 0.045 in.
1979	350 cu in LM-1 Automatic (high altitude)	1103353	0°@1000 12° @ 1600 16° @ 2400 22° @ 4600	0° @ 4 in Hg 20° @ 11 in Hg	8° BTDC @ 600 rpm	AC R45TS 0.045 in.

Where no idling speed is given, refer to Chapter 3 (also refer to the Vehicle Emission Control Sticker.)
** These figures are nominal. Allow +/- 2°.

General

Distributor directional rotation	Clockwise
Firing order	1-8-4-3-6-5-7-2
Point dwell	29 to 31-degrees
Coil (pre-1975)	
Primary resistance..	1.77 to 2.05 ohms
Secondary resistance..	3000 to 20,000 ohms
Ignition coil resistor (fixed in wiring harness)	1.35 ohms

Torque specifications

	Ft-lbs
Spark plugs..	15

1   Ignition system - general information and precautions

General information

1   In order for the engine to run correctly, it is necessary for an electrical spark to ignite the fuel/air mixture in the combustion chamber at exactly the right moment in relation to engine speed and load. The ignition coil converts low tension (LT) voltage from the battery into high tension (HT) voltage, powerful enough to jump the spark plug gap in the cylinder, providing that the system is in good condition and that all adjustments are correct.

2   The ignition system fitted to all pre-1975 cars as standard equipment is a conventional distributor with mechanical contact breaker points. On 1975 and later models, a breakerless High Energy Ignition (HEI) system is used.

Breaker point systems

3   The ignition system is divided into the primary (low tension) circuit and the secondary (high tension) circuit.

4   The primary circuit consists of the battery lead to the starter motor, the lead to the ignition switch, the calibrated resistance wire from the ignition switch to the primary coil winding and the lead from the low tension coil windings to the contact breaker points and condenser in the distributor.

5   The secondary circuit consists of the secondary coil winding, the high tension lead from the coil to the distributor cap, the rotor and the spark plug leads and spark plugs.

6   The system functions in the following manner. Low tension voltage in the coil is converted into high tension voltage by the opening and closing of the contact breaker points in the distributor. This high tension voltage is carried via the brush in the center of the distributor cap contacts to the rotor arm of the distributor cap. Every time the rotor lines up with one of the spark plug ter-

minals on the cap, it jumps the gap from the rotor arm to the terminal and is carried by the spark plug lead to the spark plug, where it jumps the spark plug gap to ground.

7   Ignition advance is controlled by both mechanical and vacuum operated systems. The mechanical governor mechanism consists of two weights, which due to centrifugal force move out from the distributor shaft as the engine speed rises. As they move outwards they rotate the cam relative to the distributor shaft, advancing spark timing. The weights are held in position by two light springs. It is the tension of these springs which determines correct spark advancement.

8   The vacuum control system consists of a diaphragm, one side of which is connected via a vacuum line to the carburetor, the other side to the contact breaker plate. Vacuum in the intake manifold and carburetor varies with engine speed and throttle opening. As the vacuum changes, it moves the diaphragm, which rotates the contact breaker plate slightly in relation to the rotor, thus advancing or retarding the spark. Control is fine-tuned by a spring in the vacuum assembly.

HEI ignition systems

9   The high energy ignition (HEI) system is a pulse triggered, transistor controlled, inductive discharge system.

10   A magnetic pick-up inside the distributor contains a permanent magnet, pole-piece and pick-up coil. A time core, rotating inside the pole piece, induces a voltage in the pick-up coil. When the teeth on the timer and pole piece line up, a signal passes to the electronic module to open the coil primary circuit. The primary circuit current collapses and a high voltage is induced in the coil secondary winding. This high voltage is directed to the spark plugs by the distributor rotor in a manner similar to a conventional system. A capacitor suppresses radio interference.

11   The HEI system features a longer spark duration than a conventional breaker point

ignition system, and the dwell period increases automatically with engine speed. These characteristics are desirable for lean firing and EGR-diluted mixtures (see Chapter 3).

12   The ignition coil and the electronic module are both housed in the distributor cap on the HEI system. The distributor does not require routine servicing.

13   Spark timing is advanced by mechanical and vacuum devices similar to those used on conventional breaker point distributors (described above).

Precautions

14   Always observe the following precautions when working on the electrical systems:

a) *Be extremely careful when servicing engine electrical components. They are easily damaged if checked, connected or handled improperly.*

b) *The alternator is driven by an engine drivebelt which could cause serious injury if your hands, hair or clothes become entangled in it with the engine running.*

c) *Both the alternator and the starter are connected directly to the battery and could arc or even cause a fire if mishandled, overloaded or shorted out.*

d) *Never leave the ignition switch on for long periods of time with the engine off.*

e) *Don't disconnect the battery cables while the engine is running.*

f) *Maintain correct polarity when connecting a battery cable from another source, such as a vehicle, during jump starting.*

g) *Always disconnect the negative cable first and hook it up last or the battery may be shorted by the tool being used to loosen the cable clamps.*

15   It's also a good idea to review the safety-related information regarding the engine electrical systems located in the *Safety First* section near the front of this manual before beginning any operation included in this Chapter.

2.2 The ignition system should be checked with a spark tester - if the ignition system produces a spark that will jump the tester gap, it's functioning properly

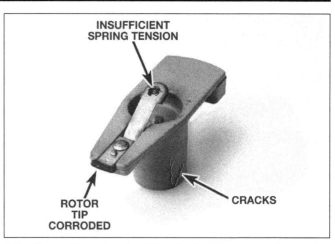

3.3a The ignition rotor should be checked for wear and corrosion as indicated here (if in doubt about its condition, buy a new one)

2   Ignition system - check

Refer to illustration 2.2

Warning: *Because of the very high secondary voltage generated by the ignition system - particularly the High Energy Ignition (HEI) system - extreme care should be taken whenever this check is performed.*

HEI system

Calibrated ignition tester method

1   If the engine turns over but won't start, disconnect the spark plug lead from any spark plug and attach it to a calibrated ignition tester for the HEI system (available at most auto parts stores).

2   Connect the clip on the tester to a ground such as a metal bracket **(see illustration)**, crank the engine and watch the end of the tester to see if bright blue, well-defined sparks occur.

3   If sparks occur, sufficient voltage is reaching the plugs to fire the engine. However, the plugs themselves may be fouled, so remove and check them as described in Section 12 or install new ones.

4   If no spark occurs, remove the distributor cap and check the cap and rotor as described in Section 3. If moisture is present, use WD-40 (or something similar) to dry out the cap and rotor, then reinstall the cap and repeat the spark test.

5   If there's still no spark, the tester should be attached to the wire from the coil (this cannot be done on vehicles with a coil-in-cap distributor) and the check should be repeated again.

6   If no spark occurs, check the primary wire connections at the coil to make sure they're clean and tight. Make any necessary repairs, then repeat the check again.

7   If sparks now occur, the distributor cap, rotor, plug wire(s) or spark plug(s) may be defective. If there's still no spark, the coil-to-cap wire may be bad. If a substitute wire doesn't make any difference, have the sys-

tem checked by a dealer service department or a repair shop.

Breaker point system

8   Remove the spark plug wire from one of the spark plugs. Using an insulated tool, hold the wire about 1/4-inch from a good ground and have an assistant crank the engine.

9   If bright blue, well-defined sparks occur, sufficient voltage is reaching the plugs to fire the engine. However, the plugs themselves may be fouled, so remove and check them as described in Section 12 or install new ones.

10   If there's no spark, check another wire in the same manner. A few sparks followed by no spark is the same condition as no spark at all.

11   If no spark occurs, remove the distributor cap and check the cap and rotor as described in Section 3. If moisture is present, use WD-40 (or something similar) to dry out the cap and rotor, then reinstall the cap and repeat the spark test.

12   If there's still no spark, disconnect the coil wire from the distributor, hold it about 1/4-inch from a good ground and crank the engine again (this cannot be done on vehicles with a coil-in-cap distributor).

13   If sparks now occur, the distributor cap, rotor or plug wire(s) may be defective.

14   If there's still no spark, check the ignition coil (see Section 11).

15   If the ignition coil checks out OK, remove the distributor cap and check the ignition points as described in Section 4. If the points appear to be in good condition and the primary wires are hooked up correctly and undamaged, adjust the points as described in Section 5, then repeat the spark check.

16   If the primary circuit is complete (battery voltage is available at the points), the points are clean and properly adjusted and the secondary ignition circuit is functioning correctly, sparks should occur at the plug wires. If there is still no spark, have the system checked by a dealer service department or a repair shop.

3   Distributor cap and rotor - check and replacement

Note: *It's common practice to install a new distributor cap and rotor whenever new spark plug wires are installed. On models that have the ignition coil mounted in the cap, the coil will have to be transferred to the new cap.*

Check

Refer to illustrations 3.3a, 3.3b, 3.6a, 3.6b and 3.12

1   Remove the air cleaner assembly.

2   Loosen the distributor cap mounting screws (note that the screws have a shoulder so they don't come completely out). On some models, the cap is held in place with latches that look like screws - to release them, push down with a screwdriver and turn them about 1/2-turn. Pull up on the cap, with the wires attached, to separate it from the distributor, then position it to one side.

3   The rotor is now visible on the end of the distributor shaft. Check it carefully for cracks and carbon tracks. Make sure the center terminal spring tension is adequate and look for corrosion and wear on the rotor tip **(see illustrations)**. If in doubt about its condition, replace it with a new one.

4   If replacement is required, detach the rotor from the shaft and install a new one. On some models, the rotor is press fit on the shaft and can be pried or pulled off. On other models, the rotor is attached to the distributor shaft with two screws.

5   The rotor is indexed to the shaft so it can only be installed one way. Press-fit rotors have an internal key that must line up with a slot in the end of the shaft (or vice versa). Rotors held in place with screws have one square and one round peg on the underside that must fit into holes with the same shape.

6   Check the distributor cap for carbon tracks, cracks and other damage. Closely examine the terminals on the inside of the cap

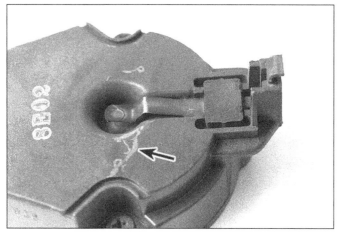

3.3b Carbon tracking on the rotor can be caused by a leaking seal between the distributor cap and the ignition coil (HEI systems); always replace both the seal and rotor when this condition is present

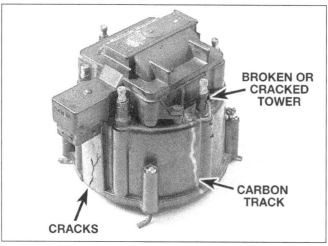

3.6a Shown here are some of the common defects to look for when inspecting the distributor cap (if in doubt about its condition, install a new one)

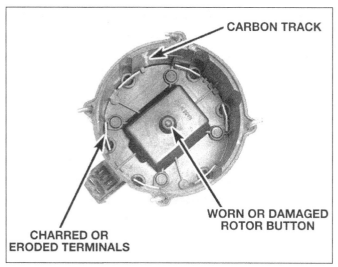

3.6b Inspect the inside of the cap, especially the metal contacts (arrows) for corrosion and wear

3.12 Remove the two screws (arrows) that retain the plastic cover to the cap

for excessive corrosion and damage **(see illustrations)**. Slight deposits are normal. Again, if in doubt about the condition of the cap, replace it with a new one. Be sure to apply a small dab of silicone lubricant to each terminal before installing the cap. Also, make sure the carbon brush (center terminal) is correctly installed in the cap - a wide gap between the brush and rotor will result in rotor burn-through and/or damage to the distributor cap.

Replacement

Conventional distributor

7   On models with a separately mounted ignition coil, simply separate the cap from the distributor and transfer the spark plug wires, one at a time, to the new cap. Be very careful not to mix up the wires!

8   Reattach the cap to the distributor, then tighten the screws or reposition the latches to hold it in place.

Coil-in-cap distributor

Refer to illustration 3.12

9   Use your thumbs to push the spark plug wire retainer latches away from the coil cover.

10   Lift the retainer ring away from the distributor cap with the spark plug wires attached to the ring. It may be necessary to work the wires off the distributor cap towers so they remain with the ring.

11   Disconnect the battery/tachometer/coil electrical connector from the distributor cap.

12   Remove the two coil cover screws and lift off the coil cover **(see illustration)**.

13   There are three small spade connectors on wires extending from the coil into the electrical connector hood at the side of the distributor cap. Note which terminals the wires are attached to, then use a small screwdriver to push them free.

14   Remove the four coil mounting screws

and lift the coil out of the cap.

15   When installing the coil in the new cap, be sure to install a new rubber arc seal in the cap.

16   Install the coil screws, the wires in the connector hood, and the coil cover.

17   Install the cap on the distributor.

18   Plug in the coil electrical connector to the distributor cap.

19   Install the spark plug wire retaining ring on the distributor cap.

4   Ignition points - replacement

Refer to illustrations 4.1, 4.2, 4.7, 4.9, 4.16, 4.17 and 4.29

1   The ignition points must be replaced at regular intervals on vehicles not equipped with electronic ignition. Occasionally the rubbing block will wear enough to require adjust-

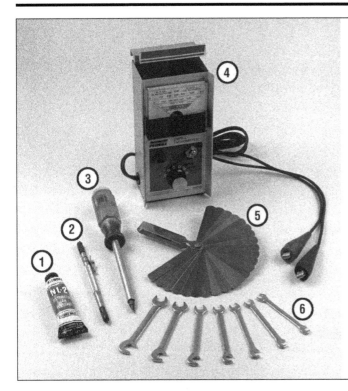

4.1 Tools and materials needed for contact point replacement and dwell angle adjustment

1   **Distributor cam lube** - *Sometimes this special lubricant comes with the new points; however, it's a good idea to buy a tube and have it on hand*

2   **Screw starter** - *This tool has special claws which hold the screw securely as it's started, which helps prevent accidental dropping of the screw*

3   **Magnetic screwdriver** - *Serves the same purpose as 2 above. If you don't have one of these special screwdrivers, you risk dropping the point mounting screws down into the distributor body*

4   **Dwell meter** - *A dwell meter is the only accurate way to determine the point setting (gap). Connect the meter according to the instructions supplied with it.*

5   **Blade-type feeler gauges** - *These are required to set the initial point gap (space between the points when they are open)*

6   **Ignition wrenches** - *These special wrenches are made to work within the tight confines of the distributor. Specifically, they are needed to loosen the nut/bolt which secures the leads to the points*

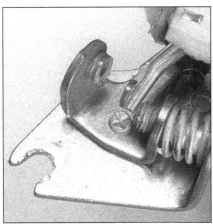

4.2 Although it is possible to restore ignition points that are pitted, burned and corroded (as shown here), they should be replaced instead

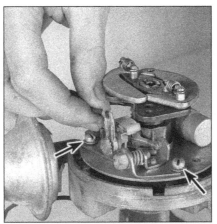
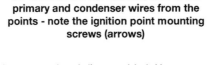

4.7 Loosen the nut and disconnect the primary and condenser wires from the points - note the ignition point mounting screws (arrows)

4.9 The condenser is attached to the breaker plate by a single screw

ment of the points. It's also possible to clean and dress them with a fine file, but replacement is recommended since they are relatively inaccessible and very inexpensive. Several special tools are required for this procedure **(see illustration)**.

2   After removing the distributor cap and rotor (Section 3), the ignition points are plainly visible. They can be examined by gently prying them open to reveal the condition of the contact surfaces **(see illustration)**. If they re rough, pitted, covered with oil or burned, they should be replaced, along with the condenser. **Caution:** *This procedure requires the removal of small screws which can easily fall down into the distributor. To retrieve them, the distributor would have to*

be removed and disassembled. Use a magnetic or spring-loaded screwdriver and be extra careful.

3   If not already done, remove the distributor cap by positioning a screwdriver in the slotted head of each latch. Press down on the latch and rotate it 1/2-turn to release the cap from the distributor body.

4   Position the cap (with the spark plug wires still attached) out of the way. Use a length of wire to hold it out of the way if necessary.

5   Remove the rotor (see Section 3 if necessary).

6   If equipped with a radio frequency interference shield (RFI), remove the mounting screws and the two-piece shield to gain

access to the ignition points.

7   Note how they are routed, then disconnect the primary and condenser wire leads from the points **(see illustration)**. The wires may be attached with a small nut (which should be loosened, but not removed) a small screw or by a spring loaded terminal. **Note:** *Some models are equipped with ignition points which include the condenser as an integral part of the point assembly. If your vehicle has this type, the condenser removal procedure below will not apply and there will be only one wire to detach from the points, rather than the two used with a separate condenser assembly.*

8   Loosen the two screws which secure the ignition points to the breaker plate, but don't completely remove the screws (most ignition point sets have slots at these locations). Slide the points out of the distributor.

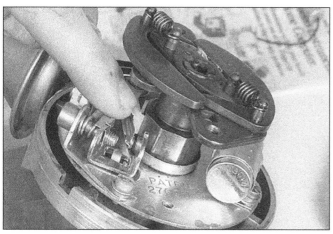

4.16 Before adjusting the point gap, the rubbing block must be resting on one of the cam lobes (which will open the points)

4.17 With the points open, insert a 0.019-inch thick feeler gauge and turn the adjustment screw with an Allen wrench

9   The condenser can now be removed from the breaker plate. Loosen the mounting strap screw and slide the condenser out or completely remove the condenser and strap **(see illustration)**. If you remove both the condenser and strap, be careful not to drop the mounting screw down into the distributor body.

10   Before installing the new points and condenser, clean the breaker plate and the cam on the distributor shaft to remove all dirt, dust and oil.

11   Apply a small amount of distributor cam lube (usually supplied with the new points, but also available separately) to the cam lobes.

12   Position the new condenser and tighten the mounting strap screw securely.

13   Slide the new point set under the mounting screw heads and make sure the protrusions on the breaker plate fit into the holes in the point base (to properly position the point set), then tighten the screws securely .

14   Attach the primary and condenser wires to the new points. Make sure the wires are routed so they don't interfere with breaker plate or advance weight movement.

15   Although the gap between the contact points (dwell angle will be adjusted later, make the initial adjustment now, which will allow the engine to be started.

16   Make sure that the point rubbing block is resting on one of the high points of the cam **(see illustration)**. If it isn't, turn the ignition switch to Start in short bursts to reposition the cam. You can also turn the crankshaft with a breaker bar and socket attached to the large bolt that holds the vibration damper in place.

17   With the rubbing block on a cam high point (points open), insert a 0.019-inch feeler gauge between the contact surfaces and use an Allen wrench to turn the adjustment screw until the point gap is equal to the thickness of the feeler gauge **(see illustration)**. The gap is correct when a slight amount of drag is felt as the feeler gauge is withdrawn.

18   If equipped, install the RFI shield.

19   Before installing the rotor, check it as described in Section 3.

20   Install the rotor. The rotor is indexed with a square peg underneath on one side and a round peg on the other side. so it will fit on the advance mechanism only one way, If so, equipped, tighten the rotor mounting screws securely.

21   Before installing the distributor cap, inspect it as described in Section 3.

22   Install the distributor cap and lock the latches under the distributor body by depressing and turning them with a screwdriver.

23   Start the engine and check the dwell angle and ignition timing.

24   Whenever new ignition points are installed or the original points are cleaned, the dwell angle must be checked and adjusted.

25   Precise adjustment of the dwell angle requires an instrument called a dwell meter. Combination tach/dwell meters are commonly available at reasonable cost from auto parts stores. An approximate setting can be obtained if a meter isn't available.

26   If a dwell meter is available, hook it up following the manufacturer's instructions.

27   Start the engine and allow it to run at idle until normal operating temperature is reached (the engine must be warm to obtain an accurate reading). Turn off the engine.

28   Raise the metal door in the distributor cap. Hold it in the open position with tape if necessary.

29   Just inside the opening is the ignition point adjustment screw. Insert a 1/8-inch Allen wrench into the adjustment screw socket **(see illustration)**.

30   Start the engine and turn the adjustment screw as required to obtain the specified dwell reading on the meter. Dwell angle specifications can be found on the tune-up decal in the engine compartment. **Note:** *When adjusting the dwell, aim for the lower end of the dwell specification range. Then, as the points wear, the d well will remain within the specified range over a longer period of*

4.29 With the door open, a 1/8-inch Allen wrench can be inserted into the adjustment screw socket and turned to adjust the point dwell

time.

31   Remove the Allen wrench and close the door on the distributor. Turn off the engine and disconnect the dwell meter, then check the ignition timing (see Section 10).

32   If a dwell meter isn't available, use the following procedure to obtain an approximate dwell setting.

33   Start the engine and allow it to idle until normal operating temperature is reached.

34   Raise the metal door in the distributor cap. Hold it in the open position with tape if necessary.

35   Just inside the opening is the ignition point adjustment screw. Insert a 1/8-inch Allen wrench into the adjustment screw socket **(see illustration 4.29)**.

36   Turn the Allen wrench clockwise until the engine begins to misfire, then turn the screw 1/2-turn counterclockwise.

37   Remove the Allen wrench and close the door. As soon as possible have the dwell angle checked and/or adjusted with a dwell meter to ensure optimum performance.

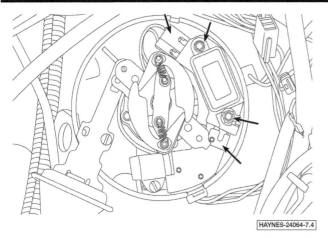

5.4 The wire connectors are attached to both ends of the module (some models) and the screws thread into the distributor body (arrows)

5.6 If you're installing a new module, apply silicone grease (included with the module) to the area where the module mounts (arrow)

5 Ignition module (HEI ignition) - replacement

Refer to illustrations 5.4 and 5.6

Note: *It's not necessary to remove the distributor from the engine to replace the ignition module.*

1 Disconnect the cable from the negative terminal of the battery. Remove the air cleaner to provide room to work around the distributor.

2 Remove the distributor cap and wires as an assembly and position them out of the way.

3 Remove the screws and detach the rotor.

4 Carefully detach the wires from the module terminals **(see illustration)**. If the wires are attached with a plastic connector, release the locking tang before pulling the connector off the terminals. If the wires are difficult to remove, you may find it easier to remove the module mounting screws first, detach the module and then pull off the wires. **Caution:** *Do not pull on the wires or the connectors may be damaged.*

5 Remove the screws and lift out the module. **Note:** *The module can only be tested with special equipment. If you suspect that the module is malfunctioning, have it checked by an auto parts store or dealer service department (auto parts stores that are equipped to handle this check will usually perform this service free of charge).*

6 Installation is the reverse of removal. Be sure to apply the silicone dielectric grease supplied with the new module to the distributor pad **(see illustration)** - DO NOT use any other type of grease! If the grease is not used, the module will overheat and destroy itself.

6 Ignition pick-up coil (HEI ignition) - check and replacement

Refer to illustrations 6.3, 6.7, 6.9 and 6.10

Check

1 Remove the distributor cap (Section 3).

2 Remove the rotor and disconnect the pick-up coil leads from the module.

3 Connect an ohmmeter to each terminal of the pick-up coil connector or wire and ground (one terminal at a time) **(see illustration)**. The ohmmeter should indicate infinite resistance. If it doesn't, the pickup coil is defective.

4 Connect the ohmmeter between both terminals or wires of the pick-up coil connector. If a vacuum advance unit is attached to the distributor, apply vacuum from an external source and watch the ohmmeter for indications of intermittent opens. The ohmmeter should indicate one steady value within the 500 to 1500 ohm range as the wires are flexed. If it doesn't, the pick-up coil is defective.

5 If the pick-up coil fails either test, replace it.

Replacement

6 Remove the distributor (Section 7). Mark the distributor shaft and gear so they can be reassembled in the same relationship.

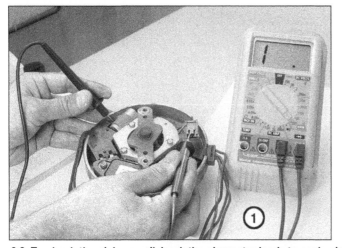

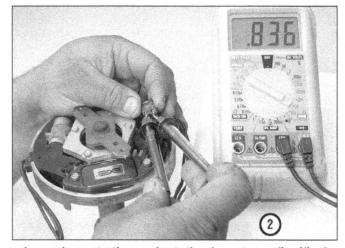

6.3 To check the pick-up coil, hook the ohmmeter leads to each wire and ground, one at a time, and note the ohmmeter reading (1) - the second check (2) requires one ohmmeter lead to be attached to each pick-up coil wire

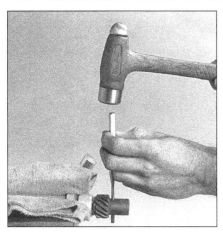

6.7 Place the distributor in a bench vise and drive out the roll pin which locks the drive gear to the bottom of the shaft

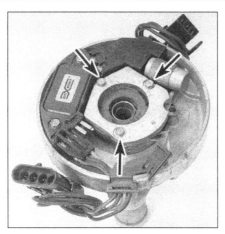

6.9 Remove the three screws (arrows) attaching the magnetic shield to the distributor base and detach the shield

6.10 The pick-up coil and pole piece assembly can be removed after carefully prying out the C-clip

7   Secure the distributor shaft housing in a bench vise and drive out the roll pin with a hammer and punch **(see illustration)**.

8   Remove the gear and tanged washer, then check the shaft for burrs. If none are found, pull the shaft from the distributor.

9   Remove the three attaching screws and separate the magnetic shield **(see illustration)**.

10   Remove the C-clip **(see illustration)** and detach the pick-up coil assembly.

11   Install the new pick-up coil assembly and make sure the C-clip is seated in the groove.

12   Install the shaft. Make sure that it's clean and lubricated.

13   Install the tanged washer (with the tangs facing up), the gear and the roll pin.

14   Spin the shaft to make sure that the teeth on the distributor shaft don't touch the teeth on the pick-up coil pole piece.

15   If the teeth touch, loosen, adjust and retighten the pole piece to eliminate contact.

16   Install the distributor (see Section 6).

7   Distributor - removal and installation

Refer to illustrations 7.5, 7.6a and 7.6b

Removal

1   After disconnecting the negative battery terminal cable, unplug the primary lead from the coil.

2   Unplug or detach all electrical leads from the distributor. To find the connectors, trace the wires from the distributor.

3   Look for a raised "1" on the distributor cap This marks the location for the number one cylinder spark plug wire terminal. If the cap does not have a mark for the number one spark plug, locate the number one spark plug and trace the wire back to its corresponding terminal on the cap.

4   Remove the distributor cap (see Section 3) and turn the engine over until the rotor is pointing toward the number one spark plug terminal.

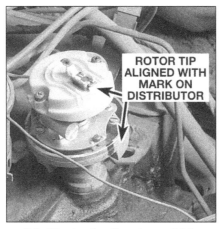

7.5 After turning the rotor until it is pointing at the terminal for the number 1 spark plug, paint or scribe a mark on the edge of the distributor base directly beneath it

5   Make a mark on the edge of the distributor base directly below the rotor tip and in line with it **(see illustration)**. Also, mark the distributor base and the engine block to ensure that the distributor is installed correctly.

6   Remove the distributor hold-down bolt and clamp **(see illustrations)**, then pull the distributor straight up to remove it. Be careful not to disturb the intermediate driveshaft. **Caution:** *DO NOT turn the engine while the distributor is removed, or the alignment marks will be useless.*

Installation

Note: *If the crankshaft has been moved while the distributor is out, locate Top Dead Center (TDC) for the number one piston and position the distributor and rotor accordingly.*

7   Insert the distributor into the engine in exactly the same relationship to the block that it was in when removed.

8   To mesh the helical gears on the camshaft and the distributor, it may be necessary to turn the rotor slightly. If the distributor

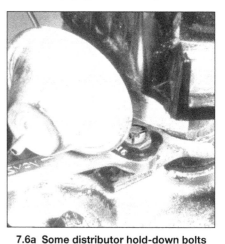

7.6a Some distributor hold-down bolts can be removed with an open-end wrench . . .

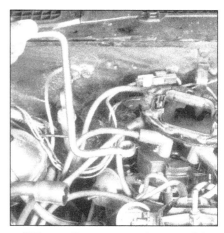

7.6b . . . others may require a special distributor wrench

doesn't seat completely, the hex shaped recess in the lower end of the distributor shaft is not mating properly with the oil pump shaft. Recheck the alignment marks between the distributor base and the block to verify that the distributor is in the same position it

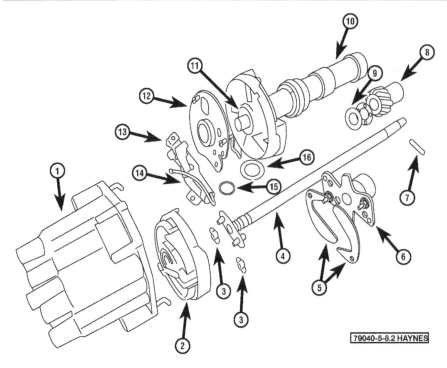

8.2 Exploded view of a typical contact breaker type distributor

1 Distributor cap	6 Cam and advance	12 Breaker plate
2 Rotor	weight base	13 Contact point
3 Advance weight	7 Drive gear roll pin	assembly
springs	8 Distributor drive gear	14 Condenser
4 Mainshaft	9 Washer and shim	15 Retaining ring
5 Advance weights	10 Distributor housing	16 Felt washer
	11 Plastic washer	

`79040-5-8.2 HAYNES`

was in before removal. Also check the rotor to see if it's aligned with the mark you made on the edge of the distributor base.
9   Place the hold-down clamp in position and loosely install the bolt.
10   Install the distributor cap and tighten the cap screws securely.
11   Plug in the module electrical connector.
12   Reattach the spark plug wires to the plugs (if removed).
13   Connect the cable to the negative terminal of the battery.
14   Check the ignition timing (see Section 10) and tighten the distributor hold-down bolt securely.

8   Distributor (breaker point type) - overhaul

Refer to illustration 8.2
1   Remove the distributor (See Section 7).
2   Remove the rotor (two screws), the advance weight springs and the weights **(see illustration)**. Where applicable, also remove the radio frequency interference (RFI) shield.
3   Drive out the roll pin retaining the gear to the shaft then pull off the gear and spacers.
4   Ensure that the shaft is not burred, then slide it from the housing.

5   Remove the cam weight base assembly.
6   Remove the screws retaining the vacuum unit and lift off the unit itself.
7   Remove the spring retainer (snap-ring) then remove the breaker plate assembly.
8   Remove the contact points and condenser, followed by the felt washer and plastic seal located beneath the breaker plate.
9   Wipe all components clean with a solvent-moistened cloth and examine them for wear, distortion and scoring. Replace parts as necessary. Pay particular attention to the rotor and distributor cap to ensure that they are not cracked.
10   Fill the lubricating cavity in the housing with general purpose grease, then fit a new plastic seal and felt washer.
11   Install the vacuum unit and the breaker plate in the housing, and the spring retainer on the upper bushing.
12   Lubricate the cam weight base and slide it on the mainshaft; install the weights and springs.
13   Insert the mainshaft in the housing then fit the shims and drivegear. Install a new roll pin.
14   Install the contact point set (see Section 4).
15   Install the rotor, aligning the round and square pilot holes.
16   Install the distributor (see Section 7).

9   Distributor (HEI type) - overhaul

Note: *To check the pick-up coil, see Section 6.*
1   Remove the distributor (see Section 7).
2   Remove the rotor (see Section 3).
3   Remove the two screws retaining the module. Move the module aside and remove the connector from the 'B' and 'C' terminals.
4   Remove the connections from the 'W' and 'G' terminals.
5   Carefully drive out the roll pin from the drive gear **(see illustration 7.7)**.
6   Remove the gear, shim and tanged washer from the distributor shaft.
7   Ensure that the shaft is not burred, then remove it from the housing.
8   Remove the washer from the upper end of the distributor housing.
9   Pry off the retaining ring and remove the pick-up coil and pole piece assembly **(see illustration 6.10)**.
10   Remove the lock ring, then take out the pick-up coil retainer, shim and felt washer.
11   Remove the vacuum unit (two screws).
12   Disconnect the capacitor lead and remove the capacitor (one screw).
13   Disconnect the wiring harness from the distributor housing.
14   Wipe all components clean with a solvent-moistened cloth and examine them for wear, distortion and other damage. Replace parts as necessary.
15   To assemble, position the vacuum unit to the housing and secure with the two screws.
16   Position the felt washer over the lubricant reservoir at the top of the housing, then position the shim on top of the felt washer.
17   Install the pick-up coil and pole-piece assembly. Make sure it engages the vacuum advance arm properly. Install the retaining ring.
18   Install the washer to the top of the housing. Install the distributor shaft, then rotate it and check for equal clearance all round between the shaft projections and pole-piece.
19   Install the tanged washer, shim and drivegear. Align the gear and install a new roll pin.
20   Loosely install the capacitor with one screw.
21   Install the connector to the 'B' and 'C' terminals on the module, with the tab at the top.
22   Apply silicone grease to the base of the module and secure it with two screws. The grease is essential to ensure good heat conduction.
23   Position the wiring harness, with the grommet in the housing notch, then connect the pink wire to the capacitor stud and the black wire to the capacitor mounting screw. Tighten the screw.
24   Connect the white wire from the pick-up coil to the module "W" terminal and the green to the 'G' terminal.
25   Install the advance weights, weight retainer (dimple downwards), and springs.

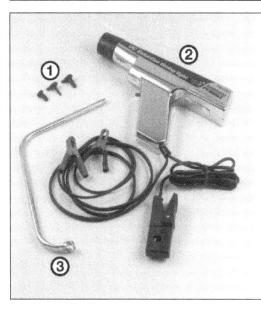

10.2 Tools needed to check and adjust the ignition timing

1   **Vacuum plugs** - *Vacuum hoses will, in most cases, have to be disconnected and plugged. Molded plugs in various shapes and sizes are available for this*
2   **Inductive pick-up timing light** - *Flashes a bright concentrated beam of light when the number one spark plug fires. Connect the leads according to the instructions supplied with the light*
3   **Distributor wrench** - *On some models, the hold-down bolt for the distributor is difficult to reach and turn with a conventional wrench or socket. A special wrench like this must be used*

7   Use chalk or paint to mark the groove in the crankshaft pulley.
8   Put a mark on the timing tab in accordance with the number of degrees called for on the VECI label or the tune-up label in the engine compartment. Each peak or notch on the timing tab represents two degrees. The word Before or the letter A indicates advance and the letter 0 indicates Top Dead Center (TDC). As an example, if your vehicle specifications call for eight degrees BTDC (Before Top Dead Center), you will make a mark on the timing tab four notches before the 0.
9   Check that the wiring for the timing light is clear of all moving engine components, then start the engine and warm it up to normal operating temperature.
10   Aim the flashing timing light at the timing mark by the crankshaft pulley, again being careful not to come in contact with moving parts. The marks should appear to be stationary. If the marks are in alignment, the timing is correct.
11   If the notch is not lining up with the correct mark, loosen the distributor hold-down bolt and rotate the distributor until the notch is lined up with the correct timing mark.
12   Retighten the hold-down bolt and recheck the timing.
13   Turn off the engine and disconnect the timing light. Reconnect the vacuum advance hose, if removed, and any other components which were disconnected.

26   Install the rotor and secure with the two screws. Ensure that the notch on the side of the rotor engages with the tab on the cam weight base.
27   Install the distributor (see Section 7).

10   Ignition timing - check and adjustment

Refer to illustrations 10.2 and 10.6
Note: *It is imperative that the procedures included on the tune-up or Vehicle Emissions Control Information (VECI) label be followed when adjusting the ignition timing. The label will include all information concerning preliminary steps to be performed before adjusting the timing, as well as the timing specifications.*

1   At the specified intervals, whenever the ignition points have been replaced, the distributor removed or a change made in the fuel type, the ignition timing should be checked and adjusted.

2   Locate the Tune-up or VECI label under the hood and read through and perform all preliminary instructions concerning ignition timing. Some special tools will be needed for this procedure **(see illustration)**.
3   Before attempting to check the timing, make sure the ignition point dwell angle is correct (Section 4), and the idle speed is as specified (Chapter 3).
4   If specified on the tune-up label, disconnect the vacuum hose from the distributor and plug the open end of the hose with a rubber plug, rod or bolt of the proper size.
5   Connect a timing light in accordance with the tool manufacturer's instructions. Generally, the light will be connected to power and ground sources and to the number one spark plug wire. The number one spark plug is the first spark plug on the right head (driver's side) as you are facing the engine from the front.
6   Locate the numbered timing tag on the front cover of the engine **(see illustration)**. It is located just behind the lower crankshaft pulley. Clean it off with solvent if necessary to read the printing and small grooves.

11   Ignition coil - check and replacement

Check

1   If the engine is hard to start (particularly when it's already hot), misses at high speed or cuts out during acceleration, the coil may be faulty. First, make sure that the battery and the distributor are in good condition, the points (pre-1975 vehicles) are properly adjusted and the plugs and plug wires are in good shape. If the problem persists, perform the following test:

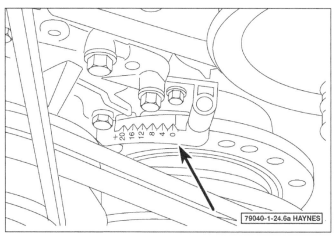

10.6   The ignition timing marks are located at the front of the engine

Breaker points ignition system

Refer to illustrations 11.3a and 11.3b

2   Before performing any of the following coil electrical checks, make sure that the coil is clean, free of any carbon tracks and that all connections are tight and free of corrosion. Also make sure that both battery terminals are clean and that the cables are securely attached (especially the ground strap at the negative terminal).

3   Detach the coil high tension cable from the distributor cap. Using an insulated tool, hold the end of the cable about 3/16-inch away from some grounded part of the engine, turn the ignition on and operate the starter. A bright blue spark should jump the gap.

   a) *If the spark is weak, yellowish or red, spark voltage is insufficient. If the points, condenser and battery are in good condition, the coil is probably weak. Check the coil primary and secondary resistance* **(see illustrations).** *Refer to the Specifications listed in the beginning of the Chapter. If the coil windings are damaged the resistance will be excessively high. Replace the coil.*

   b) *If there is no spark at all, try to locate the trouble before replacing the coil. Remove the distributor cap. Turn the engine until the points are open, or separate the points with a small piece of cardboard. Turn on the ignition switch. Using a 12-volt bulb with two test leads, attach one lead to ground somewhere on the engine and the other lead to first one of the coil's primary terminals and then the other:*

      1) *If the bulb lights when touched to the primary terminal that leads to the distributor, the coil is getting current and the primary windings are okay.*

      2) *If the bulb lights when touched to the other primary terminal but not when attached to the one leading to the distributor, the primary windings are faulty and the coil is no good.*

      3) *If the light does not go on when connected to either primary connection, the coil is not the problem. Check the ignition switch and starter solenoid.*

      4) *If the bulb lights when touched to both primary terminals (the coil is receiving current at both primary terminals), remove the high-tension cable from the center distributor cap tower and try shorting across the open distributor points with the tip of a clean (no oil) screwdriver.*

         (a) *If a spark jumps from the coil's high tension secondary wire to a grounded point on the engine as the screwdriver is removed, the points are either contaminated by oil, dirt or water, or they're burned.*

         (b) *If the screwdriver fails to produce a spark at the high tension wire, disconnect the primary wire that passes between the*

11.3a Measure the primary resistance between the ignition coil positive (+) and negative (-) terminals (points-type ignition)

11.3b Measure the secondary resistance between the positive (+) terminal and the ignition coil tower (points-type ignition)

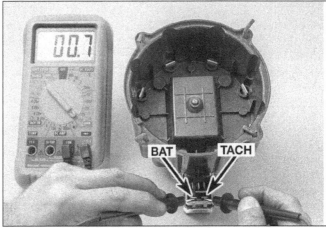

11.6 To test the HEI type coil-in-cap, attach the leads of an ohmmeter to the primary terminals and verify that the indicated resistance is zero or very near zero . . .

coil and the distributor and attach a test wire to the coil in its place. Ground the other end of the wire against the engine block, then pull it away (the test wire is simulating the points: grounding the wire is just like closing the points; pulling it away creates the same effect - producing a spark from the coil's high tension wire - as opening the points).

   (1) *If a spark jumps from the high-tension cable when the test wire is removed from the ground, the coil is okay. Either the points are grounded or the condenser is shorted.*

   (2) *If a spark does not jump during this test, the secondary windings of the coil are faulty. Replace the coil.*

4   Sometimes, a coil checks out perfectly but the engine is still hard to start and misses at higher speeds. The problem may be inadequate spark voltage caused by reversed coil polarity. If you have recently tuned up the engine or performed any service work involv-

ing the coil, it's possible that the primary leads to the coil were accidentally reversed. To check for reversed polarity, remove one of the spark plug leads and, using an insulated tool, hold it about 1/4-inch from the spark plug terminal or any ground point. Then insert the point of a pencil between the ignition lead and the plug while the engine is running (if the plug connector terminals are deeply recessed in a boot or insulating shield, straighten all but one bend in a paper clip and insert the looped end into the plug connector).

   a) *If the spark flares on the ground or spark plug side of the pencil, the polarity is correct.*

   b) *If the spark flares between the ignition lead and the pencil, however, the polarity is wrong and the primary wires should be switched at the coil.*

HEI ignition

Refer to illustrations 11.6 and 11.7

5   Remove the distributor cap (see Section 3).

6   Attach the leads of an ohmmeter across the two primary coil terminals as shown **(see illustration).** The indicated resistance should be less than one ohm.

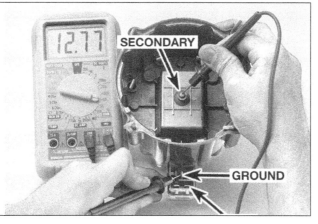

11.7 ... then, using the high scale, attach one lead to the high tension terminal and the other to the GROUND terminal and then the TACH terminal. Verify that both of the readings are not infinite - if the indicated resistance is not as specified, replace the coil

11.12 To remove a conventional coil, simply detach the high tension cable and the two primary leads and remove the mounting bracket screws

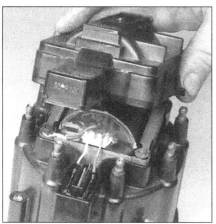

11.16a To get at the coil, remove the coil cover screws and the cover

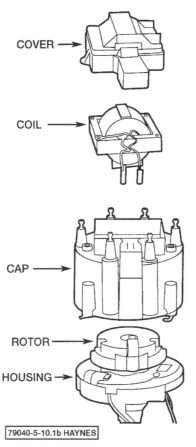

COVER →

COIL →

CAP →

ROTOR →

HOUSING →

79040-5-10.1b HAYNES

11.16b Typical HEI coil installation details

11.18 To separate the coil from the distributor cap, clearly mark the wires, detach the coil ground wire and push the leads from the underside of the connectors

7   Using the high scale, attach one lead of the ohmmeter to the SECONDARY terminal in the middle of the distributor and the other lead to the GROUND terminal **(see illustration)**.
8   Using the high scale, attach one lead of the ohmmeter to the SECONDARY terminal in the middle of the distributor and the other lead to the TACH terminal.
9   If both of the above readings indicate infinite resistance, replace the coil.

Replacement

Separately mounted coil

Refer to Illustration 11.12
10   Detach the cable from the negative terminal of the battery.
11   Disconnect the high tension cable from the coil.
12   Detach the electrical connections from the coil primary and secondary terminals. Be sure to mark the connections before removal to ensure that they are re-installed correctly. Remove the coil **(see illustration)**.
13   Installation is the reverse of removal

HEI-type coil-in-cap

Refer to Illustrations 11.16a, 11.16b, 11.18 and 11.19
14   Detach the cable from the negative terminal of the battery.

15   Disconnect the battery wire and harness connector from the distributor cap.
16   Remove the coil cover screws and the cover **(see illustrations)**.
17   Remove the coil assembly screws.
18   Note the position of each wire, marking them if necessary. Remove the coil ground wire, then push the leads from the underside of the connectors. Remove the coil from the distributor cap **(see illustration)**.
19   Installation is the reverse of the removal procedure. Be sure that the center electrode is in good shape **(see illustration)** and that the leads are connected to their original positions.

11.19 Before installing a new coil, make sure that the center electrode is in good condition

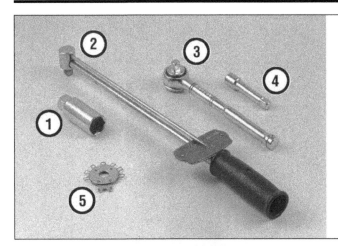

12.2 Tools necessary for changing spark plugs

1 **Spark plug socket** - This will have special padding inside to protect the spark plug's porcelain insulator
2 **Torque wrench** - Although not mandatory, using this tool is the best way to ensure the spark plugs are tightened properly
3 **Ratchet** - Standard hand tool to fit the spark plug socket
4 **Extension** - Depending on model and accessories, you may need special extensions and universal joints to reach one or more of the plugs
5 **Spark plug gap gauge** - This gauge for checking the gap comes in a variety of styles. Make sure the gap for your engine is included

12.5a Spark plug manufacturers recommend using a wire type gauge when checking the gap - if the wire does not slide between the electrodes with a slight drag, adjustment is required

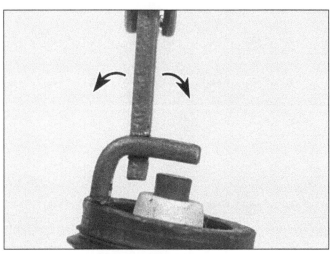

12.5b To change the gap, bend the side electrode only, as indicated by the arrows, and be very careful not to crack or chip the porcelain insulator surrounding the center electrode

12 Spark plugs - replacement

Refer to illustrations 12.2, 12.5a, 12.5b, 12.6 and 12.10

1 Open the hood.
2 In most cases, the tools necessary for spark plug replacement include a spark plug socket which fits onto a ratchet (spark plug sockets are padded inside to prevent damage to the porcelain insulators on the new plugs), various extensions and a gap gauge to check and adjust the gaps on the new plugs **(see illustration)**. A special plug wire removal tool is available for separating the wire boots from the spark plugs, but it isn't absolutely necessary. A torque wrench should be used to tighten the new plugs.
3 The best approach when replacing the spark plugs is to purchase the new ones in advance, adjust them to the proper gap and replace them one at a time. When buying the new spark plugs, be sure to obtain the correct plug type for your particular engine. This information can be found on the Emission Control Information label located under the hood and in the factory owner's manual. If differences

exist between the plug specified on the emissions label and in the owner's manual, assume that the emissions label is correct.
4 Allow the engine to cool completely before attempting to remove any of the plugs. While you're waiting for the engine to cool, check the new plugs for defects and adjust the gaps.
5 The gap is checked by inserting the proper thickness gauge between the electrodes at the tip of the plug **(see illustration)**. The gap between the electrodes should be the same as the one specified on the Emissions Control Information label. The wire should just slide between the electrodes with a slight amount of drag. If the gap is incorrect, use the adjuster on the gauge body to bend the curved side electrode slightly until the proper gap is obtained **(see illustration)**. If the side electrode is not exactly over the center electrode, bend it with the adjuster until it is. Check for cracks in the porcelain insulator (if any are found, the plug should not be used).
6 With the engine cool, remove the spark plug wire from one spark plug. Pull only on the boot at the end of the wire - do not pull

on the wire. A plug wire removal tool should be used if available **(see illustration)**.
7 If compressed air is available, use it to blow any dirt or foreign material away from the spark plug hole. The idea here is to eliminate the possibility of debris falling into the cylinder as the spark plug is removed.
8 Place the spark plug socket over the plug and remove it from the engine by turning it in a counterclockwise direction.
9 Compare the spark plug with the chart shown on the inside back cover to get an indication of the general running condition of the engine.
10 Thread one of the new plugs into the hole until you can no longer turn it with your fingers, then tighten it with a torque wrench (if available) or the ratchet. A good idea is to slip a short length of rubber hose over the end of the plug to use as a tool to thread it into place **(see illustration)**. The hose will grip the plug enough to turn it, but will start to slip if the plug begins to cross-thread in the hole - this will prevent damaged threads and the accompanying repair costs.
11 Before pushing the spark plug wire onto the end of the plug, inspect it following the

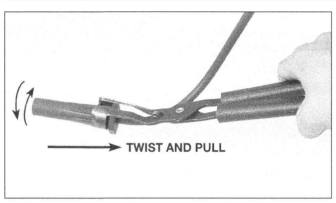

12.6 When removing a spark plug wire from a spark plug it is important to pull on the end of the boot and not on the wire itself. A slight twisting motion will also help

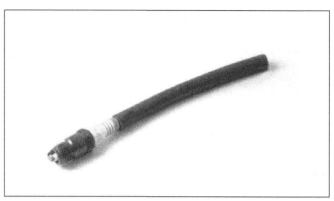

12.10 A length of rubber hose will save time and prevent damaged threads when installing the spark plugs

procedures outlined in Section 13.

12   Attach the plug wire to the new spark plug, again using a twisting motion on the boot until it's seated on the spark plug.

13   Repeat the procedure for the remaining spark plugs, replacing them one at a time to prevent mixing up the spark plug wires.

13 Spark plug wires - check and replacement

1   The spark plug wires should be checked at the recommended intervals and whenever new spark plugs are installed in the engine.

2   The wires should be inspected one at a time to prevent mixing up the order, which is essential for proper engine operation.

3   Disconnect the plug wire from one spark plug. To do this, grab the rubber boot, twist slightly and pull the wire free. Do not pull on the wire itself, only on the rubber boot **(see illustration 12.6)**.

4   Check inside the boot for corrosion, which will look like a white crusty powder. Push the wire and boot back onto the end of the spark plug. It should be a tight fit on the plug. If it isn't, remove the wire and use a pair of pliers to carefully crimp the metal connector inside the boot until it fits securely on the end of the spark plug.

5   Using a clean rag, wipe the entire length of the wire to remove any built-up dirt and grease. Once the wire is clean, check for holes, burned areas, cracks and other damage. Don't bend the wire excessively or the conductor inside might break.

6   Disconnect the wire from the distributor cap. A retaining ring at the top of the distributor may have to be removed to free the wires. Again, pull only on the rubber boot. Check for corrosion and a tight fit in the same manner as the spark plug end. Reattach the wire to the distributor cap.

7   Check the remaining spark plug wires one at a time, making sure they are securely fastened at the distributor and the spark plug when the check is complete.

8   If new spark plug wires are required, purchase a new set for your specific engine model. Wire sets are available pre-cut, with the rubber boots already installed. Remove and replace the wires one at a time to avoid mix-ups in the firing order. The wire routing is extremely important, so be sure to note exactly how each wire is situated before removing it.

Chapter 5   Clutch

Contents

Specifications

Type .... Dry plate, diaphragm spring

Operation .... Mechanical by rod linkage

Pedal free play .... 1 +/- 1/4 inch

Torque specifications   Ft-lbs

Pressure plate-to-flywheel bolts	25
Clutch housing-to-engine bolts	40
Transmission case-to-clutch housing bolts	55

1   Clutch - description and check

Refer to illustration 1.1

1   All vehicles with a manual transmission use a single dry plate, diaphragm spring type clutch **(see illustration)**. The clutch disc has a splined hub that allows it to slide along the splines of the transmission input shaft. The clutch and pressure plate are held in contact by spring pressure exerted by the diaphragm in the pressure plate.

2   The mechanical release system includes the clutch pedal, the clutch linkage which actuates the clutch release lever and the release bearing.

3   When pressure is applied to the clutch pedal to release the clutch, mechanical pressure is exerted against the outer end of the clutch release lever. As the lever pivots, the shaft fingers push against the release bearing. The bearing pushes against the fingers of the diaphragm spring of the pressure plate assembly, which in turn releases the clutch plate.

4   Terminology can be a problem when discussing the clutch components because common names are in some cases different from those used by the manufacturer. For example, the driven plate is also called the clutch plate or disc, the pressure plate is sometimes called the clutch cover, and the clutch release bearing is sometimes called a throwout bearing.

5   Other than to replace components with obvious damage, some preliminary checks

should be performed to diagnose clutch problems.

a) To check "clutch spin down time," run the engine at normal idle speed with the transmission in Neutral (clutch pedal up - engaged). Disengage the clutch (pedal down), wait several seconds and shift the transmission into Reverse. No grinding noise should be heard. A grinding noise would most likely indicate a problem in the pressure plate or the clutch disc.

b) To check for complete clutch release, run the engine (with the parking brake applied to prevent movement) and hold the clutch pedal approximately 1/2-inch from the floor. Shift the transmission between 1st gear and Reverse several times. If the shift is rough, component failure is indicated.

c) Visually inspect the pivot bushing at the top of the clutch pedal to make sure there is no binding or excessive play.

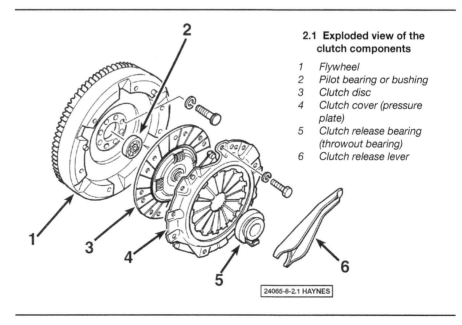

2.1 Exploded view of the clutch components

1   *Flywheel*
2   *Pilot bearing or bushing*
3   *Clutch disc*
4   *Clutch cover (pressure plate)*
5   *Clutch release bearing (throwout bearing)*
6   *Clutch release lever*

24065-8-2.1 HAYNES

d) On vehicles with mechanical release systems, a clutch pedal that is difficult to operate is most likely caused by faulty linkage. Check it for worn bushings and bent or distorted parts.

e) Crawl under the vehicle and make sure the clutch release lever is solidly mounted on the ball stud.

2   Clutch pedal free play - check and adjustment

Refer to illustration 2.3

Check

1   It is important to have the clutch free play at the proper point. Free play is the distance between the clutch pedal when it is all the way up and the point at which the clutch starts to disengage.

2   Slowly depress the clutch pedal until you can feel resistance. Do this a number of times until you can pinpoint exactly where the resistance is felt.

3   Now measure the distance the pedal travels before the resistance is felt **(see illustration)**.

4   If the distance is not as specified, the clutch pedal free play should be adjusted.

Adjustment

5   Disconnect the clutch fork return spring.

6   Hold the clutch pedal against the stop and loosen the jam nut so the clutch fork pushrod can be turned out of the swivel and back against the clutch fork. The release bearing must contact the pressure plate fingers lightly.

7   Turn the clutch fork pushrod three and a half turns.

8   Tighten the jam nut, taking care not to change the pushrod length, and reconnect the return spring.

9   Recheck the free play.

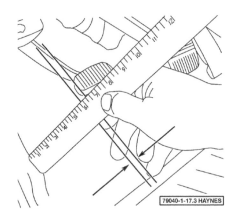

2.3 Clutch pedal free play is the distance the pedal moves before resistance is felt

3   Clutch components - removal, inspection and installation

Warning: *Dust produced by clutch wear and deposited on clutch components contains asbestos, which is hazardous to your health. DO NOT blow it out with compressed air and DO NOT inhale it. DO NOT use gasoline or petroleum-based solvents to remove the dust. Brake system cleaner should be used to flush the dust into a drain pan. After the clutch components are wiped clean with a rag, dispose of the contaminated rags and cleaner in a covered, marked container.*

Removal

Refer to illustration 3.7

1   Access to the clutch components is normally accomplished by removing the transmission, leaving the engine in the vehicle. If, of course, the engine is being removed for major overhaul, then check the clutch for wear and replace worn components as nec-

essary. However, the relatively low cost of the clutch components compared to the time and trouble spent gaining access to them warrants their replacement anytime the engine or transmission is removed, unless they are new or in near perfect condition. The following procedures are based on the assumption the engine will stay in place.

2   Referring to Chapter 6 Part A, remove the transmission from the vehicle. Support the engine while the transmission is out. Preferably, an engine hoist should be used to support it from above. However, if a jack is used underneath the engine, make sure a piece of wood is positioned between the jack and oil pan to spread the load. **Caution:** *The pickup for the oil pump is very close to the bottom of the oil pan. If the pan is bent or distorted in any way, engine oil starvation could occur.*

3   Remove the return spring and the clutch release lever pushrod.

4   Remove the clutch housing-to-engine bolts and then detach the housing. It may have to be gently pried off the alignment dowels with a screwdriver or pry bar.

5   The clutch fork and release bearing can remain attached to the housing for the time being.

6   To support the clutch disc during removal, install a clutch alignment tool through the clutch disc hub.

7   Carefully inspect the flywheel and pressure plate for indexing marks **(see illustration)**. The marks are usually an X, an O or a white letter. If they cannot be found, scribe marks yourself so the pressure plate and the flywheel will be in the same alignment during installation.

8   Turning each bolt only 1/4-turn at a time, loosen the pressure plate-to-flywheel bolts. Work in a criss-cross pattern until all spring pressure is relieved. Then hold the pressure plate securely and completely remove the bolts, followed by the pressure plate and clutch disc.

3.7 After removal of the transmission, this will be the view of the clutch components

| 1 | Pressure plate | 2 | Flywheel |

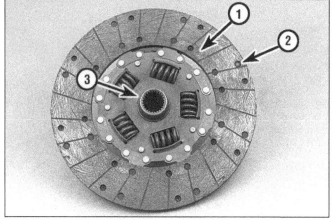

3.12 The clutch plate

1   *Lining* - this will wear down in use
2   *Rivets* - these secure the lining and will damage the flywheel or pressure plate if allowed to contact the surfaces
3   *Markings* - "Flywheel side" or something similar

 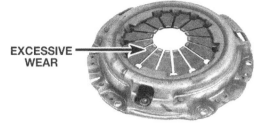 

NORMAL FINGER WEAR   EXCESSIVE → **EXCESSIVE FINGER WEAR**   **BROKEN OR BENT FINGERS**
WEAR

3.14a Replace the pressure plate if excessive wear or damage is noted

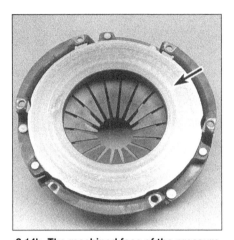

3.14b The machined face of the pressure plate must be inspected for score marks and other damage - if damage is slight, a machine shop can make the surface smooth again

3.16 A clutch alignment tool can be purchased at most auto parts stores and eliminates all guesswork when centering the clutch plate in the pressure plate

Inspection

Refer to illustrations 3.12, 3.14a and 3.14b

9   Ordinarily, when a problem occurs in the clutch, it can be attributed to wear of the clutch driven plate assembly (clutch disc). However, all components should be inspected at this time.

10   Inspect the flywheel for cracks, heat checking, grooves and other obvious defects. If the imperfections are slight, a machine shop can machine the surface flat and smooth, which is highly recommended regardless of the surface appearance. Refer to Chapter 2 for the flywheel removal and installation procedure.

11   Inspect the pilot bushing (Section 5).

12   Inspect the lining on the clutch disc. There should be at least 1/16 inch of lining above the rivet heads. Check for loose rivets, distortion, cracks, broken springs and other obvious damage **(see illustration)**. As mentioned above, ordinarily the clutch disc is routinely replaced, so if in doubt about the condition, replace it with a new one.

13   The release bearing should also be replaced along with the clutch disc (see Section 4).

14   Check the machined surfaces and the diaphragm spring fingers of the pressure plate **(see illustrations)**. If the surface is grooved or otherwise damaged, replace the pressure plate. Also check for obvious damage, distortion, cracking, etc. Light glazing can be removed with sandpaper or emery cloth. If a new pressure plate is required, new and factory-rebuilt units are available.

Installation

Refer to illustration 3.16

15   Before installation, clean the flywheel and pressure plate machined surfaces with lacquer thinner or acetone. It's important that no oil or grease is on these surfaces or the lining of the clutch disc. Handle the parts only with clean hands.

16   Position the clutch disc and pressure plate against the flywheel with the clutch held in place with an alignment tool **(see illustration)**. Make sure it's installed properly (most replacement clutch plates will be marked "flywheel side" or something similar - if not marked, install the clutch disc with the damper springs toward the transmission).

17   Tighten the pressure plate-to-flywheel bolts only finger tight, working around the pressure plate.

18   Center the clutch disc by ensuring the alignment tool extends through the splined hub and into the pilot bushing in the

crankshaft. Wiggle the tool up, down or side-to-side as needed to bottom the tool in the pilot bushing. Tighten the pressure plate-to-flywheel bolts a little at a time, working in a criss-cross pattern to prevent distorting the cover. After all of the bolts are snug, tighten them to the specified torque. Remove the alignment tool.

19   Using high-temperature grease, lubricate the inner groove of the release bearing (refer to Section 4). Also place grease on the release lever contact areas and the transmission input shaft bearing retainer.

20   Install the clutch release bearing as described in Section 4.

21   Install the clutch housing and tighten the bolts to the specified torque.

22   Install the transmission and all components removed previously. Tighten all fasteners to the proper torque specifications.

23   Refer to Section 2 for clutch pedal free play check and adjustment information.

4   Clutch release bearing and lever - removal, inspection and installation

Refer to illustrations 4.5, 4.6 and 4.7
Warning: *Dust produced by clutch wear and deposited on clutch components may contain asbestos, which is hazardous to your health. DO NOT blow it out with compressed air and DO NOT inhale it. DO NOT use gasoline or petroleum-based solvents to remove the dust. Brake system cleaner should be used to flush the dust into a drain pan. After the clutch components are wiped clean with a rag, dispose of the contaminated rags and cleaner in a covered, marked container.*

Removal

1   Disconnect the negative cable from the battery.

2   Remove the transmission (Chapter 6A).

3   Remove the clutch housing.

4   Remove the clutch release lever from the ball stud, then remove the bearing from the lever.

Inspection

5   Hold the center of the bearing and

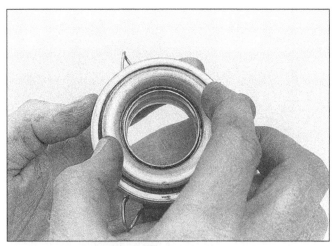

4.5 To check the bearing, hold it by the outer race and rotate the inner race while applying pressure; if the bearing doesn't turn smoothly or if it's noisy, replace the bearing

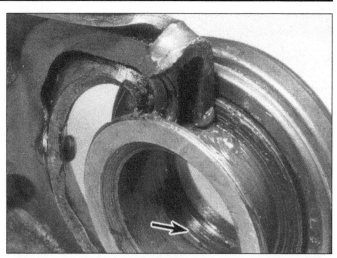

4.6 Lubricate the clutch release bearing recess (arrow) and the fork groove with high temperature grease

4.7 When installing the release bearing, make sure the fingers and the tabs fit into the bearing recess

5.5 Removing the pilot bushing with a slide hammer and internal puller attachment

rotate the outer portion while applying pressure **(see illustration)**. If the bearing doesn't turn smoothly or if it's noisy, replace it with a new one. Considering the difficulty in getting to this bearing, it is normally routinely replaced when removed. Wipe the bearing with a clean rag and inspect it for damage, wear and cracks. Don't immerse the bearing in solvent - it's sealed for life and to do so would ruin it.

Installation

6   Lightly lubricate the clutch lever crown and spring retention crown where they contact the bearing with high-temperature grease. Fill the inner groove of the bearing with the same grease **(see illustration)**.
7   Attach the release bearing to the clutch lever so that both fork tabs fit into the bearing recess **(see illustration)**.
8   Lubricate the clutch release lever ball socket with high-temperature grease and push the lever onto the ball stud until it's firmly seated.

9   Apply a light coat of high-temperature grease to the face of the release bearing, where it contacts the pressure plate diaphragm fingers.
10   Install the clutch housing and tighten the bolts to the specified torque.
11   Prior to installing the transmission, apply a light coat of grease to the transmission front bearing retainer.
12   The remainder of installation is the reverse of the removal procedure. Tighten all bolts to the specified torque.

5   Pilot bushing - inspection and replacement

Refer to illustrations 5.5 and 5.6
1   The clutch pilot bushing is pressed into the rear of the crankshaft. Its primary purpose is to support the front of the transmission input shaft. The pilot bushing should be inspected whenever the clutch components

are removed from the engine. Due to its inaccessibility, if you are in doubt as to its condition, replace it with a new one. **Note:** *If the engine has been removed from the vehicle, disregard the following steps which do not apply.*
2   Remove the transmission (refer to Chapter 6 Part A).
3   Remove the clutch components (Section 3).
4   Inspect for any excessive wear, scoring, lack of grease, dryness or obvious damage. If any of these conditions are noted, the bushing should be replaced. A flashlight will be helpful to direct light into the recess.
5   Removal requires a special puller and slide hammer **(see illustration)**. Remove the bearing with the puller and clean the crankshaft recess.
6   To install the new bearing, lightly lubricate the outside surface with lithium-based grease, then drive it into the recess with a properly sized driver and a hammer **(see illustration)**.

5.6 Install the pilot bushing with a properly sized driver or socket

and pulling lightly up a moderate incline, sudden depression of the accelerator pedal may cause the engine to increase its speed without any increase in road speed. Easing off on the accelerator will then give a definite drop in engine speed without the car slowing.

7   In extreme cases of clutch Slip the engine will race under normal acceleration conditions.

Clutch spin - diagnosis and cure

1   Clutch spin is a condition which occurs when the clutch fork travel is excessive, there is an obstruction in the clutch either on the primary gear splines, or in the operating lever itself, or the oil may have partially burnt off the clutch linings and have left a resinous deposit which is causing the clutch plate to stick to the pressure plate or flywheel.

2   The reason for clutch spin is that due to any, or a combination of, the faults just listed, the clutch pressure plate is not completely freeing from the center plate even with the clutch pedal fully depressed.

3   If clutch spin is suspected, the condition can be confirmed by extreme difficulty in engaging first gear from rest, difficulty in changing gear, and very sudden take up of the clutch drive at the fully depressed end of the clutch pedal travel as the clutch is released.

4   Check that the clutch pedal free movement is correctly adjusted and, if in order, the fault lies internally in the clutch. It will then be necessary to remove the clutch for examination.

Clutch chatter - diagnosis and cure

1   Clutch chatter is a self evident condition which occurs when the transmission or engine mountings are loose or too flexible, when there is oil on the faces of the clutch driven plate, or when the clutch pressure plate has been incorrectly adjusted during assembly.

2   The reason for clutch chatter is that due to one of the faults just listed, the clutch pressure plate is not freeing smoothly from the driven plate and is snatching,

3   Clutch chatter normally occurs when the clutch pedal is released in first or reverse gears, and the whole car shudders as it moves backwards or forwards.

11   Install the clutch components, transmission and all other components removed previously, tightening all fasteners properly.

6   Clutch pedal - removal and installation

1   Disconnect the negative cable from the battery.

2   Disconnect the clutch and brake pedal pushrods and pedal return springs.

3   Remove the pedal pivot nut and slide the pedal to the left to remove it.

4   Wipe clean all parts; however, don't use cleaning solvent on the plastic bushings. Replace all worn parts with new ones.

5   Installation is the reverse of removal. Check that the starter safety switch allows the vehicle to be started only with the clutch pedal fully depressed.

7   Troublehooting - clutch

There are four main faults to which the clutch and release mechanism are prone. They may occur by themselves or in conjunction with any of the other faults. They are clutch squeal, slip, spin and chatter.

Clutch squeal - diagnosis and cure

1   If the clutch squeals when depressing the clutch pedal, this is a sure indication of a badly worn clutch throwout bearing.

2   As well as regular wear due to normal use, wear of the clutch throwout bearing is much accentuated if the clutch is ridden, or held down for long periods in gear, with the engine running. To minimize wear of this component the car should always be taken out of gear at traffic lights and for similar hold-ups.

3   The clutch bearing is not an expensive item, but difficult to get at.

Clutch slip - diagnosis and cure

1   Clutch slip is a self-evident condition which occurs when the clutch driven plate is badly worn, oil or grease having got onto the flywheel or pressure plate faces, or the pressure plate itself is faulty.

2   The reason for clutch slip is that, due to one of the faults listed above, there is either insufficient pressure from the pressure plate, or insufficient friction from the driven plate to ensure solid drive.

3   If small amounts of oil get onto the clutch, they will be burnt off under the heat of clutch engagement, and in the process, gradually darkening the linings. Excessive oil on the clutch will burn off leaving a carbon deposit which can cause quite bad slip, or fierceness, spin and chatter.

4   The only cure for this condition is to replace the driven plate and trace and rectify the oil leak. This will either be from the transmission input shaft oil seal or the crankshaft rear main bearing oil seal.

5   If clutch slip is suspected, and confirmation of this condition is required, there are several tests which can be made.

6   With the engine in second or third gear

Chapter 6 Part A
Manual transmission

Contents

Specifications

Torque specifications

	Ft-lb
Cover-to-case bolt	17
Transmission-to-clutch housing bolt	55
Transmission-to-engine bolt	40
Extension housing-to-case bolt	46
Transmission mount bolt	30
Input shaft retainer bolt	22
Trunnion jam nut	30
Shift fork-to-shifter rail set screw	14
Shift linkage clamp screw	20
Crossmember bolt	25

1 General information

All vehicles covered in this manual come equipped with a 3-speed or 4-speed manual transmission or an automatic transmission. All information on the manual transmission is included in this Part of Chapter 6. Information on the automatic transmission can be found in Part B of this Chapter.

The manual transmissions used in these models are synchromesh units.

Due to the complexity, unavailability of replacement parts and the special tools necessary, internal repair by the home mechanic is not recommended. The information in this Chapter is limited to general information and removal and installation of the transmission.

Depending on the expense involved in having a faulty transmission overhauled, it may be a good idea to replace the unit with either a new or rebuilt one. Your local dealer or transmission shop should be able to supply you with information concerning cost, availability and exchange policy. Regardless of how you decide to remedy a transmission problem, you can still save a lot of money by removing and installing the unit yourself.

2 Column shift and backdrive linkage - adjustment

1 Shift the transmission into Reverse and turn the ignition key to the Lock position.
2 Raise the vehicle and support it securely on jackstands.
3 Loosen the swivel clamp screws at the transmission lever (Reverse and 1st gear) and at the cross shaft.
4 Place the front (2nd and 3rd gear) transmission shift lever in Neutral and the rear shift lever (Reverse and 1st gear) in Reverse.
5 Tighten swivel clamp screw securely, unlock the steering column and place the shift lever in Neutral.
6 Line up the lower shift levers and insert a 0.185 to 0.186-inch gauge pin through the hole in the levers.
7 Tighten swivel clamp screw and remove the gage pin. Check the operation of the linkage by moving it through the complete shift pattern.
8 Shift the transmission into 3rd gear and adjust the Transmission Controlled Spark (TCS) switch so that the plunger is fully depressed against the lever in 3rd and 4th gear.

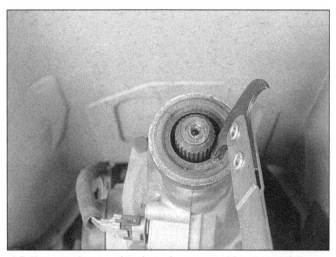

4.5 Use a seal removal tool or a long screwdriver to carefully pry the seal out of the end of the transmission

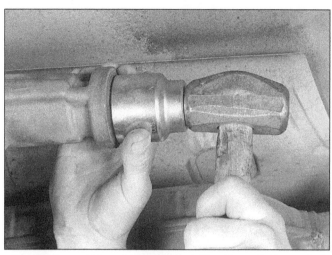

4.7 A large socket works well for installing the seal - the socket should come in contact with the outer edge of the seal

3 Transmission mount - check and replacement

1   Insert a large screwdriver or pry bar into the space between the transmission extension housing and the crossmember and pry up.
2   The transmission should not spread excessively away from the insulator.
3   To replace, remove the nuts attaching the insulator to the crossmember and the bolts attaching the insulator to the transmission.
4   Raise the transmission slightly with a jack and remove the insulator, noting which holes are used in the crossmember for proper alignment during installation.
5   Installation is the reverse of the removal procedure.

4 Transmission oil seal - replacement

Refer to illustrations 4.5 and 4.7
1   Oil leaks frequently occur due to wear of the extension housing oil seal and bushing (if equipped), and/or the speedometer drive gear oil seal and O-ring. Replacement of these seals is relatively easy, since the repairs can usually be performed without removing the transmission from the vehicle.
2   The extension housing oil seal is located at the extreme rear of the transmission, where the driveshaft is attached. If leakage at the seal is suspected, raise the vehicle and support it securely on jackstands. If the seal is leaking, transmission lubricant will be built up on the front of the driveshaft and may be dripping from the rear of the transmission.
3   Refer to Chapter 7 and remove the driveshaft.
4   Using a soft-faced hammer, carefully tap the dust shield (if equipped) to the rear and remove it from the transmission. Be careful not to distort it.

5   Using a screwdriver or pry bar, carefully pry the oil seal and bushing (if equipped) out of the rear of the transmission (**see illustration**). Take care not to damage the splines on the transmission output shaft.
6   If the oil seal and bushing cannot be removed with a screwdriver or pry bar, a special oil seal removal tool (available at auto parts stores) will be required.
7   Using a large section of pipe or a very large deep socket as a drift, install the new oil seal (**see illustration**). Drive it into the bore squarely and make sure it's completely seated. Install a new bushing using the same method.
8   Reinstall the dust shield by carefully tapping it into place. Lubricate the splines of the transmission output shaft and the outside of the driveshaft sleeve yoke with lightweight grease, then install the driveshaft. Be careful not to damage the lip of the new seal.
9   The speedometer cable and driven gear housing is located on the side of the extension housing. Look for transmission oil around the cable housing to determine if the seal and O-ring are leaking.
10   Disconnect the speedometer cable.
11   Using a hook, remove the seal.
12   Using a small socket as a drift, install the new seal.
13   Install a new O-ring in the driven gear housing and reinstall the driven gear housing and cable assembly on the extension housing.

5 Manual transmission - removal and installation

Removal

1   Disconnect the negative cable at the battery.
2   Disconnect the shift linkage.
3   Raise the vehicle and support it securely on jackstands.
4   Disconnect the speedometer cable and

wire harness connectors from the transmission.
5   Remove the driveshaft (Chapter 7). Use a plastic bag to cover the end of the transmission to prevent fluid loss and contamination.
6   Remove the exhaust system components as necessary for clearance (Chapter 3).
7   Support the engine. This can be done from above with an engine hoist, or by placing a jack (with a block of wood as an insulator) under the engine oil pan. The engine should remain supported at all times while the transmission is out of the vehicle.
8   Support the transmission with a jack - preferably a special jack made for this purpose. Safety chains will help steady the transmission on the jack.
9   Remove the rear transmission support-to-crossmember nuts/bolts.
10   Remove the nuts from the crossmember bolts. Raise the transmission slightly and remove the crossmember.
11   Remove the bolts securing the transmission to the clutch housing.
12   Make a final check that all wires and hoses have been disconnected from the transmission and then move the transmission and jack toward the rear of the vehicle until the transmission input shaft is clear of the clutch housing. Keep the transmission level as this is done.
13   Once the input shaft is clear, lower the transmission and remove it from under the vehicle.
14   The clutch components can be inspected by removing the clutch housing from the engine. In most cases, new clutch components should be routinely installed if the transmission is removed.

Installation

15   If removed, install the clutch components (Chapter 5).
16   If removed, attach the clutch housing to the engine and tighten the bolts securely.
17   With the transmission secured to the

jack, as on removal, raise the transmission into position behind the clutch housing and then carefully slide it forward, engaging the input shaft with the clutch plate hub. Do not use excessive force to install the transmission - if the input shaft does not slide into place, readjust the angle of the transmission so it is level and/or turn the input shaft so the splines engage properly with the clutch.

18   Install the transmission-to-clutch housing bolts. Tighten the bolts to the specified torque.

19   Install the crossmember and transmission support. Tighten all nuts and bolts securely.

20   Remove the jacks supporting the transmission and the engine.

21   Install the various items removed previously, referring to Chapter 7 for the installation of the driveshaft and Chapter 3 for information regarding the exhaust system components.

22   Make a final check that all wires, hoses and the speedometer cable have been connected and that the transmission has been filled with lubricant to the proper level. Lower the vehicle.

23   Connect and adjust the shift linkage.

24   Connect the negative battery cable. Road test the vehicle for proper operation and check for leakage.

6   Manual transmission overhaul - general information

Overhauling a manual transmission is a difficult job for the do-it-yourselfer. It involves the disassembly and reassembly of many small parts. Numerous clearances must be precisely measured and, if necessary, changed with select fit spacers and snap-rings. As a result, if transmission problems arise, it can be removed and installed by a competent do-it-yourselfer, but overhaul should be left to a transmission repair shop. Rebuilt transmissions may be available - check with your dealer parts department and auto parts stores. At any rate, the time and money involved in an overhaul is almost sure to exceed the cost of a rebuilt unit.

Nevertheless, it's not impossible for an inexperienced mechanic to rebuild a transmission if the special tools are available and the job is done in a deliberate step-by-step manner so nothing is overlooked.

The tools necessary for an overhaul include internal and external snap-ring pliers, a bearing puller, a slide hammer, a set of pin punches, a dial indicator and possibly a hydraulic press. In addition, a large, sturdy workbench and a vise or transmission stand will be required.

During disassembly of the transmission, make careful notes of how each piece comes off, where it fits in relation to other pieces and what holds it in place.

Before taking the transmission apart for repair, it will help if you have some idea what area of the transmission is malfunctioning. Certain problems can be closely tied to specific areas in the transmission, which can make component examination and replacement easier.

7   Troubleshooting - manual transmission

Symptom	Reason
Stiff hand control lever	Linkage out of adjustment Lack of lubrication Wear in linkage components
Gear clash on shift	Linkage out of adjustment Worn synchronizer units
Slipping out of top gear	Transmission case to clutch housing bolts loose Binding linkage Drive gear retainer broken or loose
Noisy operation (all gears)	Insufficient oil Worn countergear bearings Worn countergear anti-lash plate Worn or damaged mainshaft bearings Worn or damaged gearwheels
Noisy operation (top gear)	Damaged main drive gear bearing Damaged mainshaft bearing, Damaged top gear synchronizer unit
Noisy operation (intermediate gears)	Worn constant mesh gears Worn synchronizer unit Worn countergear bearings
Noisy operation (reverse gear)	Worn idler or bush Worn or damaged mainshaft reverse gear Worn or damaged reverse countergear
Noisy operation (in neutral, engine running)	Worn or loose pilot bushing Worn countergear anti-lash plate Worn countergear bearings Damaged main drive gear bearing
Excessive backlash in all gears	Worn countergear bearings Excessive end-play in countergear

Chapter 6 Part B
Automatic transmission

Contents

Specifications

Torque specifications

Ft-lbs (unless otherwise indicated)

Note: One foot-pound (ft-lb) of torque is equivalent to 12 inch-pounds (in-lbs) of torque. Torque values below approximately 15 ft-lbs are expressed in inch-pounds, because most foot-pound torque wrenches are not accurate at these smaller values.

Transmission-to-engine bolts	40
Floor shift lever-to-shift cable pin retaining nut	132 to 180 in-lbs
Torque converter-to-driveplate bolts	
1970 through 1985 models	25 to 35
1986 and later	46

1 General information and diagnosis

General information

1 All vehicles covered in this manual come equipped with either a 3-speed or 4-speed manual transmission or an automatic transmission. All information on the automatic transmission is included in this Part of Chapter 7. Information on the manual transmission can be found in Part A of this Chapter.

2 Due to the complexity of the automatic transmissions covered in this manual and the need for specialized equipment to perform most service operations, this Chapter contains only general diagnosis, routine maintenance, adjustment and removal and installation procedures.

3 If the transmission requires major repair work, it should be left to a dealer service department or an automotive or transmission repair shop. You can, however, remove and install the transmission yourself and save the expense, even if the repair work is done by a transmission shop.

Diagnosis (general)

Note: Automatic transmission malfunctions may be caused by five general conditions: poor engine performance, improper adjustments, hydraulic malfunctions, mechanical malfunctions or malfunctions in the computer or its signal network. Diagnosis of these problems should always begin with a check of the easily repaired items: fluid level and condition (Chapter 1), shift linkage adjustment and throttle linkage adjustment. Next, perform a road test to determine if the problem has been corrected or if more diagnosis is necessary. If the problem persists after the preliminary tests and corrections are completed, additional diagnosis should be done by a dealer service department or transmission repair shop.

Preliminary checks

4 Drive the vehicle to warm the transmission to normal operating temperature.

5 Check the fluid level as described in Section 3:

a) If the fluid level is unusually low, add enough fluid to bring the level within the designated area of the dipstick, then check for external leaks (see below).

b) If the fluid level is abnormally high, drain off the excess, then check the drained fluid for contamination by coolant. The presence of engine coolant in the automatic transmission fluid indicates that a failure has occurred in the internal radiator walls that separate the coolant from the transmission fluid.

c) If the fluid is foaming, drain it and refill the transmission, then check for coolant in the fluid or a high fluid level.

6 Check the engine idle speed. **Note:** If the engine is malfunctioning, do not proceed

A   **B**   **C**

2.1 The fluid pan gasket shape can help you determine which transmission your vehicle is equipped with

A   THM400   *B   THM250/350*   *C   Powerglide*

3.3 The automatic transmission dipstick is located at the rear of the engine compartment, usually on the right side

with the preliminary checks until it has been repaired and runs normally.

7   Check the kickdown cable for freedom of movement. Adjust it if necessary (Section 9).

8   Inspect the shift control linkage. Make sure that it's properly adjusted and that the linkage operates smoothly.

Fluid leak diagnosis

9   Most fluid leaks are easy to locate visually. Repair usually consists of replacing a seal or gasket. If a leak is difficult to find, the following procedure may help.

10   Identify the fluid. Make sure it's transmission fluid and not engine fluid or brake fluid (automatic transmission fluid is a deep red color).

11   Try to pinpoint the source of the leak. Drive the vehicle several miles, then park it over a large sheet of cardboard. After a minute or two, you should be able to locate the leak by determining the source of the fluid dripping onto the cardboard.

12   Make a careful visual inspection of the suspected component and the area immediately around it. Pay particular attention to gasket mating surfaces. A mirror is often helpful for finding leaks in areas that are hard to see.

13   If the leak still cannot be found, clean the suspected area thoroughly with a degreaser or solvent, then dry it.

14   Drive the vehicle for several miles at normal operating temperature and varying speeds. After driving the vehicle, visually inspect the suspected component again.

15   Once the leak has been located, the cause must be determined before it can be properly repaired. If a gasket is replaced but the sealing flange is bent, the new gasket will not stop the leak. The bent flange must be straightened.

16   Before attempting to repair a leak, check to make sure that the following conditions are corrected or they may cause another leak. **Note:** *Some of the following conditions cannot be fixed without highly specialized tools and expertise. Such problems must be referred to a transmission repair shop or a dealer service department.*

Gasket leaks

17   Check the pan periodically. Make sure the bolts are tight, no bolts are missing, the gasket is in good condition and the pan is flat

(dents in the pan may indicate damage to the valve body inside).

18   If the pan gasket is leaking, the fluid level or the fluid pressure may be too high, the vent may be plugged, the pan bolts may be too tight, the pan sealing flange may be warped, the sealing surface of the transmission housing may be damaged, the gasket may be damaged or the transmission casting may be cracked or porous. If sealant instead of gasket material has been used to form a seal between the pan and the transmission housing, it may be the wrong sealant.

Seal leaks

19   If a transmission seal is leaking, the fluid level or pressure may be too high, the vent may be plugged, the seal bore may be damaged, the seal itself may be damaged or improperly installed, the surface of the shaft protruding through the seal may be damaged or a loose bearing may be causing excessive shaft movement.

20   Make sure the dipstick tube seal is in good condition and the tube is properly seated. Periodically check the area around the speedometer gear or sensor for leakage. If transmission fluid is evident, check the O-ring for damage.

Case leaks

21   If the case itself appears to be leaking, the casting is porous and will have to be repaired or replaced.

22   Make sure the fluid cooler hose fittings are tight and in good condition.

Fluid comes out vent pipe or fill tube

23   If this condition occurs, the transmission is overfilled, there is coolant in the fluid, the case is porous, the dipstick is incorrect, the vent is plugged or the drain back holes are plugged.

2   Transmission identification

Refer to illustration 2.1

1   Besides checking the transmission serial number, there is a quick way to determine which of the several automatic transmissions a particular vehicle is equipped with. Read the following transmission fluid pan descriptions and refer to accompanying **illustration 2.1** to identify the various models.

Powerglide

2   This two-speed transmission case is made of either cast iron or aluminum. The word 'Powerglide' is plainly stamped on the case. The shift quadrant indicator is set up in one of two ways: P-N-D-L-R or P-R-N-D-L.

Turbo Hydra-Matic 250/350

3   This closely resembles the THM 200 but the fluid pan has 13 bolts.

Turbo Hydra-Matic 400

4   The case of the three-speed THM400 is also two-piece, but the downshifting is electrically controlled from a switch at the carburetor to the left side of the transmission. The fluid pan also has 13 bolts. The shape of the pan is elongated and irregular.

3   Automatic transmission fluid - level check

Refer to illustrations 3.3 and 3.4

1   Drive the vehicle for a minimum distance of 15 miles to ensure that the transmission fluid is at normal operating temperature.

2   Park the car on a level surface, place selector lever in 'Neutral' (Powerglide) or 'Park' (Turbo Hydra-Matic), apply the parking brake fully and let the engine idle.

3   Withdraw the dipstick (located at right rear of the engine compartment), wipe it clean and re-insert it **(see illustration)**.

4   Withdraw the dipstick for the second time and check the fluid level **(see illustration)**. If necessary, pour in fluid (into the dipstick guide tube) of the correct grade to bring it up to the 'Full' mark. **Caution:** *Add fluid a little at a time to avoid overfilling.*

4   Automatic transmission fluid - change

Refer to illustrations 4.7, 4.10 and 4.11

1   Every 24,000 miles change the fluid in the transmission unit. Should the vehicle be

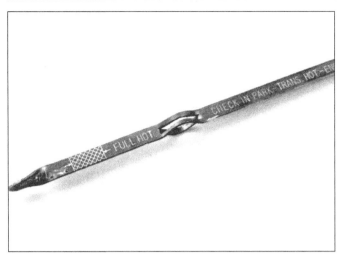

3.4 When checking the automatic transmission fluid level it is important to note the fluid temperature

4.7 With the rear bolts in place but loose, pull the front of the pan down to let the fluid drain

operating under severe conditions such as hauling a trailer or stop/start delivery operations then have the fluid change frequency to 12,000 miles.

2 Before beginning work, purchase the specified transmission fluid and a new filter.

3 Other tools necessary for this job include jackstands to support the vehicle in a raised position, a drain pan capable of holding at least eight pints, newspapers and clean rags.

4 Raise the vehicle and support it securely on jackstands.

5 With a drain pan in place, remove the front and side pan mounting bolts.

6 Loosen the rear pan bolts approximately four turns.

7 Carefully pry the transmission pan loose with a screwdriver, allowing the fluid to drain **(see illustration)**.

8 Remove the remaining bolts, pan and gasket. Carefully clean the gasket surface of the transmission to remove all traces of the old gasket and sealant.

9 Drain the fluid from the transmission pan, clean it with solvent and dry it with

compressed air.

10 Remove the filter from the mount inside the transmission **(see illustration)**.

11 Install a new filter and (if equipped) O-ring **(see illustration)**.

12 Make sure the gasket surface on the transmission pan is clean, then install a new gasket. Put the pan in place against the transmission and, working around the pan, tighten each bolt a little at a time until the final torque figure is reached.

13 Lower the vehicle and add two quarts (Powerglide)or 2-1/2 quarts (Turbo Hydra-Matic) of the specified transmission fluid through the filler tube.

14 With the transmission in Park and the parking brake set, run the engine at a fast idle, but don't race it.

15 Move the gear selector through each range and back to Park. Check the fluid level, adding fluid a little at a time until it is within the desired range on the dipstick. **Caution:** *Do not overfill!*

16 Check under the vehicle for leaks during the first few trips.

5 Column shift linkage - checking and adjustment

1 The selector linkage will be in need of adjustment if at any time 'Low' or 'Reverse' can be obtained without first having to lift the shift control lever to enable it to pass over the mechanical stop.

1973 and earlier models

2 If adjustment is required, release the control rod swivel or clamp and set the lever on the side of the transmission in the 'Drive' or L2 detent. This can be clearly defined by placing the lever in L or L1 and moving the lever back one detent (click).

3 Position the shift control lever up against the "Drive" stop and then tighten the swivel or clamp on the control rod.

4 Check all selector positions, especially "Park." In some cases, especially with worn linkage, it may be necessary to readjust slightly in order to ensure that the "Park" detent is fully engaged.

4.10 After removing the bolts or screws, lower the filter from the transmission

4.11 Some models have an O-ring on the filter spout; if it does not come out with the filter, reach up into the opening with your finger to retrieve it

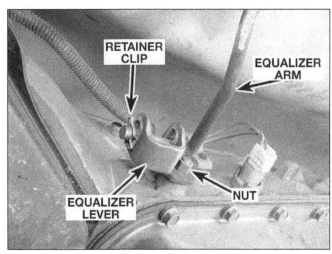 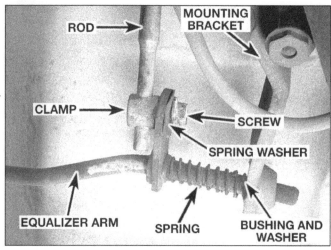

5.7 Shift linkage adjustment details (1974 and later models)

1974 and later models

Refer to illustration 5.7

5   Place the shift lever in the Neutral position of the shift indicator.

6   Position the transmission shift lever in the Neutral detent.

7   Install the clamp spring and screw assembly on the equalizer lever control rod **(see illustration)**.

8   Hold the clamp flush against the equalizer lever and tighten the clamp screw finger tight. Make sure that no force is exerted in either direction on the rod or equalizer lever while the screw is tightened.

9   Tighten the screw securely.

10   Check that the ignition key can be moved freely to the "Lock' position when the shift lever is in "Park' and not in any other position.

6   Floor shift linkage - check and adjustment

1   If the engine can be started in any of the Drive positions and the Neutral start switch is properly adjusted (Section 11), the shift linkage must be adjusted.

Powerglide

2   Set the shift lever in "Drive."

3   Working under the vehicle, disconnect the selector cable from the lever on the side of the transmission.

4   Move the lever on the side of the transmission to the "Drive' detent.

5   Measure the distance from the rear face of the cable mounting bracket to the center of the cable pivot stud. This should be 5-1/2 inches. Adjust the position of the stud if necessary to achieve this measurement.

6   Adjust the cable in its mounting bracket so that the cable end fits freely onto the pivot stud. Install the cable retaining clip.

7   Working inside the vehicle, remove the shift quadrant cover, plate and illumination bulbs.

8   Remove the selector cable clip and disconnect the cable from the shift lever.

9   Insert a gauge (0.07-inch [5/64-inch] thick) between the pawl and the detent plate, then measure the distance between the front face of the shifter assembly bracket and the center of the cable pivot pin. This should be 6-1/4 inches. If it is not, loosen the bolt and move the lever as necessary.

10   Adjust the cable mounting to the shifter bracket until the cable eye freely enters over the pivot pin. If at any time depressing the handle button does not clear the cut-outs in the detent plate, or conversely if the handle can be moved to P and R positions without depressing the button, raise or lower the detent plate after loosening the retaining bolt.

Turbo Hydra-Matic

Note: *Apply the parking brake and block the wheels to prevent the vehicle from rolling.*

11   Working under the vehicle, loosen the nut attaching the shift lever to the pin on the shift cable assembly.

12   Place the console shifter (inside the vehicle) in Park.

13   Place the manual shifting shaft on the transmission in Park.

14   Move the pin to give a 'free pin' fit, making sure that the console shifter is still in the Neutral position, and tighten the pin-to-lever retaining nut to the specified torque.

15   Make sure the engine will start in the Park and Neutral positions only.

16   If the engine can be started in any of the drive positions (as indicated by the shifter inside the vehicle), repeat the steps above or have the vehicle examined by a dealer, because improper linkage adjustment can lead to band or clutch failure and possible personal injury.

7   Powerglide - on-vehicle adjustments

Low band adjustment

1   This adjustment should normally be carried out at the time of the first fluid change

and thereafter only when unsatisfactory performance indicates it to be necessary (see Section 14).

2   Raise the vehicle to provide access to the transmission, making sure to secure the vehicle on jackstands.

3   Place the selector lever in the Neutral position.

4   Remove the protective cap from the transmission band adjusting screw.

5   Release and unscrew the adjusting screw locknut one-quarter turn and hold it in this position with a wrench.

6   Using a torque wrench and adapter, tighten the adjusting screw to 70 in-lbs., then back off the screw the exact number of turns as follows:

   a) *For a band which has been in operation for less than 6000 miles - three complete turns.*

   b) *For a band which has been in operation for more than 6000 miles - four complete turns.*

7   Tighten the adjusting screw locknut.

Throttle valve/linkage adjustment

8   Remove the air cleaner. Disconnect the accelerator linkage at the carburetor, and the accelerator and throttle valve return springs.

9   Pull the throttle valve upper rod forward with the right hand then open the carburetor throttle side with the left hand.

10   Adjust the swivel on the end of the upper throttle valve rod so that the ball stud contacts the end of the slot in the upper throttle valve rod as the carburetor reaches the wide open throttle (WOT) position.

11   Connect and adjust the accelerator linkage.

Neutral start switch alignment

12   This operation will only be required if a new switch is being installed.

13   Set the shift lever in Neutral and locate the lever tang against the transmission selector plate.

9.1 Pry out the release tab with a screwdriver

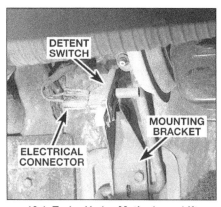

10.1 Turbo Hydra-Matic downshift (detent) switch details

14   Align the slot in the contact support with the hole in the switch by inserting a 3/32-inch diameter pin.
15   Place the contact support drive slot over the shifter tube drive tang and tighten the screws. Withdraw the pin.
16   Connect the switch wires and check that the operation of the switch is correct when the ignition is switched on.

8   Turbo Hydra-Matic 250 - on-vehicle adjustments

Intermediate band adjustment

1   This adjustment should be carried out every 24,000 miles or if the performance of the transmission indicates the need for it.
2   Raise the vehicle to gain access to the transmission, making sure to secure the vehicle on jackstands.
3   Place the speed selector lever in Neutral.
4   The adjusting screw and locknut for the intermediate band is located on the right-hand side of the transmission case.
5   Loosen the locknut 1/4-turn using a wrench or special tool, available at most auto parts stores. Hold the locknut in this position and tighten the adjusting screw to a torque of

30 in-lbs. Now back off the screw three complete turns exactly.
6   Without moving the adjusting screw, tighten the locknut to 15 ft-lbs.

Downshift (detent) cable adjustment

7   The cable will normally only require adjustment if a new one has been installed.
8   Depress the accelerator pedal fully. The ball will slide into the cable sleeve and automatically pre-set the cable tension.

9   Turbo Hydra-Matic 350 downshift (detent) cable - adjustment

Refer to illustration 9.1
1   Insert a screwdriver on each side of the snap-lock and pry out to release **(see illustration)**.
2   Compress the locking tabs and disconnect the snap-lock assembly from its bracket.
3   Manually set the carburetor in the fully open position with the throttle lever fully against its stop.
4   With the carburetor in the fully open position, push the snap-lock on the detent cable into the locked position and release the throttle lever.

10   Turbo Hydra-Matic 400 downshift (detent) switch - adjustment

Refer to illustration 10.1
1   The switch is mounted on the pedal bracket **(see illustration)**.
2   The switch is set by pushing the plunger as far forward as possible. At the first full depression of the accelerator pedal, the switch is automatically adjusted.

11   Neutral start switch - check and adjustment

Refer to illustrations 11.6a and 11.6b
1   When the switch is operating properly, the engine should crank over with the selector lever in Park or Neutral only. Also, the backup lights should come on when the lever is in Reverse.
2   If a new switch is being installed, set the shift lever against the 'Neutral' gate by rotating the lower lever on the shift tube in a counterclockwise direction as viewed from the driver's seat.
3   Locate the switch actuating tang in the shifter tube slot and then tighten the securing screws.
4   Connect the wiring harness and switch on the ignition and check that the starter motor will actuate.
5   If the switch operates correctly, move the shift lever out of Neutral, which will cause the alignment pin (installed during production of the switch) to shear.
6   If an old switch is being installed or readjusted, use a pin or drill bit (of approximately 3/32-inch diameter) to align the hole in the switch with the actuating tang. Insert the pin to a depth of 1/4-inch. Remove the pin before moving the shift lever out of Neutral **(see illustration)**. On later models the switch automatically ratchets to the proper adjustment when the shift lever is moved to Park. To readjust a later model switch, move the switch assembly housing to the Low gear position and then shift into Park **(see illustration)**.

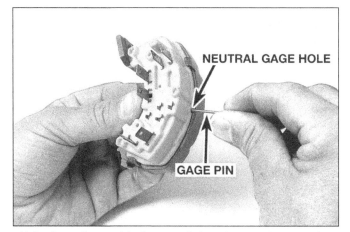

11.6a On earlier models, insert a pin into the gage hole to adjust the neutral start switch

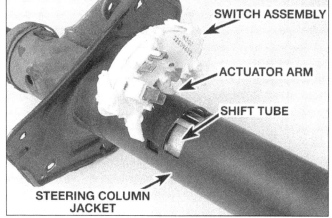

11.6b Later model neutral start switches are adjusted by moving the assembly housing to the Low gear position and then shifting into Park

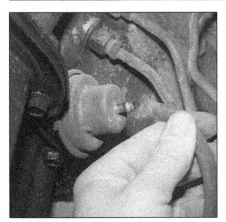

12.2 Detach the vacuum line from the modulator and inspect it for cracks, breaks and tears; then inspect the vacuum pipe on the modulator and the inside of the hose for transmission fluid

12.7 Remove the modulator hold-down bolt and clamp, then pull out the modulator

12.9 Make sure the O-ring is lubricated and the surface is clean before installing the modulator

12 Vacuum modulator - check and replacement

1   The vacuum modulator, used on some models, senses engine load and varies fluid pressure in the transmission to control shift points. Symptoms of a faulty modulator include late, harsh shifts, and/or white smoke from the exhaust pipe (if the diaphragm in the modulator fails).

Check

Refer to illustration 12.2

2   To check the modulator, raise the vehicle and support it securely on jackstands. Detach the vacuum hose from the modulator and look for the presence of transmission fluid in the hose or in the port on the modulator **(see illustration)**. If there is fluid in the hose, replace the modulator. **Note:** *Even if there's no fluid evident on the pipe or in the vacuum line, it's a good idea to insert a pipe cleaner or cotton swab into the vacuum pipe to see if there's fluid inside.*

3   Connect a hand-held vacuum pump to the port on the modulator and apply a vacuum of approximately 20 in-Hg; the modulator should hold vacuum for at least 30 seconds. If it doesn't, replace the modulator.

4   Also check the vacuum hose for cracks and general deterioration, replacing it if necessary. Be sure to check the vacuum hose at the other end of the vacuum pipe, too (at the intake manifold).

Replacement

Refer to illustrations 12.7 and 12.9

5   Raise the vehicle and support it securely on jackstands.

6   Disconnect the vacuum line at the modulator **(see illustration 12.2)**.

7   Remove the modulator hold-down bolt and clamp and remove the modulator **(see illustration)**.

8   Remove the modulator O-ring and replace it with a new one (it the old modulator is to be replaced).

9   Lubricate the O-ring with clean automatic transmission fluid and install it on the modulator **(see illustration)**.

10   Install the modulator, hold-down clamp and bolt, tightening the bolt securely.

11   Attach the vacuum hose.

12   Lower the vehicle and check the fluid level in the transmission (Section 3), adding some if necessary.

13 Automatic transmission - removal and installation

Refer to illustrations 13.5, 13.6 and 13.21

Removal

1   Disconnect the negative cable at the battery.

2   Raise the vehicle and support it securely.

3   Drain the transmission fluid (Chapter 1).

4   Remove the torque converter cover.

5   Mark the relation of the driveplate and the torque converter so they can be reinstalled in the same position **(see illustration)**.

6   Remove the torque converter-to-driveplate nuts/bolts **(see illustration)**. Turn the crankshaft bolt for access to each nut in turn.

7   Remove the starter motor (Chapter 10).

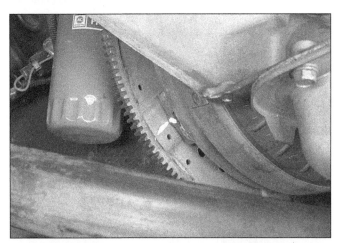

13.5 Mark the relationship of the torque converter to the driveplate

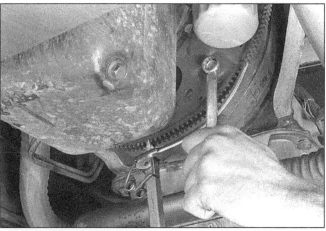

13.6 Remove the torque converter-to-driveplate retaining bolts (here, a driveplate wrench is being used to prevent the driveplate from turning, but a wrench on the crankshaft vibration damper will also work)

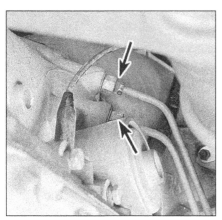

13.21 After lowering the transmission, use a flare-nut wrench to disconnect the cooler lines (arrows) and plug them to prevent fluid loss. It may be necessary to use a back-up wrench on the transmission-side fittings to prevent twisting the lines

8   Remove the driveshaft (Chapter 7).
9   Disconnect the speedometer cable.
10   Disconnect the electrical connectors from the transmission.
11   On models so equipped, disconnect the vacuum hose(s).
12   Remove any exhaust component which will interfere with transmission removal (Chapter 3).
13   Disconnect the downshift cable (if equipped).
14   Disconnect the shift linkage.

15   Support the engine using a jack and a block of wood under the fluid pan to spread the load.
16   Support the transmission with a jack - preferably a jack made for this purpose. Safety chains will help steady the transmission on the jack.
17   Remove the rear mount-to-crossmember attaching bolts and the crossmember-to-frame attaching bolts.
18   Remove the two engine rear support-to-transmission extension housing attaching bolts.
19   Raise the transmission sufficiently to allow removal of the crossmember.
20   Remove the bolts securing the transmission to the engine.
21   Lower the transmission slightly and disconnect and plug the transmission cooler lines (see illustration).
22   Remove the transmission fluid filler tube.
23   Move the transmission to the rear to disengage it from the engine block dowel pins and make sure the torque converter is detached from the driveplate. Secure the torque converter to the transmission so that it will not fall out during removal. Lower the transmission from the vehicle.

Installation

24   Make sure prior to installation that the torque converter hub is securely engaged in the pump.
25   With the transmission secured to the jack, raise the transmission into position, making sure to keep it level so the torque

converter does not slide forward. Connect the transmission cooler lines.
26   Turn the torque converter to line up the torque converter and driveplate bolt holes. The white paint mark on the torque converter and the stud made during Step 5 must line up.
27   Move the transmission carefully forward until the dowel pins are engaged and the torque converter is engaged.
28   Install the transmission housing-to-engine bolts and nuts. Tighten the bolts and nuts to the specified torque.
29   Install the torque converter-to-driveplate nuts. Tighten the nuts to the specified torque.
30   Install the transmission mount crossmember and through-bolts. Tighten the bolts and nuts securely.
31   Remove the jacks supporting the transmission and the engine.
32   Install the fluid filler tube.
33   Install the starter.
34   Connect the vacuum hose(s) (if equipped).
35   Connect the shift and TV linkage.
36   Plug in the transmission electrical connectors.
37   Install the torque converter cover.
38   Connect the driveshaft.
39   Connect the speedometer.
40   Adjust the shift linkage.
41   Install any exhaust system components which were removed.
42   Lower the vehicle.
43   Fill the transmission with the specified fluid (Section 3), run the vehicle and check for fluid leaks.

14 Troubleshooting - automatic transmission

Symptom	Reason
Powerglide	
No drive in any selector position	Low fluid level
	Clogged fluid filter screen
	Internal fault
Engine races as drive taken up but lack of acceleration	Low fluid level
	Incorrect band adjustment
	Internal fault
Engine races on upshift	Low fluid level
	Incorrect band adjustment
	Obstructed vacuum modulator line
	Clogged fluid filter screen
No upshift at all	Low band not releasing due to:
	Stuck throttle valve
	Incorrectly adjusted manual valve lever
	Internal fault
No downshift	Internal fault
Harsh jerky) upshift	Incorrect carburetor to transmission throttle valve rod adjustment
	Incorrect band adjustment
	Leaking vacuum modulator line
	Leaking vacuum modulator diaphragm
	Internal fault

Symptom	Reason

Powerglide (continued)

Symptom	Reason
Harsh jerky) downshift	High engine idle speed Incorrect band adjustment Faulty downshift valve
No drive in reverse	Incorrect linkage adjustment Internal fault
Incorrect shift points	Incorrect carburetor to transmission linkage adjustment Incorrect throttle valve adjustment
Excessive creep in Drive range	Engine idle speed too high
Creep in neutral	Incorrect linkage adjustment Low band not releasing

Turbo Hydra-Matic

Symptom	Reason
No drive in Drive range	Low fluid level Incorrect linkage adjustment
1 to 2 shift on full throttle only	Detent valve cable incorrectly set Detent valve sticking Leak in vacuum line Internal fault
No upshift from 1 to 2	Detent cable binding Incorrectly adjusted intermediate band Internal fault
No upshift from 2 to 3	Internal fault
Moves off in second speed	Intermediate band adjustment too tight
Drive in neutral	Incorrectly adjusted linkage
No drive in reverse	Low fluid level Incorrectly adjusted linkage Internal fault
Slip in all ranges and upshifts	Low fluid level Incorrectly adjusted intermediate band Internal fault
No engine braking	Incorrectly adjusted intermediate band Internal fault
No part throttle downshift	Detent valve cable broken or incorrectly adjusted
No full throttle downshift	Detent valve cable broken or incorrectly adjusted
Shift points too high or too low	Fault in vacuum line Faulty vacuum modulator assembly
Won't hold in 'P'	Incorrectly adjusted linkage Internal fault in parking pawl mechanism
White smoke from exhaust pipe, combined with late, harsh shifts	Faulty vacuum modulator

Chapter 7   Driveshaft

Contents

Specifications

Type	Tubular steel with front sliding yoke. Early models fitted with two universal joints, later models fitted with a universal joint at the transmission end and a constant velocity joint at the axle end

Torque specifications

	Ft-lbs
Universal joint strap bolts	15
Universal joint U-bolt nuts	15
Universal joint companion flange bolts (1978 and 1979 models)............	70

1   Driveshaft and universal joints - description and check

1   The driveshaft is a tube running between the transmission and the rear end. Universal joints are located at either end of the driveshaft and permit power to be transmitted to the rear wheels at varying angles. The rear U-joint on later models is a "double Cardan" constant velocity type, which is really just two U-joints joined together.

2   The driveshaft features a splined yoke at the front, which slips into the extension housing of the transmission. This arrangement allows the driveshaft to slide back and forth within the transmission as the vehicle is in operation.

3   An oil seal is used to prevent leakage of fluid at this point and to keep dirt and contaminants from entering the transmission. If leakage is evident at the front of the driveshaft, replace the oil seal, referring to the procedures in Chapter 6.

4   The driveshaft assembly requires very little service. The universal joints are lubricated for life and must be replaced if problems develop. The driveshaft must be removed from the vehicle for this procedure.

5   Since the driveshaft is a balanced unit, it's important that no undercoating, mud, etc., be allowed to stay on it. When the vehicle is raised for service it's a good idea to clean the driveshaft and inspect it for any obvious damage. Also check that the small weights used to originally balance the driveshaft are in place and securely attached. Whenever the driveshaft is removed it's important that it be reinstalled in the same relative position to preserve this balance.

6   Problems with the driveshaft are usually indicated by a noise or vibration while driving the vehicle. A road test should verify if the problem is the driveshaft or another vehicle component:

a)   *On an open road, free of traffic, drive the vehicle and note the engine speed (rpm) at which the problem is most evident.*

b)   *With this noted, drive the vehicle again, this time manually keeping the transmission in 1st, then 2nd, then 3rd gear ranges and running the engine up to the engine speed noted.*

c)   *If the noise or vibration occurs at the same engine speed regardless of which gear the transmission is in, the driveshaft*

is not at fault because the speed of the driveshaft varies in each gear.

d)   *If the noise or vibration decreased or was eliminated, visually inspect the driveshaft for damage, material on the shaft which would effect balance, missing weights and damaged universal joints. Another possibility for this condition would be tires which are out of balance.*

7   To check for worn universal joints:

a)   *On an open road, free of traffic, drive the vehicle slowly until the transmission is in High gear. Let off on the accelerator, allowing the vehicle to coast, then accelerate. A clunking or knocking noise will indicate worn universal joints.*

b)   *Drive the vehicle at a speed of about 10 to 15 mph and then place the transmission in Neutral, allowing the vehicle to coast. Listen for abnormal driveline noises.*

c)   *Raise the vehicle and support it securely on jackstands. With the transmission in Neutral, manually turn the driveshaft, watching the universal joints for excessive play.*

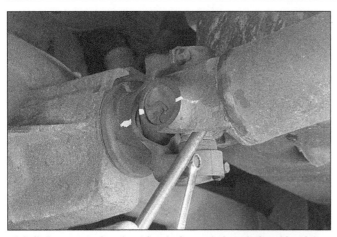

2.3 Before removing the driveshaft, mark the relationship of the driveshaft yoke to the differential flange - to prevent the driveshaft from turning when loosening the bolts (or nuts), insert a screwdriver through the yoke

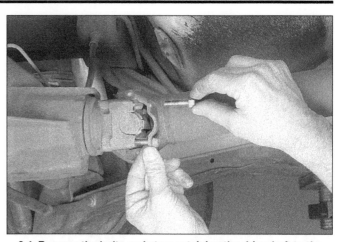

2.4 Remove the bolts and strap retaining the driveshaft to the differential yoke

2   Driveshaft - removal and installation

Refer to illustrations 2.3 and 2.4

Removal

1   Disconnect the negative cable from the battery.
2   Raise the vehicle and support it securely on jackstands. Place the transmission in Neutral with the parking brake off.
3   Using a sharp scribe, white paint or a hammer and punch, place marks on the driveshaft and the differential flange in line with each other **(see illustration)**. This is to make sure the driveshaft is reinstalled in the same position to preserve the balance.
4   Disconnect the rear universal joint by unscrewing and removing the nuts from the U-bolts, the strap retaining bolts or, on later models, the companion flange bolts **(see illustration)**. Turn the driveshaft (or tires) as necessary to bring the nuts or bolts into the most accessible position.

5   Tape the bearing caps to the spider to prevent the caps from coming off during removal.
6   Lower the rear of the driveshaft and slide the front out of the transmission.
7   To prevent loss of fluid and to protect against contamination while the driveshaft is out, wrap a plastic bag over the transmission housing and hold it in place with a rubber band.

Installation

8   Remove the plastic bag on the transmission and wipe the area clean. Inspect the oil seal carefully. Procedures for replacement of this seal can be found in Chapter 6.
9   Slide the front of the driveshaft into the transmission.
10   Raise the rear of the driveshaft into position, checking to be sure that the marks are in alignment. If not, turn the rear wheels to match the pinion flange and the driveshaft.
11   Remove the tape securing the bearing caps and install the U-bolts and nuts, straps and bolts or companion flange bolts. Tighten the nuts/bolts to the specified torque.

3   Universal joints - replacement

Note: *A press or large vise will be required for this procedure. It may be advisable to take the driveshaft to a local dealer or machine shop where the universal joints can be replaced for you, normally at a reasonable charge.*

Cleveland type joint

Refer to illustrations 3.1 and 3.2

1   Clean away all dirt from the ends of the bearings on the yokes so that the snap-rings can be removed using a pair of snap-ring pliers. If the snap-rings are very tight, tap the end of the bearing cup (inside the snap-ring) to relieve the pressure **(see illustration)**.
2   Support the trunnion yoke on a short piece of tube or the open end of a socket, then use a suitably sized socket to press out the cross (trunnion) by means of a vise **(see illustration)**.
3   Press the trunnion through as far as possible, then grip the bearing cup in the jaws of a

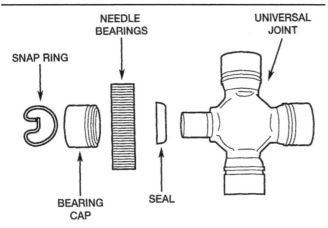

3.1 Cleveland type universal joint repair kit

3.2 To press the universal joint out of the driveshaft, set it up in a vise with the small socket (on the left) pushing the joint and bearing cap into the large socket

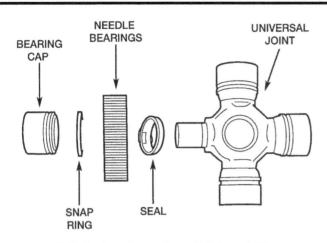

3.16 Saginaw type universal joint repair kit

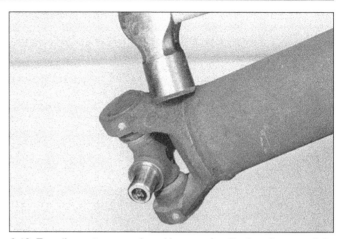

3.18 To relieve stress produced by pressing the bearing caps into the yokes, strike the yoke in the area shown

vise to fully remove it. Repeat the procedure for the remaining cups.

4 On some models, the slip yoke at the transmission end has a vent hole. When dismantling, ensure that this vent hole is not blocked.

5 A universal joint repair kit will contain a new trunnion, seals, bearings, cups and snap-rings. Some replacement universal joints are equipped with a grease fitting. Be sure it is offset in the proper direction (toward the driveshaft).

6 Commence reassembly by packing each of the reservoirs at the trunnion ends with lubricant.

7 Make sure that the dust seals are correctly located on the trunnion, so that the cavities in the seals are nearer the trunnion.

8 Using a vise, press one bearing cup into the yoke so that it enters not more than 1/4-inch.

9 Using a thick grease, stick each of the needle rollers inside the cup.

10 Insert the trunnion into the partially fitted bearing cup, taking care not to displace the needle rollers.

11 Stick the needle bearings into the oppo-site cup and then, holding the trunnion in correct alignment, press both cups fully home in the jaws of the vise.

12 Install the new snap-rings.

13 Repeat the operations on the other two bearing cups.

14 In extreme cases of wear or neglect, it is possible that the bearing cup housings in the yoke will have worn so much that the cups are a loose fit in the yokes. In such cases, replace the complete driveshaft assembly.

15 Always check the wear in the sliding sleeve splines and replace the sleeve if worn.

Saginaw type joint

Refer to illustrations 3.16 and 3.18

16 Where a Saginaw joint is to be disassembled, the procedure given in the previous Section for pressing out the bearing cup is applicable. If the joint has been previously repaired it will be necessary to remove the snap-rings inboard of the yokes; if this is to be the first time that servicing has been carried out, there are no snap-rings to remove, but the pressing operation in the vise will shear the plastic molding material **(see illustration)**.

17 Having removed the cross (trunnion), remove the remains of the plastic material from the yoke. Use a small punch to remove the material from the injection holes.

18 Reassembly is similar to that given for the Cleveland type joint except that the snap-rings are installed inside the yoke. If difficulty is encountered, strike the yoke firmly with a hammer to assist in seating **(see illustration)**.

Double Cardan type constant velocity joint

Refer to illustrations 3.19 and 3.21

19 An inspection kit containing two bearing cups and two retainers is available to permit the joint to be dismantled to the stage where the joint can be inspected. Before any dismantling is started, mark the flange yoke and coupling yoke so that they can be reassembled in the same relative position **(see illustration)**.

20 Dismantle the joint by removing the bearing cups in a similar way to that described in the preceding Section, according to type.

21 Disengage the flange yoke and trunnion from the centering ball. Pry the seal from the ball socket and remove the washers, springs and three ball seats **(see illustration)**.

3.19 CV joint alignment marks made before disassembly

3.21 Place the companion flange in a vise as shown and remove the seal and centering ball

22 Clean the ball seat insert bushing and inspect for wear. If evident, the flange yoke and trunnion assembly must be replaced.
23 Clean the seal, ball seats, spring and washers and inspect for wear. If excessive wear is evident or parts are broken, a replacement service kit must be used.
24 Remove all plastic material from the groove of the coupling yoke (if this type of joint is used).
25 Inspect the centering ball; if damaged, it must be replaced.

26 Withdraw the centering ball from the stud using a suitable extractor. Provided that the ball is not to be re-used, it will not matter if it is damaged.
27 Press a new ball onto the stud until it seats firmly on the stud shoulder. It is extremely important that no damage to the ball occurs during this stage, and suitable protection must be given to it.
28 Using the grease provided in the repair kit, lubricate all the parts and insert them into the ball seat cavity in the following order:

spring, washer (small OD), three ball seats (largest opening outwards to receive the ball), washer (large OD) and the seal.
29 Lubricate the seal lips and press it (lip inwards) into the cavity. Fill the cavity with the grease provided.
30 Install the flange yoke to the centering ball, ensuring that the alignment marks are correctly positioned.
31 Install the trunnion caps as described previously for the Cleveland or Saginaw types.

4 Troubleshooting - driveshaft

Symptom

Vibration * ....

Reason

Wear in sliding sleeve splines
Loose bolts on rear universal joint
Worn universal joint bearings
Out of balance driveshaft
Distorted driveshaft

Knock or 'clunk' when taking up drive or shifting gear ....

Loose bolts on rear universal joint
Worn universal joint bearings
Worn drive pinion splines causing looseness in companion flange
Excessive backlash in differential gears

* It is sometimes possible to reduce driveshaft vibration by rotating the shaft through 180 degrees with respect to the drive flange of the differential unit. Before this is attempted, ensure that there is no mud, undercoating, etc., on the shaft, which could be affecting its balance

Chapter 8   Rear axle

Contents

Specifications

Type .... Hypoid, semi-floating

Ratios

Axle ratios have varied according to the year of manufacture and engine capacity.
The ratios given are typical, and apply to 1975 models in particular.

Standard ratio	2.73 : 1
Highway ratio	2.56 : 1
High altitude ratio (standard ratio with 350 cu in. 4 barrel carburetor)	3.08 : 1
Axle identification code	G

(This is the third letter of the axle code number which appears on the front
of the right-hand axle tube, about three inches from the differential cover).

Pinion bearing preload

New	15 to 30 in-lbs
Used	5 to 10 in-lbs
Oil capacity (drain and refill)	3.5 US pints (1969 models)
	4 US pints (1969 models with 8.875 in. ring gear)
	3.75 US pints (1970/71 models)
	4.25 US pints (1970/71 models with 8.875 in. ring gear, and 1972 and later models)

Torque specifications

	Ft-lbs (unless otherwise indicated)
Differential cover bolts	25
Pinion shaft lock bolt	25
Filler plug	25

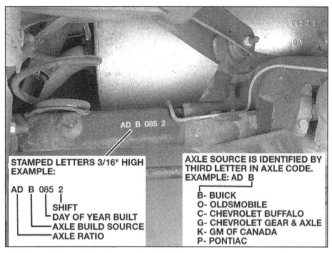

1.3 Location of the rear axle identification number - early models
have only a three-letter code

2.3a Remove the pinion shaft lock bolt . . .

1   Rear axle - description and check

Refer to illustration 1.3

Description

1   The rear axle assembly is a hypoid, semi-floating type (the centerline of the pinion gear is below the centerline of the ring gear). The differential carrier is a casting with a pressed steel cover, and the axle tubes are made of steel, pressed and welded into the carrier.

2   An optional locking rear axle is also available. This differential allows for normal differential operation until one wheel loses traction. The unit utilizes multi-disc clutch packs and a speed sensitive engagement mechanism which locks both axleshafts together, applying equal rotational power to both wheels.

3   In order to undertake certain operations, particularly replacement of the axleshafts, it's important to know the axle identification number. It's located on the front face of the right

side axle tube. The third letter of the code identifies the manufacturer of the axle. This is important, as axle design varies slightly among manufacturers **(see illustration)**.

Check

4   Many times a fault is suspected in the rear axle area when, in fact, the problem lies elsewhere. For this reason, a thorough check should be performed before assuming a rear axle problem.

5   The following noises are those commonly associated with rear axle diagnosis procedures:

a) *Road noise is often mistaken for mechanical faults. Driving the vehicle on different surfaces will show whether the road surface is the cause of the noise. Road noise will remain the same if the vehicle is under power or coasting.*

b) *Tire noise is sometimes mistaken for mechanical problems. Tires which are worn or low on pressure are particularly susceptible to emitting vibrations and*

noises. Tire noise will remain about the same during varying driving situations, where rear axle noise will change during coasting, acceleration, etc.

c) *Engine and transmission noise can be deceiving because it will travel along the driveline. To isolate engine and transmission noises, make a note of the engine speed at which the noise is most pronounced. Stop the vehicle and place the transmission in Neutral and run the engine to the same speed. If the noise is the same, the rear axle is not at fault.*

6   Overhaul and general repair of the rear axle is beyond the scope of the home mechanic due to the many special tools and critical measurements required. Thus, the procedures listed here will involve axleshaft removal and installation, axleshaft oil seal replacement, axleshaft bearing replacement and removal of the entire unit for repair or replacement. For differential overhaul procedures, refer to the Haynes *Suspension, Steering and Driveline Manual*.

2.3b . . . then carefully remove the pinion shaft from the
differential carrier (don't turn the wheels or the carrier after the
shaft has been removed, or the spider gears may fall out)

2.4 Push the axle flange in, then remove the C-lock from
the inner end of the axleshaft

2.5 Carefully pull the axleshaft from the housing to avoid damaging the seal

3.2 The axleshaft oil seal can sometimes be pried out with the end of the axle

2   Axleshaft - removal and installation

Refer to illustrations 2.3a, 2.3b, 2.4 and 2.5

1   Raise the rear of the vehicle, support it securely and remove the wheel and brake drum. See Chapter 9 for information if difficulty is experienced in removing the brake drums.

2   Remove the cover from the differential and allow the oil to drain into a container.

3   Remove the lock bolt from the differential pinion shaft. Remove the pinion shaft **(see illustrations)**.

4   Push the outer (flanged) end of the axleshaft in and remove the C-lock from the inner end of the shaft **(see illustration)**.

5   Withdraw the axleshaft, taking care not to damage the oil seal in the end of the axle housing as the splined end of the axleshaft passes through it **(see illustration)**.

6   Installation is the reverse of removal. Tighten the lock bolt to the specified torque.

7   Always use a new cover gasket and tighten the cover bolts to the specified torque.

8   Refill the axle with the correct quantity and grade of lubricant.

3   Axleshaft oil seal - replacement

Refer to illustrations 3.2

1   Remove the axleshaft as described in the preceding Section.

2   Pry the old oil seal out of the end of the axle housing, using a large screwdriver or the inner end of the axleshaft itself as a lever **(see illustration)**.

3   Using a large socket as a seal driver, tap the seal into position so that the lips are facing in and the metal face is visible from the end of the axle housing. When correctly installed, the face of the oil seal should be flush with the end of the axle housing. Lubricate the lips of the seal with gear oil.

4   Installation of the axleshaft is described in the preceding Section.

4   Axleshaft bearing - replacement

Refer to illustrations 4.3 and 4.4

1   Remove the axleshaft (refer to Section 2) and the oil seal (refer to Section 3).

2   A bearing puller will be required, or a tool which will engage behind the bearing will have to be fabricated.

3   Attach a slide hammer and pull the bearing from the axle housing **(see illustration)**.

4   Clean out the bearing recess and drive in the new bearing using a special tool available at most auto parts stores **(see illustration)**. **Caution:** *Failure to use this tool could result in bearing damage. Lubricate the new bearing with gear lubricant. Make sure that the bearing is tapped into the full depth of its recess and that the numbers on the bearing are visible from the outer end of the housing.*

5   Discard the old oil seal and install a new one, then install the axleshaft.

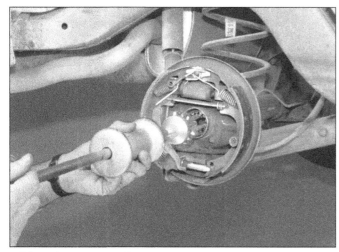

4.3 A slide hammer with a special bearing puller attachment is required to pull the axleshaft bearing from the axle housing

4.4 A special bearing driver, available at auto parts stores, is needed to install the axleshaft bearing without damaging it

5.3 Check the torque required to rotate the pinion with an in-lb torque wrench

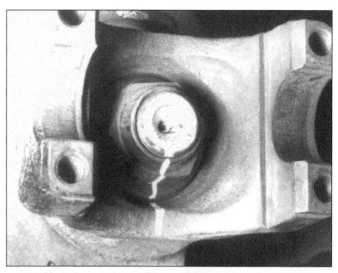

5.4 Mark the relative positions of the pinion, nut and flange before removing the nut

5   Pinion oil seal - replacement

Refer to illustrations 5.3, 5.4, 5.6 and 5.10

1   Raise the rear of the vehicle and support it securely on jackstands placed under the axle. Block the front wheels to keep the vehicle from rolling off the stands.

2   Disconnect the driveshaft and fasten it out of the way (see Chapter 7).

3   Use an inch-pound torque wrench to check the torque required to rotate the pinion. Record it for use later **(see illustration)**.

4   Scribe or punch alignment marks on the pinion shaft, nut and flange **(see illustration)**.

5   Count the number of threads visible between the end of the nut and the end of the pinion shaft and record it for use later.

6   Keep the companion flange from moving while the self-locking pinion nut is loosened **(see illustration)**.

7   Remove the pinion nut.

8   Withdraw the companion flange. It may be necessary to use a two or three-jaw puller engaged behind the flange to draw it out. Do not attempt to pry behind the flange or hammer on the end of the pinion shaft.

9   Pry out the old seal and discard it.

10   Lubricate the lips of the new seal with high-temperature grease and tap it evenly into position with a seal installation tool or a large socket. Make sure it enters the housing squarely and is tapped in to its full depth **(see illustration)**.

11   Align the mating marks made before disassembly and install the companion flange. If necessary, tighten the pinion nut to draw the flange into place. Do not try to hammer the flange into position.

12   Apply non-hardening sealant to the ends of the splines visible in the center of the flange so oil will be sealed in.

13   Install the washer (if equipped) and pinion nut. Tighten the nut carefully until the original number of threads are exposed and the marks line up.

14   Measure the torque required to rotate the pinion and tighten the nut in small increments until it matches the figure recorded in Step 5. In order to compensate for the drag of the new oil seal, the nut should be tightened more until the rotational torque of the pinion slightly exceeds what was recorded earlier, but not by more than 5 in-lbs. **Caution:** *Don't exceed the recommended preload torque; if you do, the collapsible spacer on the pinion shaft will have to be replaced. Also, don't loosen the pinion nut in an attempt to reduce preload.*

15   Connect the driveshaft (Chapter 7) and lower the vehicle.

6   Rear axle assembly - removal and installation

1   Raise the rear of the vehicle and support it securely on jackstands placed under the body frame rails.

2   Remove the rear wheels and brake drums. See Chapter 9 for information if difficulty is experienced in removing the brake drums.

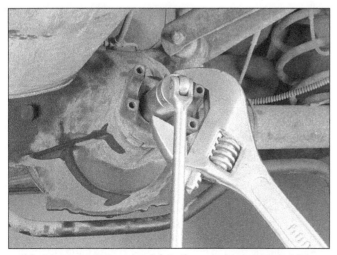

5.6 A large wrench or arc-joint pliers can be used to hold the companion flange while the pinion nut is loosened

5.10 Use a large socket or piece of pipe to tap the pinion oil seal into place

3   Disconnect the driveshaft from the differential (see Chapter 7), then hang the back end of the driveshaft up, out of the way, using a wire hung from underneath the floorpan.
4   Disconnect the axle vent hose from the top of the axle housing.
5   Disconnect the brake hose from the junction block on top of the axle housing and plug the hose to prevent leaks.
6   Disconnect the rear parking brake cables from the equalizer (see Chapter 9).
7   Position a floor jack under the differential housing and just take up the weight. Do not raise the jack sufficiently to take the weight of the vehicle from the jackstands.
8   Disconnect the lower shock absorber mounts (see Chapter 11).
9   Remove the nuts from the leaf spring U-bolts, then remove the spring plates and U-bolts (see Chapter 11).
10   Remove the shackles from the rear ends of the leaf springs and allow the leaf springs to pivot down to the floor (see Chapter 11).
11   Carefully lower the axle assembly, then pull the floor jack to the rear and remove the axle assembly from under the vehicle.
12   Installation is a reversal of removal, but tighten all suspension bolts and nuts to the specified torque (see Chapter 11).
13   Bleed the brakes (see Chapter 9).

7   Troubleshooting - rear axle

Symptom	Reason
Vibration	Worn axle shaft bearing
	Loose 'U' bolts or strap bolts (driveshaft to pinion flange)
	Tires require balancing
	Driveshaft out of balance
Noise on turns	Worn differential gear
Noise on drive or coasting...	Worn or incorrectly adjusted ring and pinion gear
'Clunk' on acceleration or deceleration...................................	Worn differential gear cross shaft
	Worn driveshaft universal joints
	Loose flange 'U' bolts or strap bolts

Positraction type axle

Chatter on turns..	Incorrect lubricant in differential
	Brake cones worn

It must be noted that tire noise, wear in the rear suspension bushings and worn or loose shock absorber mounts can all mislead the mechanic into thinking that components of the rear axle are the source of trouble.

Chapter 9 Brakes

Contents

Specifications

Brake fluid type .... DOT 3 brake fluid

Disc brakes

Minimum brake pad thickness	1/8 inch
Disc minimum thickness	Refer to minimum thickness casting on disc
Lateral runout	0.004 inch maximum
Disc thickness variation (parallelism)	0.0005 inch
Caliper-to-knuckle clearance	0.005 to 0.012 inch

Drum brakes

Minimum brake lining thickness	1/16 inch
Drum maximum diameter	Refer to maximum diameter casting on drum
Out-of-round	0.006 inch maximum
Taper	0.003 inch maximum

Torque specifications   Ft-lbs (unless otherwise indicated)

Note: One foot-pound (ft-lb) of torque is equivalent to 12 inch-pounds (in-lbs) of torque. Torque values below approximately 15 ft-lbs are expressed in inch-pounds, because most foot-pound torque wrenches are not accurate at these smaller values.

Master cylinder mounting nuts	24
Power booster mounting nuts	24
Caliper mounting bolts	35
Caliper support plate-to-steering knuckle (upper bolt)	140
Caliper support plate (steering arm-to-knuckle bolts	70
Wheel cylinder mounting bolts	144 in-lbs
Brake hose-to-caliper inlet fitting bolt	32
Wheel lug nuts	70

Typical rear drum brake assembly

1   Return spring
2   Return spring
3   Hold-down spring
4   Primary brake shoe
5   Adjuster screw assembly
6   Actuator lever
7   Lever return spring
8   Actuator spring (not all models)
9   Hold-down spring
10 Secondary brake shoe

1   General information

The vehicles covered by this manual are equipped with hydraulically operated front and rear brake systems. The front brakes are disc or drum type and the rear brakes are drum type. Both the front and rear brakes are self adjusting. The front disc brakes automatically compensate for pad wear, while the drum brakes incorporate an adjustment mechanism which is activated as the brakes are applied when the vehicle is driven in Reverse.

Hydraulic system

The hydraulic system consists of two separate circuits. The master cylinder has separate reservoirs for the two circuits and in the event of a leak or failure in one hydraulic circuit, the other circuit will remain operative. A visual warning of circuit failure or air in the system is given by a warning light activated by displacement of the piston in the pressure differential switch portion of the combination valve from its normal "in balance" position.

Combination valve

A combination valve, located in the engine compartment below the master cylinder, consists of three sections providing the following functions. The metering section limits pressure to the front brakes until a predetermined front input pressure is reached and

until the rear brakes are activated. There is no restriction at inlet pressures below three psi, allowing pressure equalization during non-braking periods. The proportioning section of the valve reduces pressure to the rear brakes during heavy braking, reducing the tendency for the rear wheels to lock up. The valve is also designed to ensure full pressure to one brake system should the other system fail. The pressure differential warning switch incorporated into the combination valve is designed to continuously compare the front and rear brake pressure from the master cylinder and energize the dash warning light in the event of either a front or rear brake system failure. The design of the switch and valve are such that the switch will stay in the "warning" position once a failure has occurred. The only way to turn the light off is to repair the cause of the failure and apply a brake pedal force of 450 psi.

Power brake booster

The power brake booster, utilizing engine manifold vacuum and atmospheric pressure to provide assistance to the hydraulically operated brakes, is mounted on the firewall in the engine compartment.

Parking brake

The parking brake operates the rear brakes only, through cable actuation. It's activated by a pedal mounted on the left side kick panel.

Service

After completing any operation involving disassembly of any part of the brake system, always test drive the vehicle to check for proper braking performance before resuming normal driving. When testing the brakes, perform the tests on a clean, dry flat surface. Conditions other than these can lead to inaccurate test results.

Test the brakes at various speeds with both light and heavy pedal pressure. The vehicle should stop evenly without pulling to one side or the other. Avoid locking the brakes, because this slides the tires and diminishes braking efficiency and control of the vehicle.

Tires, vehicle load and front-end alignment are factors which also affect braking performance.

2   Disc brake pads - replacement

Refer to illustrations 2.5 and 2.6a through 2.6g

Warning: *Disc brake pads must be replaced on both front wheels at the same time - never replace the pads on only one wheel. Also, the dust created by the brake system may contain asbestos, which is harmful to your health. Never blow it out with compressed air and don't inhale any of it. An approved filtering mask should be worn when working on the brakes. Do not, under*

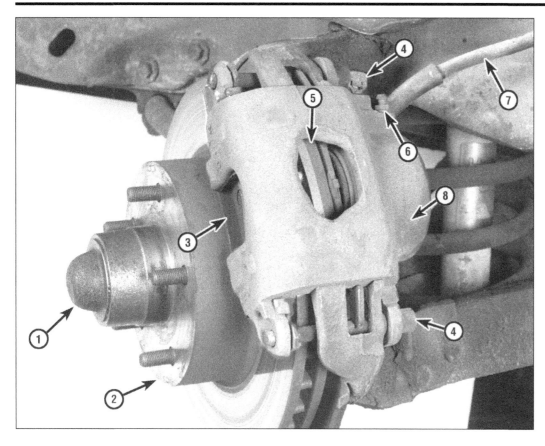

Typical front disc brake assembly

1   Dust cap
2   Brake disc/hub assembly
3   Outer brake pad
4   Caliper mounting bolts
5   Inner brake pad
6   Bleed screw
7   Brake hose
8   Caliper

any circumstances, use petroleum-based solvents to clean brake parts. Use brake system cleaner only!

1   Remove the cover from the brake fluid reservoir.

2   Loosen the wheel lug nuts, raise the front of the vehicle and support it securely on jackstands.

3   Remove the front wheels. Work on one brake assembly at a time, using the assembled brake for reference if necessary. Before beginning work, wash the brake assembly

with brake system cleaner.

4   Inspect the brake disc carefully as outlined in Section 4. If machining is necessary, follow the information in that Section to remove the disc, at which time the pads can be removed from the calipers as well.

5   Push the piston back into the bore to provide room for the new brake pads. A C-clamp can be used to accomplish this **(see illustration)**. As the piston is depressed to the bottom of the caliper bore, the fluid in the master cylinder will rise. Make sure it doesn't

overflow. If necessary, siphon off some of the fluid.

6   Follow the accompanying illustrations, beginning with **illustration 2.6a**, for the actual pad replacement procedure. Be sure to stay in order and read the caption under each illustration.

7   When reinstalling the caliper, be sure to tighten the mounting bolts to the specified torque. After the job has been completed, firmly depress the brake pedal a few times to bring the pads into contact with the disc.

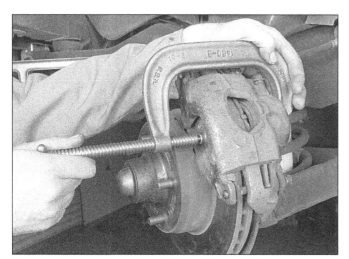

2.5 Using a large C-clamp, push the piston back into the caliper bore - note that one end of the clamp is on the flat area near the brake hose fitting and the other end (screw end) is pressing against the outer brake pad

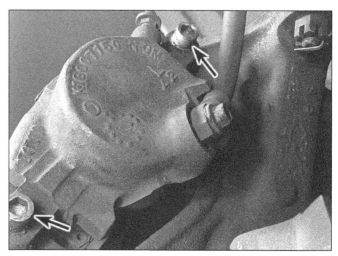

2.6a Remove the two caliper-to-steering knuckle mounting bolts (arrows) (this will require the use of an Allen wrench)

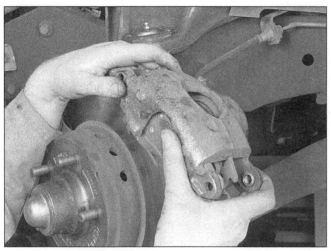

2.6b Slide the caliper up and off the disc

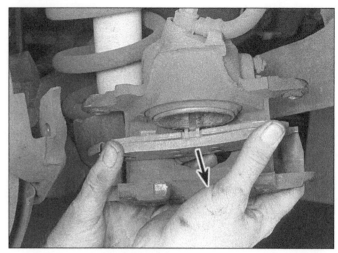

2.6c Pull the inner pad straight out, disengaging the retainer spring from the caliper piston

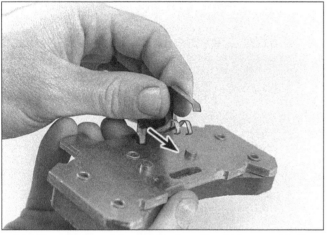

2.6d Transfer the retainer spring from the old inner pad to the new one - hook the end of the spring in the hole at the top of the pad, then insert the two prongs of the spring into the slot on the pad backing plate

2.6e Lubricate the caliper mounting ears with multi-purpose grease

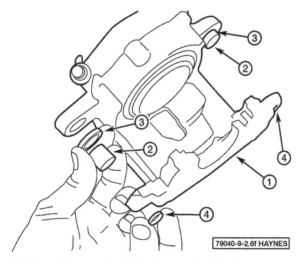

2.6f Push the mounting bolt sleeves out of the bores, remove the old bushings and install the new ones supplied with the brake pads

1   Caliper   3   Bushings
2   Sleeves   4   Bushings

2.6g Slide the caliper assembly over the disc, install the mounting bolts, then insert a screwdriver between the disc and outer brake pad, pry up, then strike the pad ears with a hammer to eliminate all play between the pad and caliper

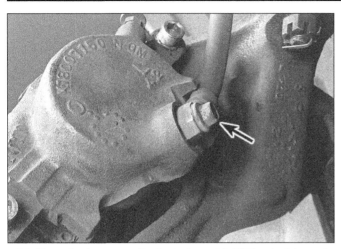

3.4 It's easier to remove the brake hose inlet fitting bolt (arrow)
before removing the caliper mounting bolts

3.8 With the caliper padded to catch the piston, use compressed
air to force the piston out of the bore - make sure your hands or
fingers are not between the piston and caliper!

3   Disc brake caliper - removal, overhaul and installation

Warning: *Dust created by the brake system may contain asbestos, which is harmful to your health. Never blow it out with compressed air and don't inhale any of it. An approved filtering mask should be worn when working on the brakes. Do not, under any circumstances, use petroleum-based solvents to clean brake parts. Use brake system cleaner only!*

Note: *If an overhaul is indicated (usually because of fluid leakage) explore all options before beginning the job. New and factory rebuilt calipers are available on an exchange basis, which makes this job quite easy. If it's decided to rebuild the calipers, make sure a rebuild kit is available before proceeding. Always rebuild the calipers in pairs - never rebuild just one of them.*

Removal

Refer to illustration 3.4

1   Remove the cover from the brake fluid reservoir, siphon off two-thirds of the fluid into a container and discard it.

2   Loosen the wheel lug nuts, raise the front of the vehicle and support it securely on jackstands. Remove the front wheels.

3   Bottom the piston in the caliper bore **(see illustration 2.5)**.

4   **Note:** *Do not remove the brake hose from the caliper if you are only removing the caliper.* Remove the brake hose inlet fitting bolt and detach the hose **(see illustration)**. On models with a hydraulic fitting (tube nut), use a flare nut wrench. Have a rag handy to catch spilled fluid and wrap a plastic bag tightly around the end of the hose to prevent fluid loss and contamination.

5   Remove the two mounting bolts and detach the caliper from the vehicle (refer to Section 2 if necessary).

Overhaul

Refer to illustrations 3.8, 3.9, 3.10, 3.11, 3.15, 3.16, 3.17 and 3.18

6   Refer to Section 2 and remove the brake pads from the caliper.

7   Clean the exterior of the caliper with brake cleaner or denatured alcohol. Never use gasoline, kerosene or petroleum-based cleaning solvents. Place the caliper on a clean workbench.

8   Position a wooden block or several shop rags in the caliper as a cushion, then use compressed air to remove the piston from the caliper **(see illustration)**. Use only enough air pressure to ease the piston out of the bore. If the piston is blown out, even with the cushion in place, it may be damaged. **Warning:** *Never place your fingers in front of the piston in an attempt to catch or protect it when applying compressed air, as serious injury could occur.*

9   Carefully pry the dust boot out of the caliper bore **(see illustration)**.

10   Using a wood or plastic tool, remove the piston seal from the groove in the caliper bore **(see illustration)**. Metal tools may cause bore damage.

11   Remove the caliper bleeder screw, then remove and discard the sleeves and bushings from the caliper ears. Discard all rubber parts **(see illustration)**.

12   Clean the remaining parts with brake system cleaner.

3.9 When prying the dust boot out of the caliper, be very careful
not to scratch the bore surface

3.10 Remove the piston seal with a wooden or plastic tool to
avoid scratching the bore and seal groove

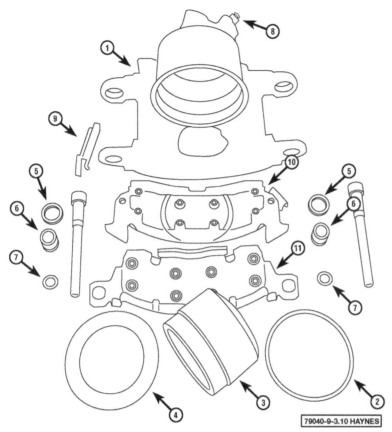

3.15 Position the seal in the caliper bore groove, making sure it isn't twisted

3.16 Install the new dust boot in the piston groove (note that the folds are at the open end of the piston)

3.11 Exploded view of the disc brake caliper components

1	Caliper housing	5	Rubber bushing	9	Clip
2	Piston seal	6	Sleeve	10	Inner brake pad
3	Piston	7	Rubber bushing	11	Outer brake pad
4	Boot	8	Bleeder screw		

3.17 Install the piston squarely in the caliper bore then bottom it by pushing down evenly

13   Carefully examine the piston for nicks and burrs and loss of plating. If surface defects are present, the parts must be replaced.

14   Check the caliper bore in a similar way. Light polishing with crocus cloth is permissible to remove light corrosion and stains. Discard the mounting bolts if they're corroded or damaged.

15   When assembling, lubricate the piston bores and seal with clean brake fluid. Posi-

tion the seal in the caliper bore groove **(see illustration)**.

16   Lubricate the piston with clean brake fluid, then install a new boot in the piston groove with the fold toward the open end of the piston **(see illustration)**.

17   Insert the piston squarely into the caliper bore, then apply force to bottom it **(see illustration)**.

3.18 Seat the boot in the counterbore (a seal driver is being used in this photo, but a drift punch will work if care is exercised)

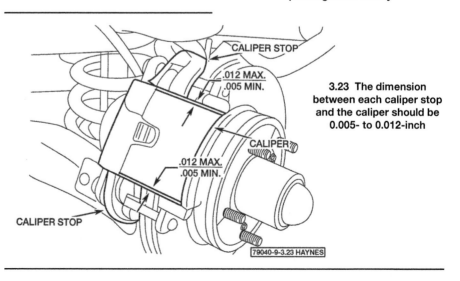

3.23 The dimension between each caliper stop and the caliper should be 0.005- to 0.012-inch

4.3 The brake pads on this vehicle were obviously neglected, as they wore down to the rivets and cut deep grooves into the disc - wear this severe means the disc must be replaced

4.4a Check disc runout with a dial indicator - if the reading exceeds the maximum allowable runout, the disc will have to be resurfaced or replaced

18   Position the dust boot in the caliper counterbore, then use a drift to drive it into position **(see illustration)**. Make sure the boot is recessed evenly below the caliper face.
19   Install the bleeder screw.
20   Install new bushings in the mounting bolt holes and fill the area between the bushings with the silicone grease supplied in the rebuild kit. Push the sleeves into the mounting bolt holes.

Installation

Refer to illustration 3.23
21   Inspect the mounting bolts for excessive corrosion.
22   Place the caliper in position over the disc and mounting bracket, install the bolts and tighten them to the specified torque.
23   Check to make sure the total clearance between the caliper and the bracket stops is between 0.005- and 0.012-inch **(see illustration)**.
24   Install the brake hose and inlet fitting bolt, using new copper washers, then tighten the bolt to the specified torque.
25   If the line was disconnected, be sure to bleed the brakes (Section 10).
26   Install the wheels and lower the vehicle.
27   After the job has been completed, firmly depress the brake pedal a few times to bring the pads into contact with the disc.
28   Check brake operation before driving the vehicle in traffic.

4   Brake disc - inspection, removal and installation

Inspection

Refer to illustrations 4.3, 4.4a, 4.4b and 4.5
1   Loosen the wheel lug nuts, raise vehicle and support it securely on jackstands. Remove the wheel.
2   Remove the brake caliper as outlined in Section 3. Remove caliper mounting bracket

4.4b Using a swirling motion, remove the glaze from the disc with sandpaper or emery cloth

on models so equipped. It's not necessary to disconnect the brake hose. After removing the caliper bolts, suspend the caliper out of the way with a piece of wire. Don't let the caliper hang by the hose and don't stretch or twist the hose.
3   Visually check the disc surface for score marks and other damage. Light scratches and shallow grooves are normal after use and may not always be detrimental to brake operation, but deep score marks - over 0.015-inch (0.38 mm) - require disc removal and refinishing by an automotive machine shop. Be sure to check both sides of the disc **(see illustration)**. If pulsating has been noticed during application of the brakes, suspect disc runout. Be sure to check the wheel bearings to make sure they're properly adjusted.
4   To check disc runout, place a dial indicator at a point about 1/2-inch from the outer edge of the disc **(see illustration)**. Set the indicator to zero and turn the disc. The indicator reading should not exceed the specified allowable runout limit. If it does, the disc should be refinished by an automotive machine shop. **Note:** *Professionals recommend resurfacing of brake discs regardless of*

4.5 A micrometer is used to measure disc thickness (the minimum thickness is cast into the inside of the disc)

the dial indicator reading (to produce a smooth, flat surface that will eliminate brake pedal pulsations and other undesirable symptoms related to questionable discs). At the very least, if you elect not to have the discs resurfaced, deglaze them with sandpaper or emery cloth (use a swirling motion to ensure a nondirectional finish) **(see illustration)**.
5   The disc must not be machined to a thickness less than the specified minimum refinish thickness. The minimum wear (or discard) thickness is cast into the inside of the disc. The disc thickness can be checked with a micrometer **(see illustration)**.

Removal

6   Refer to Chapter 11, *Front wheel bearing check, repack and adjustment* for the hub/disc removal procedure. Remove the two lug nuts that were installed to hold the disc in place and pull the disc from the hub.

Installation

7   Install the disc and hub assembly and adjust the wheel bearing (Chapter 11).
8   Install the caliper and brake pad assembly over the disc and position it on the steer-

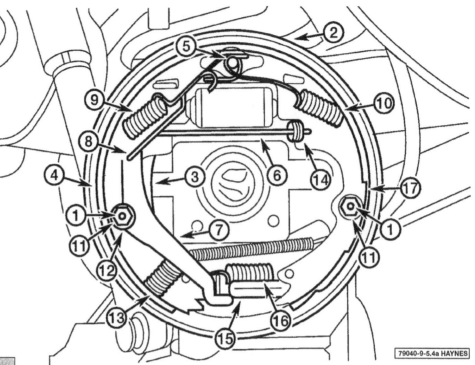

5.4a Drum brake components

1   *Hold-down pins*
2   *Backing plate*
3   *Parking brake lever (rear only)*
4   *Secondary shoe*
5   *Shoe guide*
6   *Parking brake strut (rear only)*
7   *Actuator lever*
8   *Actuator link*
9   *Return spring*
10 *Return spring*
11 *Hold-down spring*
12 *Lever pivot*
13 *Lever return spring*
14 *Strut spring (rear only)*
15 *Adjusting screw assembly*
16 *Adjusting screw spring*
17 *Primary shoe*

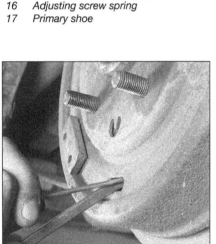

**5.4b If the drum won't come off, knock
out the plug from the drum (or backing
plate on some models), then disengage
the self-adjuster lever from the star wheel
with one tool and turn the star wheel with
another tool . . .**

**5.4c . . . this is how it looks with the drum
removed for clarity**

ing knuckle (refer to Section 3 for the caliper installation procedure, if necessary). Tighten the caliper bolts to the specified torque.

9   Install the wheel, then lower the vehicle to the ground. Depress the brake pedal a few times to bring the brake pads into contact with the disc. Bleeding of the system will not be necessary unless the brake hose was disconnected from the caliper. Check the operation of the brakes carefully before placing the vehicle into normal service.

5   Drum brake shoes (front and rear) - replacement

Refer to illustrations 5.4a through 5.4x
Warning: *Drum brake shoes must be replaced on both wheels at the same time - never*

**5.4d Remove the shoe return springs -
the spring tool shown here is available at
most auto parts stores and makes this job
much easier and safer**

replace the shoes on only one wheel. Also, the dust created by the brake system may contain asbestos, which is harmful to your health. Never blow it out with compressed air and don't inhale any of it. An approved filtering mask should be worn when working on the brakes. Do not, under any circumstances, use petroleum-based solvents to clean brake parts. Use brake system cleaner only!
Caution: *Whenever the brake shoes are replaced, the retractor and hold-down springs should also be replaced. Due to the continuous heating/cooling cycle that the springs are subjected to, they lose their tension over a period of time and may allow the shoes to drag on the drum and wear at a much faster rate than normal.*

1   Loosen the wheel lug nuts, raise the rear of the vehicle and support it securely on jackstands. Block the front wheels to keep the vehicle from rolling.
2   Release the parking brake.
3   Remove the wheel. **Note:** *All four rear brake shoes must be replaced at the same time, but to avoid mixing up parts, work on only one brake assembly at a time.*
4   Follow the accompanying photos **(illustrations 5.4a through 5.4x)** for the inspection and replacement of the brake shoes. Be sure to stay in order and read the caption under each illustration. After removing the drum and before beginning work, wash the brake components with brake system cleaner. **Note:** *If the brake drum cannot be easily pulled off the axle and shoe assembly, make sure that the parking brake is completely released, then apply some penetrating oil at the hub-to-drum joint. Allow the oil to soak in and try to pull the drum off. If the drum still cannot be pulled off, the brake shoes will have to be retracted. This*

5.4e Pull the bottom of the actuator lever toward the secondary brake shoe, compressing the lever return spring - the actuator link can now be removed

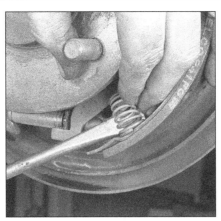

5.4f Pry the actuator lever spring out with a large screwdriver

5.4g Slide the parking brake strut out from between the axle flange and primary shoe (rear drum brakes only)

5.4h Remove the hold-down springs and pins - the hold-down spring tool shown here is available at most auto parts stores

5.4i Remove the actuator lever and pivot - be careful not to let the pivot fall out of the lever. On some models, the lever is two pieces, attached with a spring - you don't normally need to disassemble the lever

5.4j Spread the top of the shoes apart and slide the assembly around the axle

is accomplished by first removing the plug from the backing plate or brake drum. With the plug removed, pull the lever off the adjusting star wheel with one small screwdriver

while turning the adjusting wheel with another small screwdriver, moving the shoes away from the drum **(see illustrations 5.4b and 5.4c)**. *The drum should now come off.*

5   Before reinstalling the drum it should be checked for cracks, score marks, deep scratches and hard spots, which will appear as small discolored areas. If the hard spots

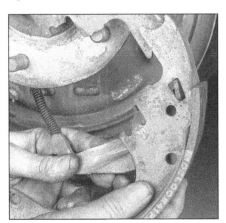

5.4k Unhook the parking brake lever from the secondary shoe (rear drum brake only)

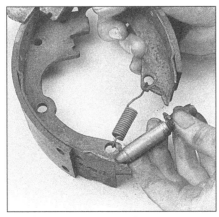

5.4l Spread the bottom of the shoes apart and remove the adjusting screw assembly

5.4m Clean the adjusting screw with solvent, dry it off and lubricate the threads and end with multi-purpose grease, then reinstall the adjusting screw assembly between the new brake shoes

5.4n Lubricate the shoe contact points on the backing plate with high temperature brake grease

5.4o Insert the parking brake lever into the opening in the secondary brake shoe (rear drum brake only)

5.4p Spread the shoes apart and slide them into position on the backing plate

5.4q Install the hold-down pin through the backing plate and primary shoe, then install the spring and retainer

5.4r Insert the lever pivot into the actuator lever, place the lever over the secondary shoe hold-down pin and install the hold-down spring and retainer

5.4s Guide the parking brake strut behind the axle flange and engage the rear end of it in the slot on the parking brake lever - spread the shoes enough to allow the other end to seat against the primary shoe (rear drum brake only)

cannot be removed with fine emery cloth or if any of the other conditions listed above exist, the drum must be taken to an automotive machine shop to have it resurfaced. **Note:** *Professionals recommend resurfacing the drums whenever a brake job is done. Resurfacing will eliminate the possibility of out-of-round drums. If the drums are worn so much that they can't be resurfaced without exceeding the maximum allowable diameter (stamped into the*

drum), then new ones will be required. At the very least, if you elect not to have the drums resurfaced, remove the glazing from the surface with medium-grit emery cloth using a swirling motion. If it was necessary to knock out the plug from the drum (in Step 4), install a new *metal* plug in the drum. If the plug was removed from the backing plate, install a rubber plug in the backing plate.

6   Install the brake drum on the axle flange.
7   Mount the wheel, install the lug nuts, then lower the vehicle.
8   Make a number of forward and reverse stops to adjust the brakes until satisfactory pedal action is obtained.
9   Check brake operation before driving the vehicle in traffic.

5.4t Place the shoe guide over the anchor pin

5.4u Hook the lower end of the actuator link to the actuator lever, then loop the top end over the anchor pin

5.4v Install the lever return spring over the tab on the actuator lever, then push the spring up onto the brake shoe

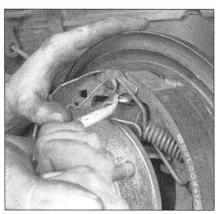

5.4w Install the primary and secondary shoe return springs

6   Wheel cylinder - removal, overhaul and installation

Note: *If an overhaul is indicated (usually because of fluid leakage or sticky operation) explore all options before beginning the job. New wheel cylinders are available, which makes this job quite easy. If it's decided to rebuild the wheel cylinder, make sure that a rebuild kit is available before proceeding. Never overhaul only one wheel cylinder - always rebuild both of them at the same time.*

Removal

Refer to illustrations 6.4, 6.5a and 6.5b

1   Raise the rear of the vehicle and support it securely on jackstands. Block the front wheels to keep the vehicle from rolling.

2   Remove the brake shoe assembly (Section 5).

3   Remove all dirt and foreign material from around the wheel cylinder.

4   Unscrew the brake line fitting **(see illustration)**. Don't pull the brake line away from the wheel cylinder.

5.4x Pull out on the actuator lever to disengage it from the adjusting screw wheel, turn the wheel to adjust the shoes in or out as necessary - the brake drum should slide over the shoes and turn with a very slight amount of drag

5   Remove the wheel cylinder mounting bolts. On models using retaining clips, use curved needle-nose pliers to remove the wheel cylinder retaining clip **(see illustration)**, from the rear of the wheel cylinder. If curved needle-nose pliers are not available, insert two awls into the access slots between the cylinder pilot and the retainer locking tabs and bend both tabs away at the same time **(see illustration)**.

6   Detach the wheel cylinder from the brake backing plate and place it on a clean workbench. Immediately plug the brake line to prevent fluid loss and contamination. **Note:** *If the brake shoe linings are contaminated with brake fluid, install new brake shoes.*

Overhaul

Refer to illustration 6.7

7   Remove the bleeder screw, cups, pistons, boots and spring assembly from the wheel cylinder body **(see illustration)**.

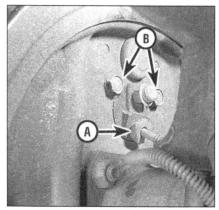

6.4 Completely loosen the brake line fitting (A) then remove the two wheel cylinder mounting bolts (B) (if used)

6.5a Using a pair of needle-nose pliers, squeeze the retaining clip together to release it from the wheel cylinder

8   Clean the wheel cylinder with brake fluid, denatured alcohol or brake system cleaner. **Warning:** *Do not, under any circumstances, use petroleum based solvents to clean brake parts!*

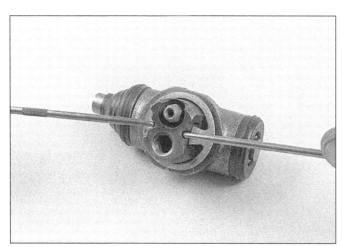

6.5b Insert two awls in between the wheel cylinder and the retaining clip tangs and pull back to release the wheel cylinder (cylinder removed for clarity)

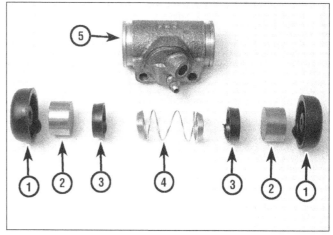

6.7 Exploded view of the wheel cylinder components

1 Dust boot	4 Spring
2 Piston	5 Wheel cylinder housing
3 Cup	

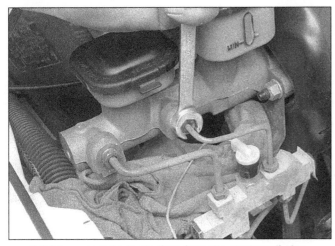

**7.2 Disconnect the brake lines from the master cylinder -
a flare-nut wrench should be used**

7.9 Removing the secondary stop bolt

9   Use compressed air to remove excess fluid from the wheel cylinder and to blow out the passages.
10   Check the cylinder bore for corrosion and score marks. Crocus cloth can be used to remove light corrosion and stains, but the cylinder must be replaced with a new one if the defects cannot be removed easily, or if the bore is scored.
11   Lubricate the new cups with brake fluid.
12   Assemble the wheel cylinder components. Make sure the cup lips face in.

Installation

13   Place the wheel cylinder in position and install the bolts, or, on models that use a retaining clip, install the clip.
14   Connect the brake line and tighten the fitting. Install the brake shoe assembly.
15   Bleed the brakes (Section 10).
16   Check brake operation before driving the vehicle in traffic.

7   Master cylinder - removal, overhaul and installation

Note: *Before deciding to overhaul the master cylinder, investigate the availability and cost of a new or factory rebuilt unit and also the availability of a rebuild kit.*

7.10a Removing retaining ring

Removal

Refer to illustration 7.2

1   Place rags under the brake line fittings and prepare caps or plastic bags to cover the ends of the lines once they are disconnected. **Caution:** *Brake fluid will damage paint. Cover all body parts and be careful not to spill fluid during this procedure.*
2   Loosen the tube nuts at the ends of the brake lines where they enter the master cylinder. To prevent rounding off the flats on these nuts, a flare-nut wrench, which wraps around the nut, should be used **(see illustration)**.
3   Pull the brake lines away from the master cylinder slightly and plug the ends to prevent contamination.
4   On manual (non-power) brakes, disconnect the pushrod at the brake pedal inside the car.
5   Remove the two master cylinder mounting nuts. If there is a bracket retaining the combination valve, move it forward slightly, taking care not to bend the hydraulic lines running to the combination valve, and remove the master cylinder from the vehicle **(see illustration)**.

6   Remove the reservoir cover and reservoir diaphragm, then discard any fluid remaining in the reservoir.

Overhaul

Refer to illustrations 7.9, 7.10a, 7.10b, 7.13a, 7.13b, 7.14, 7.16a, 7.16b and 7.17

7   Drain all fluid from the master cylinder and place the unit in a vise. Use wood blocks to cushion the jaws of the vise.
7   On manual brakes remove the pushrod retaining ring.
8   Remove the secondary stop bolt from the bottom of the front fluid reservoir (Delco Moraine) or from the base of the master cylinder body (Bendix) **(see illustration)**.
10   Remove the retaining ring from the groove and take out the primary piston assembly **(see illustrations)**. Following the primary piston out of the bore will be the secondary piston, spring and retainer. You may have to invert the master cylinder and tap it on a block of wood to eject the pistons.
11   Examine the inside surface of the master cylinder and the secondary piston. If there is evidence of scoring or "bright' wear areas, the

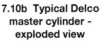

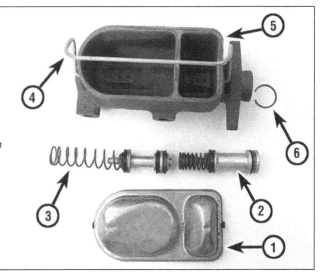

**7.10b Typical Delco
master cylinder -
exploded view**

1   *Reservoir cover*
2   *Primary piston
     and spring*
3   *Secondary piston
     and spring*
4   *Bail*
5   *Master cylinder
     body*
6   *Retaining ring*

7.13a Removing line seats

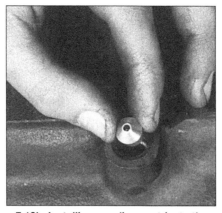

7.13b Installing new line seat (note the cone-shaped end is installed facing up)

7.14 Seals correctly installed on piston

entire master cylinder should be replaced with a new one. If the bore looks good, light honing is permitted to remove any glazing (this applies to cast iron master cylinders only).

12   If the components are in good condition, wash them with brake system cleaner. Discard all the rubber components and the primary piston. Purchase a rebuild kit which will contain all the necessary parts for the overhaul.

13   Inspect the line seats which are located in the master cylinder body where the lines connect. If they appear damaged they should be replaced with new ones, which come in the overhaul kit. They are forced out of the body by threading a screw into the tube and then prying outwards **(see illustration)**. The new ones are forced into place using a spare brake line nut **(see illustration)**. All parts necessary for this should be included in the rebuild kit.

14   Place the new secondary seals in the grooves of the secondary piston **(see illustration)**.

15   Assemble the primary seals and seal protector over the end of the secondary piston.

16   Lubricate the cylinder bore and secondary piston with clean brake fluid. Insert the spring retainer into the spring then place the retainer and spring over the end of the secondary piston **(see illustrations)**. The retainer should locate inside the primary seal lips.

17   With the master cylinder vertical, push the secondary piston into the bore completely **(see illustration)**.

18   Coat the seals of the primary piston with brake fluid and fit it into the cylinder bore. Hold it down while the retaining ring is installed in the cylinder groove.

19   Continue to hold the piston down while the stop screw is installed.

20   Install the reservoir diaphragm into the reservoir cover plate, making sure it is fully collapsed inside the recess lid.

21   **Note:** *Whenever the master cylinder is removed, the complete hydraulic system must be bled. The time required to bleed the system can be reduced if the master cylinder is filled with fluid and bench bled (refer to Steps 22 through 26) before the master cylinder is installed on the vehicle.*

Bench bleeding master cylinder

22   Insert threaded plugs of the correct size into the cylinder outlet holes and fill the reservoirs with brake fluid. The master cylinder should be supported in such a manner that brake fluid will not spill during the bench bleeding procedure.

23   Loosen one plug at a time and push the piston assembly into the bore to force air from the master cylinder. To prevent air from being drawn back into the cylinder, the appropriate plug must be replaced before allowing the piston to return to its original position.

24   Stroke the piston three or four times for each outlet to ensure that all air has been expelled.

25   Since high pressure is not involved in the bench bleeding procedure, an alternative to the removal and replacement of the plugs with each stroke of the piston assembly is available. Before pushing in on the piston assembly, remove one of the plugs completely. Before releasing the piston, however, instead of replacing the plug, simply put your finger tightly over the hole to keep air from being drawn back into the master cylinder. Wait several seconds for the brake fluid to be drawn from the reservoir to the piston bore, then repeat the procedure. When you push down on the piston it will force your finger off the hole, allowing the air inside to be expelled. When only brake fluid is being ejected from the hole, replace the plug and go on to the other port.

26   Refill the master cylinder reservoirs and install the diaphragm and cover assembly. **Note:** *The reservoirs should only be filled to the top of the reservoir divider to prevent overflowing when the cover is installed.*

Installation

27   Carefully install the master cylinder by reversing the removal steps, then bleed the brakes (refer to Section 10).

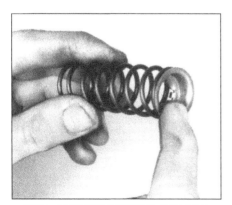

7.16a Assembling retainer and spring

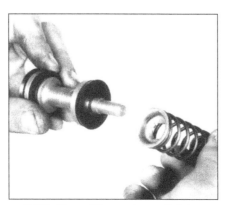

7.16b Assembling spring and piston

7.17 Pushing piston into master cylinder bore

8 Combination valve - check and replacement

Check

1   Disconnect the wire from the pressure differential switch. **Note:** *When unplugging the connector, squeeze the side lock releases, moving the inside tabs away from the switch, then pull up. Pliers may be used as an aid if necessary.*
2   Using a jumper wire, connect the switch wire to a good ground, such as the engine block.
3   Turn the ignition key to the On position. The warning light in the instrument panel should light.
4   If the warning light does not light, either the bulb is burned out or the electrical circuit is defective. Replace the bulb or repair the electrical circuit as necessary.
5   When the warning light functions correctly, turn the ignition switch off, disconnect the jumper wire and reconnect the wire to the switch terminal.
6   Make sure the master cylinder reservoirs are full, then attach a bleeder hose to one of the rear wheel bleeder valves and immerse the other end of the hose in a container partially filled with clean brake fluid.
7   Turn the ignition switch On.
8   Open the bleeder valve while a helper applies moderate pressure to the brake pedal. The brake warning light on the instrument panel should light.
9   Close the bleeder valve before the helper releases the brake pedal.
10   Reapply the brake pedal with moderate to heavy pressure. The brake warning light should go out.
11   Attach the bleeder hose to one of the front brake bleeder valves and repeat Steps 8 through 10. The warning light should react in the same manner as in Steps 8 and 10.
12   Turn the ignition switch Off.
13   If the warning light did not come on in Steps 8 and 11, but does light when a jumper is connected to ground, the warning light

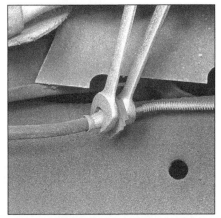

9.2 Place a wrench on the hose fitting to prevent it from turning and disconnect the line with a flare-nut wrench

switch portion of the combination valve is defective and the combination valve must be replaced with a new one since the components of the combination valve are not individually serviceable.

Replacement

14   Place a container under the combination valve and protect all painted surfaces with newspapers or rags.
15   Disconnect the hydraulic lines at the combination valve, then plug the lines to prevent further loss of fluid and to protect the lines from contamination.
16   Disconnect the electrical connector from the pressure differential switch.
17   Remove the bolt holding the valve to the mounting bracket and remove the valve from the vehicle.
18   Installation is the reverse of the removal procedure.
19   Bleed the entire brake system.

9 Brake hoses and lines - inspection and replacement

Inspection

1   About every six months, with the vehicle raised and supported securely on jackstands, the rubber hoses which connect the steel brake lines with the front and rear brake assemblies should be inspected for cracks, chafing of the outer cover, leaks, blisters and other damage. These are important and vulnerable parts of the brake system and inspection should be complete. A light and mirror will be helpful for a thorough check. If a hose exhibits any of the above conditions, replace it with a new one.

Replacement
Front brake hose
Refer to illustration 9.2

2   Using a back-up wrench, disconnect the brake line from the hose fitting, being careful not to bend the frame bracket or brake line **(see illustration)**.
3   Use a pair of pliers to remove the U-clip from the female fitting at the bracket, then detach the hose from the bracket.
4   Unscrew the brake hose from the caliper. At the caliper end of the hose, remove the bolt from the fitting block, then remove the hose and the copper washers on either side of the fitting block.
5   To install the hose, first connect it to the caliper, using new copper washers on either side of the fitting block. Lubricate the inlet fitting bolt threads with clean brake fluid before installation.
6   With the fitting engaged with the caliper locating ledge, attach the hose to the caliper, tightening the fitting bolt to the specified torque.
7   Without twisting the hose, install the female fitting in the hose bracket. It will fit the bracket in only one position.

8   Install the U-clip retaining the female fitting to the frame bracket.
9   Using a back-up wrench on the hose fitting, attach the brake line to the hose fitting, tightening it securely.
10   When the brake hose installation is complete, there should be no kinks in the hose. Make sure the hose doesn't contact any part of the suspension. Check this by turning the wheels to the extreme left and right positions. If the hose makes contact, remove it and correct the installation as necessary. Bleed the system (Section 10).

Rear brake hose
11   Using a back-up wrench, disconnect the hose at the frame bracket, being careful not to bend the bracket or steel lines.
12   Remove the U-clip with a pair of pliers and separate the female fitting from the bracket.
13   Disconnect the two hydraulic lines at the junction block, then unbolt and remove the hose.
14   Bolt the junction block to the axle housing and connect the lines, tightening them securely. Without twisting the hose, install the female end of the hose in the frame bracket.
15   Install the U-clips retaining the female end to the bracket.
16   Using a back-up wrench, attach the steel line fittings to the female fittings. Again, be careful not to bend the bracket or steel line.
17   Make sure the hose installation did not loosen the frame bracket. Tighten the bracket if necessary.
18   Fill the master cylinder reservoir and bleed the system (refer to Section 10).

Metal brake lines
19   When replacing brake lines be sure to use the correct parts. Don't use copper tubing for any brake system components. Purchase steel brake lines from a dealer or auto parts store.
20   Prefabricated brake line, with the tube ends already flared and fittings installed, is available at auto parts stores and dealers.
21   If necessary, carefully bend the line to the proper shape. A tube bender is recommended for this **Warning:** *Do not crimp or damage the line.*
22   When installing the new line make sure it's securely supported in the brackets and has plenty of clearance between moving or hot components.
23   After installation, check the master cylinder fluid level and add fluid as necessary. Bleed the brake system as outlined in the next Section and test the brakes carefully before driving the vehicle in traffic.

10 Brake system bleeding

Refer to illustration 10.8
Warning: *Wear eye protection when bleeding the brake system. If the fluid comes in con-*

10.8 When bleeding the brakes, a hose is connected to the bleeder valve and then submerged in brake fluid - air will be seen as bubbles in the container and the hose

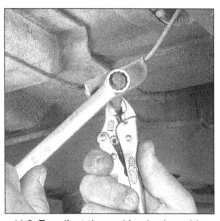

11.3 To adjust the parking brake cable, turn the adjusting nut on the equalizer while preventing the cable from turning with a pair of locking pliers clamped to the end of the adjuster rod

tact with your eyes, immediately rinse them with water and seek medical attention.
Note: *Bleeding the hydraulic system is necessary to remove any air that manages to find its way into the system when it's been opened during removal and installation of a hose, line, caliper or master cylinder.*

1   It will probably be necessary to bleed the system at all four wheels if air has entered the system due to low fluid level, or if the brake lines have been disconnected at the master cylinder.

2   If a brake line was disconnected only at a wheel, then only that caliper or wheel cylinder must be bled.

3   If a brake line is disconnected at a fitting located between the master cylinder and any of the calipers or wheel cylinders, that part of the system served by the disconnected line must be bled.

4   Remove any residual vacuum from the brake power booster by applying the brake several times with the engine off.

5   Remove the master cylinder reservoir cover and fill the reservoir with brake fluid. Reinstall the cover. **Note:** *Check the fluid level often during the bleeding operation and add fluid as necessary to prevent the fluid level from falling low enough to allow air bubbles into the master cylinder.*

6   Have an assistant on hand, as well as a supply of new brake fluid, a clear container partially filled with clean brake fluid, a length of clear tubing to fit over the bleeder valve and a wrench to open and close the bleeder valve.

7   Beginning at the right rear wheel, loosen the bleeder valve slightly, then tighten it to a point where it is snug but can still be loosened quickly and easily.

8   Place one end of the tubing over the bleeder valve and submerge the other end in brake fluid in the container **(see illustration)**.

9   Have the assistant pump the brakes slowly a few times to build pressure in the system, then hold the pedal firmly depressed.

10   While the pedal is held depressed, open the bleeder valve just enough to allow a flow of fluid to leave the valve. Watch for air bubbles to exit the submerged end of the tube. When the fluid flow slows after a couple of seconds, close the valve and have your assistant release the pedal.

11   Repeat Steps 9 and 10 until no more air is seen leaving the tube, then tighten the bleeder valve and proceed to the left rear wheel, the right front wheel and the left front wheel, in that order, and perform the same procedure. Be sure to check the fluid in the master cylinder reservoir frequently.

12   Never use old brake fluid. It contains moisture which can boil, rendering the brake system useless.

13   Refill the master cylinder with fluid at the end of the operation.

14   Check the operation of the brakes. The pedal should feel solid when depressed, with no sponginess. If necessary, repeat the entire process. **Warning:** *Do not operate the vehicle if you are in doubt about the effectiveness of the brake system.*

11 Parking brake - adjustment

Refer to illustration 11.3

1   The adjustment of the parking brake cable may be necessary whenever the rear brake cables have been disconnected or the parking brake cables have stretched due to age and stress.

2   Depress the parking brake pedal exactly three ratchet clicks and then raise the car for access underneath, supporting it securely on jackstands.

3   Tighten the adjusting nut until the left rear wheel can just barely be turned in a rearward motion **(see illustration)**. The wheel should be completely locked from moving in a forward rotation.

4   Carefully release the parking brake pedal and check that the wheel is able to rotate freely in either direction. It is important that there is no drag on the rear brakes with the pedal released.

12 Parking brake cables - replacement

Rear cables

Refer to illustrations 12.4 and 12.5

1   Loosen the wheel lug nuts, raise the rear of the vehicle and support it securely on jackstands. Remove the wheel(s).

2   Loosen the equalizer nut to slacken the cables, then disconnect the cable to be replaced from the equalizer.

3   Remove the brake drum from the axle flange. Refer to Section 5 if any difficulty is encountered.

4   Remove the brake shoe assembly far enough to disconnect the cable end from the parking brake lever **(see illustration)**.

5   Depress the tangs on the cable housing retainer and push the housing and cable through the backing plate **(see illustration)**.

6   To install the cable, reverse the removal procedure and adjust the cable as described in the preceding Section.

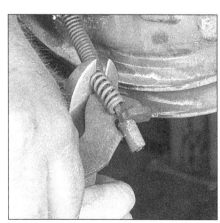

12.4 To disconnect the cable end from the parking brake lever, pull back on the return spring and maneuver the cable out of the slot in the lever

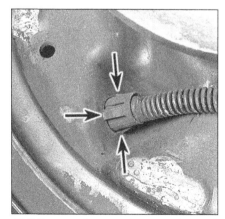

12.5 Depress the retention tangs (arrows) to free the cable and housing from the backing plate

Front cable

7   Raise the vehicle and support it securely on jackstands.
8   Loosen the equalizer assembly to provide slack in the cable.
9   Disconnect the front cable from the cable joiner near the equalizer.
10   Disconnect the cable from the pedal assembly.
11   Free the cable from the routing clips and push the cable and grommet through the firewall.
12   To install the cable, reverse the removal procedure and adjust the cable as described in the preceding Section.

13   Parking brake pedal - removal and installation

1   Disconnect the battery ground cable and the parking brake warning switch wire.
2   Remove the clip and ball from the clevis (if necessary, the equalizer nut can be loosened) **(see illustration 11.3)**.
3   Remove the pedal rear mounting bolt and the nuts from the mounting studs at the front of the dash panel (under the hood).
4   Remove the pedal assembly.
5   Installation is the reverse of the removal procedure.

14   Power brake booster - check, removal and installation

Operating check

1   Depress the brake pedal several times with the engine off and make sure that there is no change in the pedal reserve distance.
2   Depress the pedal and start the engine. If the pedal goes down slightly, operation is normal.

Airtightness check

3   Start the engine and turn it off after one or two minutes. Depress the brake pedal several times slowly. If the pedal goes down farther the first time but gradually rises after the second or third depression, the booster is airtight.

4   Depress the brake pedal while the engine is running, then stop the engine with the pedal depressed. If there is no change in the pedal reserve travel after holding the pedal for 30 seconds, the booster is airtight.

Removal and installation

Note: *Dismantling of the power brake unit requires special tools. If a problem develops, it is recommended that a new or factory-exchange unit be installed rather than trying to overhaul the original booster.*

5   Remove the mounting nuts which hold the master cylinder to the power brake unit (see Section 7). Move the master cylinder forward, but be careful not to bend or kink the lines leading to the master cylinder. If there is any strain on the lines, disconnect them at the master cylinder and plug the ends.
6   Disconnect the vacuum hose leading to the front of the power brake booster. Cover the end of the hose.
7   Disconnect the power brake pushrod from the brake pedal. Do not force the pushrod to the side when disconnecting it.
8   Remove the four booster mounting nuts and carefully lift the unit out of the engine compartment.
9   When installing, loosely install the four mounting nuts, then connect the pushrod to the brake pedal. If the old retaining clip doesn't snap into place, replace it with a new clip. Tighten the booster retaining nuts to the torque listed in this Chapter's Specifications. Reconnect the vacuum hose. Install the master cylinder (see Section 7). If the brake lines were disconnected from the master cylinder, bleed the brake system to eliminate any air which has entered the system (see Section 10).

15   Brake light switch - check, adjustment and replacement

Refer to illustration 15.1

1   The brake light switch **(see illustration)** is located on a flange or bracket protruding from the brake pedal support.
2   With the brake pedal in the fully released position, the plunger of the switch should be fully pressed in. When the pedal

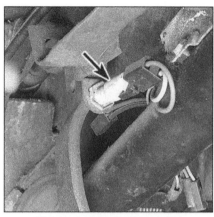

15.1 Typical brake light switch

is depressed, the plunger releases and sends electrical current to the brake lights at the rear of the car.
3   To check the brake light switch, simply note whether the brake lights come on when the pedal is depressed and go off when the pedal is released. If they don't, adjust the switch. To do this, loosen the locknut and thread the switch inward until the body (not the plunger) of the switch contacts the brake pedal, then release the pedal.
4   If the lights still don't come on, either the switch is not getting voltage, the switch itself is defective, or the circuit between the switch and the lights is defective. There is always the remote possibility that all of the brake light bulbs are burned out, but this is not very likely.
5   Use a voltmeter or test light to verify that there's voltage present at one side of the switch connector. If no voltage is present, troubleshoot the circuit from the switch to the fuse box. If there is voltage present, check for voltage on the other terminal when the brake pedal is depressed. If no voltage is present, replace the switch. If there is voltage present, troubleshoot the circuit from the switch to the brake lights.
6   To replace a faulty switch, simply unplug the electrical connector(s) (one for the switch circuit, one for cruise control) and pull the switch out of its bracket. Installation is the reverse of removal. Be sure to adjust the switch as described in Step 2.

16   Troubleshooting - brake system

Symptom

Pedal travels almost to floor before brakes operate ....

Reason

Brake fluid level too low
Caliper or wheel cylinder leaking
Master cylinder leaking (bubbles in master cylinder fluid)
Brake flexible hose leaking
Brake line fractured
Brake system unions loose
Pad or shoe linings over 75% worn
Brake adjusters not functioning

Symptom	Reason
Brake pedal feels springy	New linings not yet bedded-in Brake discs or drums badly worn or cracked Master cylinder securing nuts loose Power brake booster loose (where fitted)
Brake pedal feels 'spongy' and 'soggy'	Air in hydraulic system Caliper or wheel cylinder leaking Master cylinder leaking (bubbles in master cylinder reservoir) Brake pipe line or flexible hose leaking Unions in brake system loose
Excessive effort required to brake car	Pad or shoe linings badly worn New pads or shoes recently fitted - not yet bedded in Harder linings fitted than standard causing increase in pedal pressure Linings and brake drums contaminated with oil, grease or brake fluid Brake booster malfunction (where fitted) or faulty vacuum hose
Brakes uneven and pulling to one side	Linings and discs or drums contaminated with oil, grease or brake fluid Tire pressures unequal Brake caliper loose Brake pads or shoes fitted incorrectly Different type of linings fitted at each wheel Anchorages for front suspension or rear suspension loose Brake discs or drums badly worn, cracked or distorted
Brakes tend to bind, drag or lock-on	Rear brakes over-adjusted (release automatic adjuster) Internally split brake hose

Notes

Chapter 10
Electrical system

Contents

Specifications

Torque specifications

	Ft-lbs
Starter motor mounting bolts	25 to 35
Alternator pulley nut	40 to 50

1 General description

The electrical system is a 12-volt, negative ground type. Power for the lights and all electrical accessories is supplied by a lead/acid-type battery which is charged by the alternator.

This Chapter covers repair and service procedures for the various electrical components not associated with the ignition system. Information on the ignition system can be found in Chapter 4.

It should be noted that when portions of the electrical system are serviced, the negative battery cable should be disconnected from the battery to prevent electrical shorts and/or fires.

2 Battery - removal and installation

1   **Caution:** *Always disconnect the negative cable first and hook it up last or the battery may be shorted by the tool being used to loosen the cable clamps. Disconnect both cables from the battery terminals.*
2   Remove the battery hold-down clamp.
3   Lift out the battery. Use the proper lifting technique - the battery is heavy.
4   While the battery is out, inspect the battery carrier (tray) for corrosion. Neutralize any corrosion with a water and baking soda mixture, then dry off the tray and paint it with an anti-rust paint.
5   If you are replacing the battery, make sure that you get an identical battery, with the same dimensions, amperage rating, "cold cranking" rating, etc.
6   Installation is the reverse of removal.

3 Battery - check and maintenance

Refer to illustrations 3.1, 3.4, 3.6a and 3.6b

Warning: *Certain precautions must be followed when checking and servicing the battery. Hydrogen gas, which is highly flammable, is always present in the battery cells, so keep lighted tobacco and all other open flames and sparks away from the battery. The electrolyte inside the battery is actually dilute sulfuric acid, which will cause injury if splashed on your skin or in your eyes. It will also ruin clothes and*

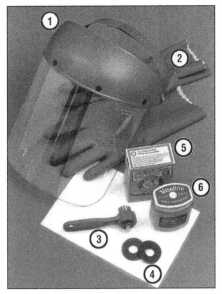

3.1 Tools and materials required for battery maintenance

1   *Face shield/safety goggles* - *When removing corrosion with a brush, the acidic particles can easily fly up into your eyes*
2   *Rubber gloves* - *Another safety item to consider when servicing the battery - remember that's acid inside the battery!*
3   *Battery terminal/cable cleaner* - *This wire brush cleaning tool will remove all traces of corrosion from the battery cable*
4   *Treated felt washers* - *Placing one of these on each terminal, directly under the cable end, will help prevent corrosion (be sure to get the correct type for side-terminal batteries)*
5   *Baking soda* - *A solution of baking soda and water can be used to neutralize corrosion*
6   *Petroleum jelly* - *A layer of this on the battery terminal bolts will help prevent corrosion*

painted surfaces. When removing the battery cables, always detach the negative cable first and hook it up last!

1   Carry out the regular weekly maintenance described in the Routine maintenance Section at the front of this manual. Battery maintenance is an important procedure which will help ensure that you are not stranded because of a dead battery. Several tools are required for this procedure **(see illustration)**.

2   When checking/servicing the battery, always turn the engine and all accessories off.

3   A sealed (sometimes called maintenance-free), side-terminal battery is standard equipment on most of these vehicles. The cell caps cannot be removed, no electrolyte checks are required and water cannot be added to the cells. However, if a standard top-terminal aftermarket battery has been

3.4 Remove the cell caps to check the water level in the battery - if the level is low, add distilled water only

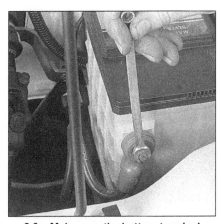

3.6a Make sure the battery terminal bolts are tight

installed, the following maintenance procedure can be used.

4   Remove the caps and check the electrolyte level in each of the battery cells **(see illustration)**. It must be above the plates. There's usually a split-ring indicator in each cell to indicate the correct level. If the level is low, add distilled water only, then reinstall the cell caps. **Caution:** *Overfilling the cells may cause electrolyte to spill over during periods of heavy charging, causing corrosion and damage to nearby components.*

5   The external condition of the battery should be checked periodically. Look for damage such as a cracked case.

6   Check the tightness of the battery cable bolts **(see illustration)** to ensure good electrical connections. Inspect the entire length of each cable, looking for cracked or abraded insulation and frayed conductors.

7   If corrosion (visible as white, fluffy deposits) is evident, remove the cables from the terminals, clean them with a battery brush and reinstall them. Corrosion can be kept to a minimum by applying a layer of petroleum jelly or grease to the bolt threads.

8   Make sure the battery carrier is in good condition and the hold-down clamp is tight. If the battery is removed, make sure that no parts remain in the bottom of the carrier when

Terminal end corrosion or damage.

Insulation cracks.

Chafed insulation or exposed wires.

Burned or melted insulation.

3.6b Typical battery cable problems

it's reinstalled. When reinstalling the hold-down clamp, don't overtighten the bolt.

9   Corrosion on the carrier, battery case and surrounding areas can be removed with a solution of water and baking soda. Apply the mixture with a small brush, let it work, then rinse it off with plenty of clean water.

10   Any metal parts of the vehicle damaged by corrosion should be coated with a zinc-based primer, then painted.

---

4   Battery cables - check and replacement

---

1   Periodically inspect the entire length of each battery cable for damage, cracked or burned insulation and corrosion. Poor battery cable connections can cause starting problems and decreased engine performance.

2   Check the cable-to-terminal connections at the ends of the cables for cracks, loose wire strands and corrosion. The presence of white, fluffy deposits under the insulation at the cable terminal connection is a sign that the cable is corroded and should be replaced. Check the terminals for distortion, missing mounting bolts and corrosion.

3   When removing the cables, always disconnect the negative cable first and hook it

up last or the battery may be shorted by the tool used to loosen the cable clamps. Even if only the positive cable is being replaced, be sure to disconnect the negative cable from the battery first.

4   Disconnect the old cables from the battery, then trace each of them to their opposite ends and detach them from the starter solenoid and ground. Note the routing of each cable to insure correct installation.

5   If you are replacing either or both of the old cables, take them with you when buying new cables. It is vitally important that you replace the cables with identical parts. Cables have characteristics that make them easy to identify: positive cables are usually red, larger in cross-section and have a larger diameter battery post clamp; ground cables are usually black, smaller in cross-section and have a slightly smaller diameter clamp for the negative post.

6   Clean the threads of the solenoid or ground connection with a wire brush to remove rust and corrosion. Apply a light coat of battery terminal corrosion inhibitor, or petroleum jelly, to the threads to prevent future corrosion.

7   Attach the cable to the solenoid or ground connection and tighten the mounting nut/bolt securely.

8   Before connecting a new cable to the battery, make sure that it reaches the battery post without having to be stretched.

9   Connect the positive cable first, followed by the negative cable.

5   Battery - emergency jump starting

Refer to the *Booster battery (jump) starting* procedure at the front of this manual.

6   Alternator - maintenance and special precautions

1   The alternator fitted to Nova models is a Delco-Remy Delcotron. Three types have been used, the 5.5 inch Series 1D, 10DN Series 100B and the Series 10SI. The first two types require a separate voltage regulator; the last type has an integral regulator.

2   Alternator maintenance consists of occasionally wiping away any dirt or oil which may have collected.

3   Check the tension of the driving belt every 6000 miles. Refer to Chapter 2, Section 7.

4   No lubrication is required as alternator bearings are grease sealed for the life of the unit.

5   Take extreme care when making circuit connections to a vehicle fitted with an alternator and observe the following. When making connections to the alternator from a battery always match correct polarity. Before using electric-arc welding equipment to repair any part of the vehicle, disconnect the connector from the alternator and disconnect the posi-

tive battery terminal. Never start the car with a battery charger connected. Always disconnect both battery leads before using a battery charger. If boosting from another battery, always connect in parallel using heavy cable.

7   Charging system - check

Refer to illustration 7.4

1   If a malfunction occurs in the charging circuit, don't automatically assume that the alternator is causing the problem. First check the following items:

a) *Check the drivebelt tension and its condition. Replace it if worn or deteriorated.*
b) *Make sure that the alternator mounting and adjustment bolts are tight.*
c) *Inspect the alternator wiring harness and the connectors at the alternator and voltage regulator. They must be in good condition and tight.*
d) *Check the fusible link (if equipped) located between the starter solenoid and the alternator. If it's burned, determine the cause, repair the circuit and replace the link (the vehicle won't start and/or the accessories won't work if the fusible link blows).*
e) *Start the engine and check the alternator for abnormal noises (a shrieking or squealing sound indicates a bad bushing).*
f) *Check the specific gravity of the battery electrolyte. If it's low, charge the battery (doesn't apply to maintenance free batteries).*
g) *Make sure that the battery is fully charged (one bad cell in a battery can cause overcharging by the alternator).*
h) *Disconnect the battery cables (negative first, then positive). Inspect the battery posts and the cable clamps for corrosion. Clean them thoroughly if necessary. Reconnect the cable to the positive terminal.*
i) *With the key off, insert a test light between the negative battery post and the disconnected negative cable clamp.*
   1) *If the test light does not come on, reattach the clamp and proceed to the next Step.*
   2) *If the test light comes on, there is a short in the electrical system of the vehicle. The short must be repaired before the charging system can be checked.*
   3) *Disconnect the alternator wiring harness.*
      a) *If the light goes out, the alternator is bad.*
      b) *If the light stays on, pull each fuse until the light goes out (this will tell you which component is shorted).*

2   Using a voltmeter, check the battery voltage with the engine off. It should be approximately 12 volts.

3   Start the engine and check the battery voltage again. It should now be approximately 13.5 to 14.5 volts.

4   Locate the D-shaped test hole in the back of the alternator. Ground the tab inside the hole to the alternator body by inserting a screwdriver blade through the hole and touching the tab and the body at the same time **(see illustration)**. **Caution:** *Don't run the engine with the tab grounded any longer than necessary to obtain a voltmeter reading. If the alternator is charging, it's running unregulated during this test, which may overload the electrical system and cause damage to the components.*

5   The reading on the voltmeter should be 15-volts or higher with the tab grounded.

6   If the voltmeter indicates low voltage, the alternator is faulty and should be rebuilt or replaced with a new one.

7   If the voltage reading is 15-volts or higher and a "no charge" condition exists, the regulator or field circuit is the problem. Remove the alternator and have it load tested and repaired or replaced.

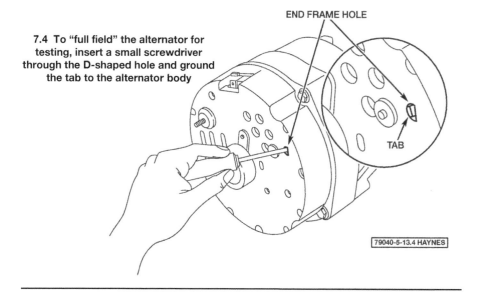

7.4 To "full field" the alternator for testing, insert a small screwdriver through the D-shaped hole and ground the tab to the alternator body

END FRAME HOLE

TAB

79040-5-13.4 HAYNES

8.2 The first step in alternator removal is disconnecting all electrical leads (which will vary somewhat with alternator type and year of manufacture)

8.3a To remove the drivebelt, loosen the alternator mounting bolt . . .

8.3b . . . then loosen the adjustment bolt and slip off the belt - finally, remove the mounting and adjusting bolts and remove the alternator

9.1 An exploded view of the 10SI alternator

1   Nut
2   Washer
3   Pulley
4   Washer
5   Collar
6   Fan
7   Drive end frame
8   Bearing
9   Plate
10 Stator
11 Rotor
12 Slip-ring end frame
13 Bearing
14 Terminal component stud
15 Voltage regulator
16 Rectifier bridge
17 Diode trio
18 Brushes
19 Capacitor
20 Brush springs
21 Brush holder

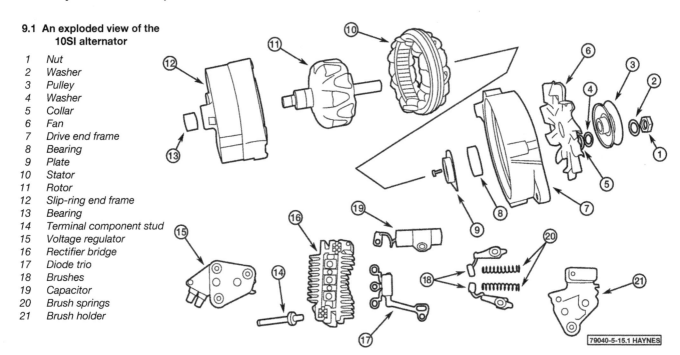

8   Alternator - removal and installation

Refer to illustrations 8.2, 8.3a, 8.3b

1   Detach the cable from the negative terminal of the battery.
2   Detach the electrical connectors from the alternator and the voltage regulator **(see illustration)**.
3   Loosen the alternator adjustment and pivot bolts and detach the drivebelt **(see illustrations)**.
4   Remove the adjustment and pivot bolts and separate the alternator from the engine.
5   If you are replacing the alternator, take the old alternator with you when purchasing a replacement unit. Make sure that the new/rebuilt unit is identical to the old alternator. Look at the terminals - they should

be the same in number, size and location as the terminals on the old alternator. Finally, look at the identification markings - they will be stamped in the housing or printed on a tag or plaque affixed to the housing. Make sure that these numbers are the same on both alternators.
6   Many new/rebuilt alternators do not have a pulley installed, so you may have to switch the pulley from the old unit to the new/rebuilt one. When buying an alternator, find out the shop's policy regarding pulleys - some shops will perform this service free of charge.
7   Installation is the reverse of removal.
8   After the alternator is installed, adjust the drivebelt tension (see Chapter 2, Section 7).
9   Check the charging voltage to verify proper operation of the alternator (see Section 7).

9.2 Paint an alignment mark on the alternator body between each end frame

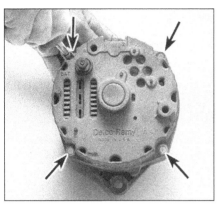

9.3 Remove the bolts (arrows) securing the rear end frame to the drive end frame

9.4 Separate the two halves

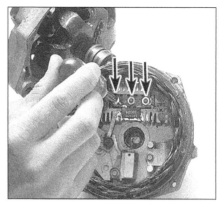

9.5 Remove the three nuts (arrows) retaining the stator windings to the rectifier bridge

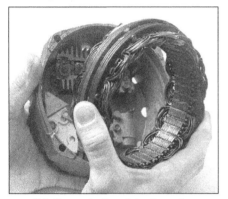

9.6 Separate the stator from the end frame

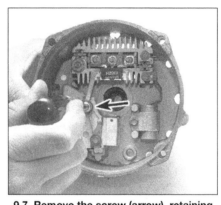

9.7 Remove the screw (arrow) retaining the diode trio to the regulator/brush assembly

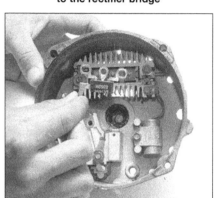

9.8 Lift the diode trio from the alternator

9   Alternator components - check and replacement

Refer to illustrations 9.1, 9.2, 9.3, 9.4, 9.5, 9.6, 9.7, 9.8, 9.9, 9.10, 9.13, 9.14a, 9.14b, 9.14c, 9.14d, 9.14e, 9.15a, 9.15b, 9.16a, 9.16b, 9.17a, 9.17b and 9.17c

1   Remove the alternator (see Section 8) and place it on a clean workbench **(see illustration)**.
2   Paint an alignment mark on each side of the end frame to insure correct reassembly **(see illustration)**.

3   Remove the four bolts that secure the two end frames together **(see illustration)**.
4   Separate the two halves. The rotor will remain in one half while the stator will remain in the other half **(see illustration)**.
5   Remove the three nuts retaining the stator windings to the rectifier bridge **(see illustration)**.
6   Remove the stator from the end frame **(see illustration)**.
7   Remove the screw retaining the diode trio to the brush assembly **(see illustration)**
8   Lift the diode trio from the alternator

assembly **(see illustration)**.
9   Remove the brush assembly and voltage regulator **(see illustration)**.
10   Remove the nut and washer from the output stud. Remove the screws retaining the rectifier bridge to the rear end frame and lift the assembly out **(see illustration)**.
11   Remove the nut and washer and separate the front pulley and fan from the drive end frame.
12   Remove the rotor from the drive end frame.
13   Remove the condenser mounting bolt and condenser **(see illustration)**.

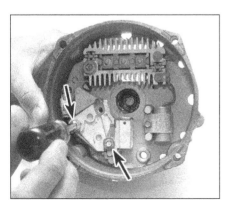

9.9 Remove the two remaining screws (arrows) and remove the brush assembly and voltage regulator

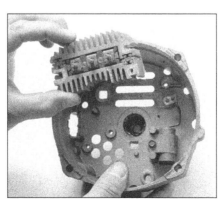

9.10 Remove the rectifier assembly from the end frame

9.13 Remove the condenser mounting bolt

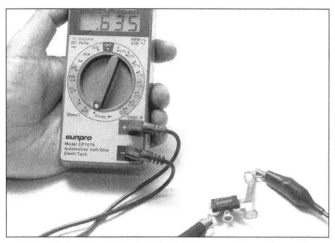

9.14a To test the diode trio, install the negative probe on the extension leg and the positive probe onto each diode (in turn). Continuity should exist through each diode

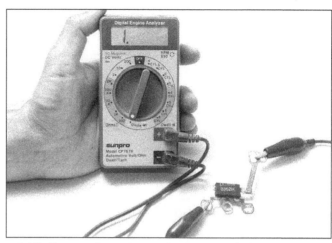

9.14b Reverse the probes to check the continuity in the other direction. The diode should allow continuity in one direction only

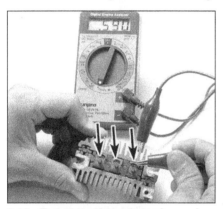

9.14c To test the three diodes (arrows) on the top side of the rectifier bridge, install the positive probe onto the cooling fin and touch the negative probe to each metal tab. Continuity should exist through each diode

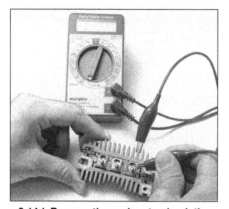

9.14d Reverse the probes to check the continuity in the other direction. Continuity should NOT exist. The rectifier bridge diodes should allow continuity in one direction only

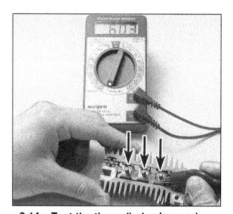

9.14e Test the three diodes (arrows) on the bottom row in the same manner

14   Test the diodes as follows:

a) *Test the diode trio by installing the negative probe of the multimeter onto the leg and the positive probe onto each terminal of the diode trio. Using the diode testing function on the multimeter, continuity should exist in one position only* **(see illustrations)**. *Reverse the polarity of the test by changing the position of the probes. Continuity should NOT exist. This checks the unidirectional capability of the operating diode. Next, repeat this test for each of the other two diodes (total six checks). Each diode should have the exact same characteristics. It will be necessary to use a multimeter that includes a diode check function. This function allows slight current application to the diode to assist in opening the voltage gate.*

b) *To test the rectifier assembly, follow the exact same procedure as detailed above. Be sure to use a multimeter with a diode check function. Use a screwdriver to lift the tabs off the contact sur-*

face to separate the negative and positive diode rectifiers. Make sure the tabs do not touch the posts, the heat sinks or each other. Install the multimeter probes on the tabs and the cooling fins on one side of the rectifier, then the other, to

obtain results **(see illustrations)**. *If the test results are incorrect for any one of the diodes, replace the complete unit.*

15   Test the rotor as follows:

a) *Check for an open between the two slip rings* **(see illustration)**. *There should be 1 to 5 ohms resistance between the slip rings.*

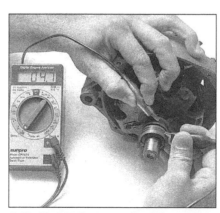

9.15a To test the rotor, check for an open between the two slip rings. There should be 1 to 5 ohms resistance

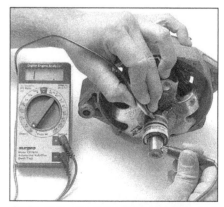

9.15b Also, check the rotor slip rings and the rotor for possible grounds. Continuity should NOT exist between the slip ring and the shaft or frame

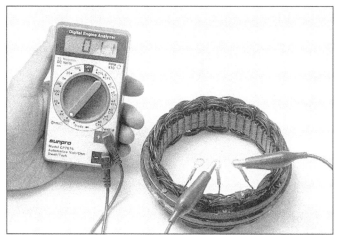

9.16a To test the stator, check for opens between each end terminal of the stator windings. Continuity should exist between all the terminals

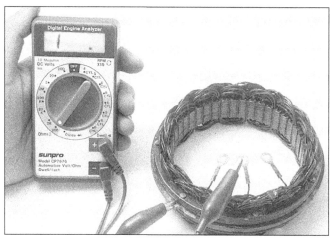

9.16b Also, check for a grounded stator winding to the end frame. Continuity should NOT exist between any terminal and the frame

9.17a Apply a small amount of grease to the bearing surface

b) *Check for grounds between each slip ring and the rotor* **(see illustration)**. *There should be no continuity (infinite resistance) between the rotor and either slip ring. If the rotor fails either test, or if the slip rings are excessively worn, the rotor is defective.*

16   Test the stator as follows:
a) *Check for opens between each end terminal of the stator windings* **(see illustration)**. *If either reading is high (infinite resistance), the stator is defective.*
b) *Check for a grounded stator winding between each stator terminal and the frame* **(see illustration)**. *If there's continuity between any stator winding and the frame, the stator is defective.*

17   Reassembly is a reversal of disassembly. Observe the following points:
a) *Take great care to position the insulating washers and sleeves correctly on the rectifier bridge and brush assembly screws.*
b) *Place a small amount of grease onto the bearing surface before installing the alternator halves together* **(see illustration)**.
c) *Press the brushes into the holder and insert a stiff wire (such as a straightened-out paper clip) through the small hole in the back of the alternator to hold the brushes in a retracted position* **(see illustration)**. *This will prevent them from*

catching on the slip rings as the alternator halves are assembled.
d) *Clean the brush contact surfaces on the slip rings before installing the end-frame.*
e) *Make sure that the marks on the rear end frame and drive end frame (which were made before dismantling) are in alignment.*
f) *Tighten the pulley nut securely.*
g) *Remove the wire or paper clip from the end-frame to permit the brushes to release onto the slip rings* **(see illustration)**.

10   External voltage regulator - check and replacement

Refer to illustration 10.4

1   A discharged battery will normally be due to a fault in the voltage regulator, but before testing the unit, do the following:
a) *Check the drivebelt tension (see Chapter 2, Section 7).*
b) *Test the condition of the battery.*
c) *Check the charging circuit for loose connections and broken wires.*
d) *Made sure that lights or other electrical accessories have not been switched on inadvertently.*
e) *Check the generator indicator lamp for normal illumination with the ignition switched on and off, and with the engine idling and stationary.*

2   Disconnect the battery ground cable. Disconnect the harness connector from the regulator.

3   Under no circumstances should the voltage regulator or field relay contacts be cleaned since any abrasive materials will destroy the contact material.

4   Voltage regulator point gap and air gap adjustments can be checked with a feeler gauge of the specified thickness (point gap should be 0.014-inch and the air gap should be 0.067-inch). Check the voltage regulator

9.17b Press the two brushes into the holder and slide a paper clip through the eyelet to keep the brushes in the holder assembly

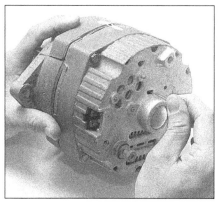

9.17c After the alternator is completely assembled remove the wire or paper clip to release the brushes

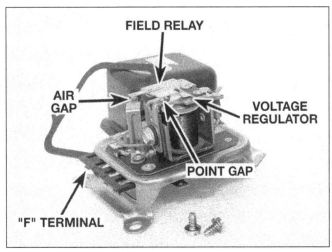

10.4 The voltage regulator point and air gap adjustments can be checked with a feeler gauge of the specified thickness (point gap = 0.014-inch; air gap = 0.067-inch)

13.4a Working from underneath the vehicle (supported securely on jackstands), remove the starter motor mounting bolts . . .

point opening of the upper contacts with the lower contacts just touching. Adjustments are made by carefully bending the upper contact arm. Check the voltage regulator air gap with the lower contacts touching and adjust it, if necessary **(see illustration)**.

5   The field relay point opening may be adjusted by bending the armature stop. The air gap is checked with the points just touching and is adjusted by bending the flat contact support spring. **Note:** *The field relay will normally operate satisfactorily even if the air gap is outside the specified limits, and should not be adjusted when the system is functioning satisfactorily.*

6   If the regulator must be replaced, simply remove the mounting screws.

7   Installation is the reverse of the removal procedure. Ensure that the rubber gasket is in place on the regulator base.

11   Starting system - general information and precautions

The sole function of the starting system is to turn over the engine quickly enough to allow it to start.

The starting system consists of the battery, the starter motor, the starter solenoid and the wires connecting them. The solenoid is mounted directly on the starter motor or is a separate component located in the engine compartment.

The solenoid/starter motor assembly is installed on the lower part of the engine, next to the transmission bellhousing.

When the ignition key is turned to the Start position, the starter solenoid is actuated through the starter control circuit. The starter solenoid then connects the battery to the starter. The battery supplies the electrical energy to the starter motor, which does the actual work of cranking the engine.

The starter motor on a vehicle equipped with a manual transmission can only be oper-

ated when the clutch pedal is depressed; the starter on a vehicle equipped with an automatic transmission can only be operated when the transmission selector lever is in Park or Neutral.

Always observe the following precautions when working on the starting system:

a) *Excessive cranking of the starter motor can overheat it and cause serious damage. Never operate the starter motor for more than 15 seconds at a time without pausing to allow it to cool for at least two minutes.*

b) *The starter is connected directly to the battery and could arc or cause a fire if mishandled, overloaded or shorted out.*

c) *Always detach the cable from the negative terminal of the battery before working on the starting system.*

12   Starter motor - testing in vehicle

Note: *Before diagnosing starter problems, make sure that the battery is fully charged.*

1   If the starter motor does not turn at all when the ignition switch is operated, make sure that the shift lever is in Neutral or Park (automatic transaxle) or that the clutch pedal is depressed (manual transaxle).

2   Make sure that the battery is charged and that all cables, both at the battery and starter solenoid terminals, are clean and secure.

3   If the starter motor spins but the engine is not cranking, the overrunning clutch in the starter motor is slipping and the starter motor must be replaced.

4   If, when the switch is actuated, the starter motor does not operate at all but the solenoid clicks, then the problem lies with either the battery, the main solenoid contacts or the starter motor itself, or the engine is seized.

5   If the solenoid plunger cannot be heard when the switch is actuated, the battery is bad, the fusible link is burned (the circuit is open) or the solenoid itself is defective.

6   To check the solenoid, connect a jumper lead between the battery (positive terminal) and the ignition switch terminal (the small terminal) on the solenoid. If the starter motor now operates, the solenoid is OK and the problem is in the ignition switch, neutral start switch or in the wiring.

7   If the starter motor still does not operate, remove the starter/solenoid assembly for disassembly, testing and repair.

8   If the starter motor cranks the engine at an abnormally slow speed, first make sure that the battery is charged and that all terminal connections are tight. If the engine is partially seized, or has the wrong viscosity oil in it, it will crank slowly.

9   Run the engine until normal operating temperature is reached, then disconnect the coil wire from the distributor cap and ground it on the engine.

10   Connect a voltmeter positive lead to the battery positive post and then connect the negative lead to the negative post.

11   Crank the engine and take the voltmeter readings as soon as a steady figure is indicated. Do not allow the starter motor to turn for more than 15 seconds at a time. A reading of nine volts or more, with the starter motor turning at normal cranking speed, is normal. If the reading is nine volts or more but the cranking speed is slow, the motor is faulty. If the reading is less than nine volts and the cranking speed is slow, the solenoid contacts are probably burned, the starter motor is bad, the battery is discharged or there is a bad connection.

13   Starter motor - removal and installation

Refer to illustrations 13.4a and 13.4b

1   Detach the cable from the negative terminal of the battery.

2   Raise the vehicle and support it securely on jackstands.

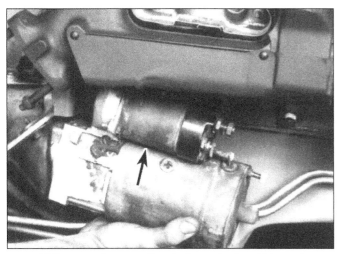

13.4b . . . and remove the starter and solenoid (arrow) as an assembly

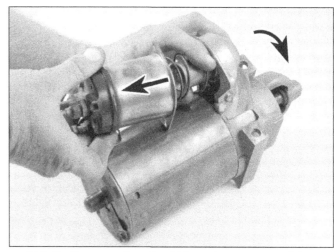

14.3 To disengage the solenoid from the starter, turn it in a clockwise direction

3   Clearly label, then disconnect the wires from the terminals on the starter motor and solenoid.
4   Remove the starter motor mounting bolts **(see illustrations)**. Remove the starter.
5   Installation is the reverse of removal.

14   Starter motor solenoid - removal, repair and installation

Refer to Illustrations 14.3 and 14.4
1   After removing the starter/solenoid unit (see Section 13), disconnect the connector strap from the solenoid MOTOR terminal.
2   Remove the two screws which secure the solenoid housing to the end-frame assembly.
3   Twist the solenoid in a clockwise direction to disengage the flange key and then withdraw the solenoid **(see illustration)**.
4   Remove the nuts and washers from the solenoid terminals and then unscrew the two solenoid end-cover retaining screws and washers and pull off the end-cover **(see illustration)**.
5   Unscrew the nut washer from the battery terminal on the end-cover and remove the terminal.
6   Remove the resistor bypass terminal and contactor.
7   Remove the motor connector strap terminal and solder a new terminal in position.
8   Use a new battery terminal and install it to the end-cover. Install the bypass terminal and contactor.
9   Install the end-cover and the remaining terminal nuts.
10   Install the solenoid to the starter motor by first checking that the return spring is in position on the plunger and then insert the solenoid body into the drive housing and turn the body counter clockwise to engage the flange key.
11   Install the two solenoid securing screws and connect the MOTOR connector strap.

15   Starter motor - overhaul

Refer to Illustration 15.2
Note: *Due to the critical nature of the disassembly and testing of the starter motor it may be advisable for the home mechanic to simply purchase a new or factory-rebuilt unit. If it is decided to overhaul the starter, check on the* availability of singular replacement components before proceeding.
1   Disconnect the starter motor field coil connectors from the solenoid terminals.
2   Unscrew and remove the through-bolts **(see illustration)**.
3   Remove the commutator end-frame, field frame assembly and the armature from the drive housing.

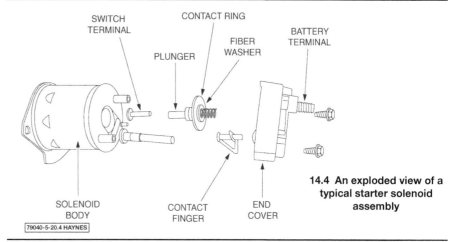

14.4 An exploded view of a typical starter solenoid assembly

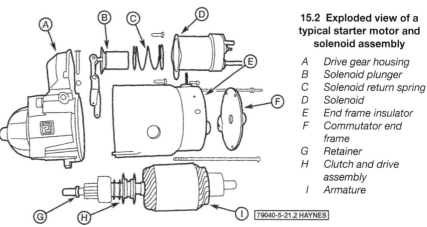

15.2 Exploded view of a typical starter motor and solenoid assembly

A   Drive gear housing
B   Solenoid plunger
C   Solenoid return spring
D   Solenoid
E   End frame insulator
F   Commutator end frame
G   Retainer
H   Clutch and drive assembly
I   Armature

4 Slide the two-section thrust collar off the end of the armature shaft and then, using a piece of suitable tube, drive the stop/retainer up the armature shaft to expose the snap-ring.

5 Extract the snap-ring from its shaft groove and then slide the stop/retainer and overrunning clutch assembly from the armature shaft.

6 Dismantle the brush components from the field frame.

7 Release the V-shaped springs from the brush holder supports.

8 Remove the brush holder support pin and then lift the complete brush assembly upwards.

9 Disconnect the leads from the brushes if they are worn down to half their original length and they are to be replaced.

10 The starter motor is now completely dismantled except for the field coils. If these are found to be defective during the tests described later in this Section, removal of the pole shoe screws is best left to a service facility which has the necessary pressure driver.

11 Clean all components and replace any obviously worn components.

12 Never attempt to undercut the insulation between the commutator segments on starter motors having the molded type commutators. On commutators of conventional type, the insulation should be undercut (below the level of the segments) by 1/32-inch. Use an old hacksaw blade to do this, and make sure that the undercut is the full width of the insulation and the groove is quite square at the bottom. When the undercutting is completed, brush away all dirt and dust.

13 Clean the commutator by spinning it while a piece of '00' sandpaper is wrapped around it. Never use any other type of abrasive material for this work.

14 If necessary, because the commutator is in such bad shape, it may be turned down in a lathe to provide a new surface. Make sure to undercut the insulation when the turning is completed.

15 To test the armature for ground: use a lamp-type circuit tester. Place one lead on the armature core or shaft and the other on a segment of the commutator. If the lamp lights, then the armature is grounded and must be replaced.

16 To test the field coils for open circuit: place one test probe on the insulated brush and the other on the field connector bar. If the lamp does not light, the coils are open and must be replaced.

17 To test the field coils for ground: place one test probe on the connector bar and the other on the grounded brush. If the lamp lights, then the field coils are grounded.

18 The overrunning clutch cannot be repaired, and if faulty, it must be replaced as a complete assembly.

19 Install the brush assembly to the field frame as follows:

20 Install the brushes to their holders.

21 Assemble the insulated and grounded brush holders together with the V-spring and then locate the unit on its support pin.

22 Push the holders and spring to the bottom of the support and then rotate the spring to engage the V in the support slot.

23 Connect the ground wire to the grounded brush and the field lead wire to the insulated brush.

24 Repeat the operations for the second set of brushes.

25 Smear silicone oil onto the drive end of the armature shaft and then slide the clutch assembly (pinion to the front) onto the shaft.

26 Slide the pinion stop/retainer onto the shaft so that its open end is facing away from the pinion.

27 Stand the armature vertically on a piece of wood and then position the snap-ring on the end of the shaft. Using a hammer and a piece of hardwood, drive the snap-ring onto the shaft.

28 Slide the snap-ring down the shaft until it drops into its groove.

29 Install the thrust collar on the shaft so that the shoulder is next to the snap-ring. Using two pairs of pliers, squeeze the thrust collar and stop/retainer together until the snap-ring fully enters the retainer.

30 Lubricate the drive housing bushing with silicone oil and after ensuring that the thrust collar is in position against the snap-ring, slide the armature and clutch assembly into the drive housing so that, at the same time, the shift lever engages with the clutch.

31 Position the field frame over the armature and apply sealing compound between the frame and the solenoid case.

32 Position the field frame against the drive housing, taking care not to damage the brushes.

33 Lubricate the bushing in the commutator end-frame using silicone oil; place the leather brake washer on the armature shaft and then slide the commutator end-frame onto the shaft.

34 Reconnect the field coil connectors to the MOTOR terminal of the solenoid.

35 Now check the pinion clearance. To do this, connect a 12-volt battery between the solenoid S terminal and ground and at the same time fix a heavy connecting cable between the MOTOR terminal and ground (to prevent any possibility of the starter motor rotating). As the solenoid is energized it will push the pinion forward into its normal cranking position and retain it there. With the fingers, push the pinion away from the stop/retainer in order to eliminate any slack, then check the clearance between the face of the pinion and the face of stop/retainer using a feeler gauge. The clearance should be between 0.010 and 0.140-inch to ensure correct engagement of the pinion with the flywheel (or driveplate - automatic transmission) ring-gear. If the clearance is incorrect, the starter will have to be dismantled again and any worn or distorted components replaced, no adjustment being provided for.

16 Electrical troubleshooting - general information

A typical electrical circuit consists of an electrical component, any switches, relays, motors, fuses, fusible links or circuit breakers related to that component and the wiring and connectors that link the component to both the battery and the chassis. To help you pinpoint an electrical circuit problem, wiring diagrams are included at the end of this book.

Before tackling any troublesome electrical circuit, first study the appropriate wiring diagrams to get a complete understanding of what makes up that individual circuit. Trouble spots, for instance, can often be narrowed down by noting if other components related to the circuit are operating properly. If several components or circuits fail at one time, chances are the problem is in a fuse or ground connection, because several circuits are often routed through the same fuse and ground connections.

Electrical problems usually stem from simple causes, such as loose or corroded connections, a blown fuse, a melted fusible link or a bad relay. Visually inspect the condition of all fuses, wires and connections in a problem circuit before troubleshooting it.

If testing instruments are going to be utilized, use the diagrams to plan ahead of time where you will make the necessary connections in order to accurately pinpoint the trouble spot.

The basic tools needed for electrical troubleshooting include a circuit tester or voltmeter (a 12-volt bulb with a set of test leads can also be used), a continuity tester, which includes a bulb, battery and set of test leads, and a jumper wire, preferably with a circuit breaker incorporated, which can be used to bypass electrical components. Before attempting to locate a problem with test instruments, use the wiring diagram(s) to decide where to make the connections.

Voltage checks

Voltage checks should be performed if a circuit is not functioning properly. Connect one lead of a circuit tester to either the negative battery terminal or a known good ground. Connect the other lead to a connector in the circuit being tested, preferably nearest to the battery or fuse. If the bulb of the tester lights, voltage is present, which means that the part of the circuit between the connector and the battery is problem free. Continue checking the rest of the circuit in the same fashion. When you reach a point at which no voltage is present, the problem lies between that point and the last test point with voltage. Most of the time the problem can be traced to a loose connection. **Note:** *Keep in mind that some circuits receive voltage only when the ignition key is in the Accessory or Run position.*

17.1 On most models the fuse block is located under the dash to the left of the driver

Finding a short

One method of finding shorts in a circuit is to remove the fuse and connect a test light or voltmeter in its place to the fuse terminals. There should be no voltage present in the circuit. Move the wiring harness from side to side while watching the test light. If the bulb goes on, there is a short to ground somewhere in that area, probably where the insulation has rubbed through. The same test can be performed on each component in the circuit, even a switch.

Ground check

Perform a ground test to check whether a component is properly grounded. Disconnect the battery and connect one lead of a self-powered test light, known as a continuity tester, to a known good ground. Connect the other lead to the wire or ground connection being tested. If the bulb goes on, the ground is good. If the bulb does not go on, the ground is not good.

Continuity check

A continuity check is done to determine if there are any breaks in a circuit - if it is passing electricity properly. With the circuit off (no power in the circuit), a self-powered continuity tester can be used to check the circuit. Connect the test leads to both ends of the circuit (or to the "power" end and a good ground), and if the test light comes on, the circuit is passing current properly. If the light doesn't come on, there is a break somewhere in the circuit. The same procedure can be used to test a switch, by connecting the continuity tester to the switch terminals. With the switch turned On, the test light should come on.

Finding an open circuit

When diagnosing for possible open circuits, it is often difficult to locate them by sight because oxidation or terminal misalignment are hidden by the connectors. Merely wiggling a connector on a sensor or in the

wiring harness may correct the open circuit condition. Remember this when an open circuit is indicated when troubleshooting a circuit. Intermittent problems may also be caused by oxidized or loose connections.

Electrical troubleshooting is simple if you keep in mind that all electrical circuits are basically electricity running from the battery, through the wires, switches, relays, fuses and fusible links to each electrical component (light bulb, motor, etc.) and to ground, from which it is passed back to the battery. Any electrical problem is an interruption in the flow of electricity to and from the battery.

17 Fuses, fusible links and circuit breakers - general information

Note: *The electrical circuits of the vehicle are protected by a combination of fuses, circuit breakers and fusible links.*

Fuses

Refer to illustration 17.1

1   The fuse block is located under the instrument panel on the left side of the dashboard **(see illustration)**.

2   Each of the fuses is designed to protect a specific circuit, and the various circuits are identified on the fuse panel itself.

3   A blown fuse can be readily identified by inspecting the element inside the glass tube. If this metal element is broken, the fuse is inoperable and must be replaced with a new one. The easiest and fastest way to check fuses, however, is with a test light. Check for power at each end of each fuse. If power is present on one side of the fuse but not the other, the fuse is blown.

4   Be sure to replace blown fuses with the correct type. Fuses of different ratings are physically interchangeable, but only fuses of the proper rating should be used. Replacing a fuse with one of a higher or lower value than specified is not recommended. Each electrical circuit needs a specific amount of protection. The amperage value of each fuse is molded into the fuse body.

5   If the replacement fuse immediately fails, don't replace it again until the cause of the problem is isolated and corrected. In most cases, the cause will be a short circuit in the wiring caused by a broken or deteriorated wire.

Fusible links

6   Some circuits are protected by fusible links. The links are used in circuits which are not ordinarily fused, such as the ignition circuit.

7   Although the fusible links appear to be of a heavier gauge than the wire they are protecting, the appearance is due to the thick insulation. All fusible links are several wire gauges smaller than the wire they are designed to protect.

8   Fusible links cannot be repaired, but a new link of the same size wire can be put in

its place. The procedure is as follows:

a) *Disconnect the negative cable from the battery.*
b) *Disconnect the fusible link from the starter solenoid.*
c) *Cut the damaged fusible link out of the wiring just behind the connector.*
d) *Strip the insulation back approximately 1/2-inch.*
e) *Position the connector on the new fusible link and crimp it into place.*
f) *Use rosin core solder at each end of the new link to obtain a good solder joint.*
g) *Use plenty of electrical tape around the soldered joint. No wires should be exposed.*
h) *Connect the fusible link to the starter solenoid. Connect the battery ground cable. Test the circuit for proper operation.*

Circuit breakers

9   Circuit breakers protect components such as power windows and headlights. Some circuit breakers are located in the fuse box.

10   On some models the circuit breaker resets itself automatically, so an electrical overload in a circuit breaker protected system will cause the circuit to fail momentarily, then come back on. If the circuit does not come back on, check it immediately. Once the condition is corrected, the circuit breaker will resume its normal function. Some circuit breakers must be reset manually.

18 Relays - general information

Several electrical accessories in the vehicle use relays to transmit the electrical signal to the component. If the relay is defective, that component will not operate properly.

The various relays are grouped together in several locations. If a faulty relay is suspected, it can be removed and tested by a dealer service department or a repair shop. Defective relays must be replaced as a unit.

19 Instrument panel - service operations

Instrument cluster - replacement

Refer to illustration 19.3

1   Disconnect the battery ground cable then lower the steering column (refer to Chapter 11 if necessary). Take care to prevent damage to the steering mast jacket.

2   Remove three screws from the front of the heater control.

3   Where applicable, remove the radio control knobs, washers, bezel nuts and front support at the lower edge of the cluster (the radio need not be removed) **(see illustration)**.

19.3 Removing the instrument cluster bezel screws

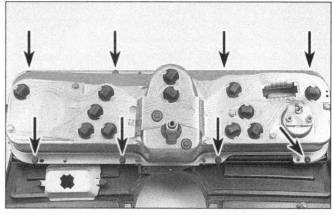

19.8 Instrument cluster-to-housing screws (arrows)

4   Remove the instrument carrier screws at the top, bottom and sides. Tilt the carrier forwards, and disconnect the speedometer cable (Step 21) and plug connectors.

5   Lift the carrier out of the panel after removing the screws.

6   installation is the reverse of the removal procedure, but take care not to kink the cable housing.

Fuel gauge (and optional tachometer) - replacement

Refer to illustration 19.8

7   Disconnect the battery ground cable then reach behind the cluster to remove the cluster lamps from the rear cover plate. Remove the gauge terminal nuts.

8   Carefully move the printed circuit away from the gauge and remove the screws retaining the gauge assembly to the cluster housing **(see illustration)**.

9   Remove the gauge assembly and unscrew the three terminal nuts which secure the gauge. Note the order of assembly of the external parts of the gauge assembly.

10   Installation is the reverse of the removal procedure.

Clock - replacement

11   Disconnect the battery ground cable.

12   Where applicable, remove the radio completely. Full details of this operation cannot be given in view of the differing types of equipment which may have been fitted. However, as a guide it will normally be necessary to remove the knobs, bezels and control shaft knobs and washers, the mounting screws (which may mean disconnection of air conditioning ducts) followed by the antenna, power feed and speaker connections. (When installing, always ensure that the speaker is connected before applying power to the radio.)

13   Remove the clock set stem at the front of the cluster. Remove the cluster lamps and circuit attaching screws from around the clock.

14   Remove the clock from the rear of the cluster housing.

15   Installation is the reverse of the removal procedure.

Speedometer- replacement

Note: *Attempts at speedometer repair are not recommended. In the event of a fault arising the services of a repair specialist should be obtained or a replacement unit fitted.*

16   Remove the instrument cluster (Steps 1 thru 6).

17   Remove the rear cover retaining screws then bend the ground strap away to enable the cover to be removed.

18   Unscrew the retaining screws and remove the speedometer from the rear cover.

19   Installation is the reverse of the removal procedure.

Speedometer cable core - replacement

20   Where applicable, remove the radio (Step 12).

21   Disconnect the speedometer cable by depressing the retaining spring on the back of the instrument cluster case, pushing the casing inwards then backwards.

22   Remove the dash panel sealing plug to allow the casing to be moved then pull out the old core. If the cable is broken, the vehicle must be raised for access to the transmission end.

23   Lubricate the replacement cable with silicone grease and push the core into the housing.

24   The remainder of the installation procedure is the reverse of removal, but road test the vehicle to ensure correct operation on completion.

Instrument panel bulbs

25   All bulbs are accessible except for the center top bulb above the hi-beam indicator, which will require removal of the instrument cluster (Steps 1 thru 6).

26   To replace bulbs on the left of the steering column reach up under the dash and twist the bulb socket holder loose. Installation is the reverse of this procedure.

27   To replace bulbs on the right of the steering column, disconnect the battery ground cable and remove the radio (where applicable - refer to paragraph 12), then refer to the procedure in paragraph 26.

Printed circuit replacement

28   Remove the instrument cluster (paragraphs 1 thru 6), then remove all the cluster illuminating and indicator lamps.

29   Remove the nuts securing the fuel gauge and clock terminals to the printed circuit and housing.

30   Remove the four hexagon head screws which retain the printed circuit to the cluster housing then remove the printed circuit.

31   Installation is the reverse of the removal procedure, but ensure that all retaining screws and terminal nuts are installed to ensure proper ground circuits for the printed circuit.

Seat separator instrument console gauges - replacement

32   Disconnect the battery ground cable.

33   Remove the three cover retaining screws and lift the cover from the cluster.

34   Remove the three screws which retain each gauge mounting plate then carefully disengage the gauge plate from the housing and remove the electrical connections. Individual gauges can now be removed from the gauge plate.

35   Installation is the reverse of the removal procedure.

---

20   Turn signal and hazard flashers - troubleshooting

---

1   Different turn signal and hazard flasher types are used in various Chevrolet vehicles and it is essential that a replacement is the same as that removed.

2   The turn signal flasher is mounted in a clip on the instrument panel lower lip, but it may be necessary to remove the cigarette lighter connection to gain access to the flasher connection.

3   Where applicable, a hazard warning flasher is mounted in the fuse panel (see Section 17). This will not normally operate when the brake pedal is depressed.

4   If the turn signals fail on one side of the vehicle, and the flasher unit cannot be heard, a faulty bulb is indicated. If the flasher unit can be heard, a short to ground is indicated.

A fault may also be attributable to a wiring defect, an incorrect type of bulb fitted or a defective switch these points should be systematically checked in the event of a fault developing.

5   If the turn signals fail on both sides, the fault may be due to a blown fuse, faulty flasher unit or switch, or a broken or loose connection. Where a fuse has blown, examine for short circuits between wires and to ground before fitting a replacement.

6   If the hazard warning lamps are inoperative the same diagnostic procedure as in paragraph 5 should be followed.

21 Horns - troubleshooting

Refer to illustration 21.1

1   A relay is incorporated in the horn circuit and in the event of the horn not operating first check the fuse and then disconnect the electrical supply lead from the horn terminal and run a jumper wire direct from the battery I+) terminal. Provided the horn ground connection is good, the horn will blow.

2   Take the disconnected horn lead and connect it to a 12 volt buzzer or lamp which is well grounded. Depress the horn control on the steering wheel and observe whether the circuit is completed. If so then the horn is faulty **(see illustration)**.

3   If the circuit is not completed, disconnect the multi-connector on the steering column and ground the horn wire terminal. If the circuit is not found to be open during this test, replace the horn relay. If the circuit is found to be open (horn control depressed) during the test then it is probable that the harness wiring is at fault.

22 Front lighting - service operations

Headlight - adjustment

Refer to illustrations 22.4a and 22.4b

1   Headlight adjustment can only be carried out accurately using beam aiming equipment.

However, the procedure given in this sub-Section is satisfactory for most practical purposes.

2   Ensure that the fuel tank is half-full, tires are correctly inflated, the vehicle is loaded in the manner used for normal driving, and the headlight lenses are clean.

3   Position the vehicle so that it is 25 feet from and square to a wall or screen.

4   Mark the screen or wall as shown **(see illustration)** and adjust the beams as necessary to obtain the correct adjustment pattern **(see illustration)**.

Headlight - replacement

Refer to illustrations 22.5, 22.6 and 22.7

5   Remove the headlight trim **(see illustration)**.

21.1 The horn is located in the left front corner of the engine compartment

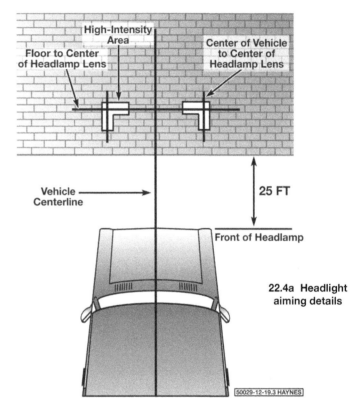

22.4a Headlight aiming details

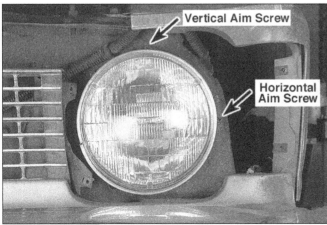

22.4b Locations of the headlight aiming screws

22.5 Remove these screws and detach the headlight trim

22.6 Remove the screws and lift off the headlight retaining ring - don't disturb the headlight aiming screws

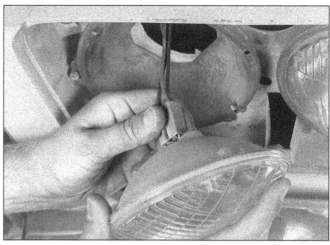

22.7 Pull the headlight forward and unplug the electrical connector

6   Remove the screws and detach the headlight retaining ring **(see illustration)**.
7   Pull the light forward and unplug the electrical connector **(see illustration)**.
8   Installation is the reverse of the removal procedure, but ensure that the numbers on the lens face are at the top. Adjust the aim on completion (Steps 1 thru 4).

Parking lamp bulb (pre-1973 models) - replacement

9   Remove the two lens retaining screws and disengage the lens to gain access to the bulb.

Parking lamp bulb (1973 and later models) - replacement

10   Twist out the socket from the rear of the housing to gain access to the bulb.

Lamp housing (pre-1973 models) - replacement

11   Disconnect the lamp wire assembly in-line connector then remove the two bolts securing the housing.
12   Remove the housing then remove the lens and bulb assembly.

22.16 Turn the socket 1/4-turn and detach it from the housing; to remove the bulb, pull it straight out of the socket

13   Installation is the reverse of the removal procedure.

Lamp housing (1973 models onwards) - replacement

14   Remove the twist-out socket from the rear of the housing then unscrew the housing assembly retaining screws from the bumper. Remove the housing assembly.
15   Installation is the reverse of the removal procedure.

Side marker lamp

Refer to illustration 22.16
16   To replace the bulb, turn the connector 1/4-turn counterclockwise to remove the plug for access to the bulb **(see illustration)**.
17   If necessary, remove the housing assembly by unscrewing the retaining nuts or screws inside the fender. Pull away the assembly and disconnect the twist socket. Installation is the reverse of the removal procedure.

---

23   Rear lighting - service operations

Side marker lamps

1   All procedures for the rear side marker lamps are similar to those for the front side marker lamps (see previous Section).

Rear lamp cluster and license plate lamp bulbs - replacement

2   On all models, access to the bulb is from the rear compartment. Bulbs are either retained in push-in or twist-in holders.

Rear lamp cluster lens - replacement

3   On 1969 models the lens is retained by two screws on the outside. For all other models it is necessary to remove the housing.

Rear lamp cluster housing - replacement

4   On 1969 to 1972 models, the housing can be removed from the outside after removing the retaining nuts or screws in the rear compartment.
5   On 1973 models onwards the housing is removed from inside after unscrewing the retaining nuts or screws. On 1975 and later models, this operation also enables the bezel to be withdrawn from the outside.

---

24   Door jamb switches - replacement and adjustment

---

1   Door jamb switches can be removed by unscrewing the locking nut and screwing out the switch.
2   After fitting a replacement, adjustment should be made by closing the door slowly to contact the switch plunger, then screwing the switch in, or out, to obtain the correct operation. The locking nut should be tightened securely.

---

25   Seat separator console lamps- bulb replacement

---

1   *Compartment lamps:* Pry up the switch assembly from the console opening to gain access to the bulb and bulb holder.
2   *Courtesy lamps:* Remove two screws and the lens or ashtray at the rear of the console to gain access to the bulb.
3   *Automatic floor shift quadrant lamp - bulb replacement:* Remove the quadrant trim plate retaining screws and lift out the trim plate to gain access to the lamp socket and bulb.

26.2 The dome lamp base is retained by two screws. When removing the bulb, pry only on the ends

26   Dome lamp - removal and installation

Refer to illustration 26.2

1   Insert a flat-bladed screwdriver between the lens and base, then press inwards and down to disengage the lens retaining tabs from the base.
2   Remove the bulb then remove the two base retaining screws **(see illustration)**.
3   To disengage the wire harness, push the clips through the back of the base.
4   Installation is the reverse of the removal procedure.

27   Lighting switch - replacement

Refer to illustration 27.2

1   Disconnect the battery ground cable.
2   Pull the switch knob to the 'On' position then reach up behind the instrument panel to depress the switch shaft retainer. Pull the knob and shaft assembly out **(see illustration)**.

3   Remove the ferrule nut then withdraw the switch assembly and disconnect the multi-way connector. (This may require prying apart with a screwdriver.)
4   When installing, first connect the multi-way connector then reverse the removal procedure.

28   Wiper/washer switch - replacement

1   Disconnect the battery ground cable then pull off the switch electrical connector.
2   Remove the three switch mounting screws then lift the switch rearwards out of the instrument panel.
3   Installation is the reverse of the removal procedure.

29   Brake light switch - replacement

    Refer to Chapter 9 for the brake light switch check, adjustment and replacement procedures.

30   Headlight dimmer switch - replacement

Refer to illustration 30.2
Note: *When the headlight dimmer switch fails, the electrical connector, wiring and terminals attached to the switch frequently burn, necessitating replacement of the connector and terminals, and splicing in new lengths of wire.*
1   Fold back the floor mat, disengage the switch connector lock fingers and disconnect the connector.
2   Remove the two switch retaining screws **(see illustration)**.
3   When installing, install the connector and check the switch operation, then reverse the removal procedure.

31   Back-up light switch - replacement

Manual transmission, mast jacket mounted (includes seat belt warning switch)

1   Disconnect the wiring at the switch terminals then remove the switch attaching screws. Remove the switch from the mast jacket.
2   Position the column switch tube in the reverse detent using the 1st/Reverse lower shift lever.
3   Position the replacement switch with the drive tang in the slot in the shift tube and the right side of the tang touching the right surface of the shift tube slot.
4   Install the two attaching screws. Do not move the selector lever, or the shear pin will break.
5   Install the wiring connector and check the operation of the switch.

Manual transmission, transmission mounted

6   Raise the vehicle then disconnect the switch wiring at the harness in-line connector.
7   Remove the wiring clip retaining bolt and the clip retaining the reverse lever rod to the transmission.
8   Remove the screws which retain the switch and shield assembly then remove the switch.
9   Installation is the reverse of the removal procedure. On completion check the switch operation.

Automatic transmission, column shift (includes neutral start and seat belt warning switches)

10   Disconnect the wiring connectors at the switch terminals.

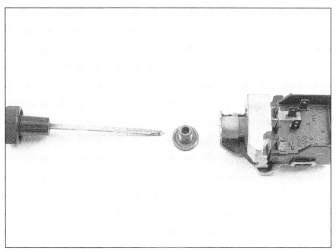

27.2 Headlight switch components

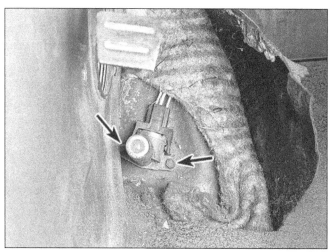

30.2 The headlight dimmer switch is retained by two screws

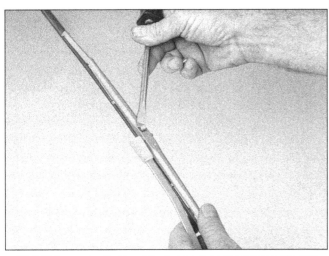

36.3 Using a small screwdriver, pry up on the release pin and detach the blade assembly from the wiper arm pin

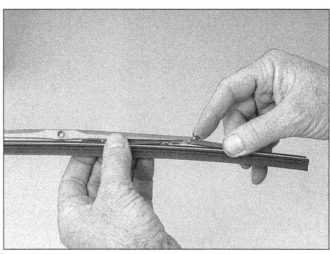

36.4 Depress the button and slide the rubber element off of the wiper blade

11   Remove the retaining screws and the switch from the mast jacket.
12   When installing, position the shift lever in 'Neutral' ('Drive' with torque drive transmission) by rotating the lower lever. **Note**: *Where the pin has already been sheared, align the switch assembly with the actuating tang and insert a suitable pin to a depth of 0.25 inch to hold the tang in place.*
13   Assemble the switch assembly to the column by inserting the tang in the shifter tube slot, then tightening the retaining screws.
14   Connect the wiring harness and check for correct switch operation.
15   Where applicable, remove the pin (Step 12). Otherwise move the selector lever out of 'Neutral' (or 'Drive') to shear the switch pin.

Automatic transmission, floor shift

16   Disconnect the shift control lever arm from the control rod and remove the control knob.
17   Remove the trim plate retaining screws and the trim plate assembly.
18   Remove the control assembly, then remove the switch retaining nuts and the switch.
19   To install, position the gearshift in 'Drive' (pre- 1972 models) or 'Park' (1972 models onwards) and assemble the switch assembly to the lever and bracket by inserting the pin in the actuating tang of the switch assembly.
20   Install the mounting screws and connect the wiring harness.
21   Check the switch operation. The remainder of the installation procedure is reverse of removal.

32   Clutch start switch (manual transmission) - replacement

1969 models

1   Unplug the switch connector then remove the retainer from the pins or link on the clutch pedal.
2   Remove the switch (1 screw) from the pedal support.
3   When installing, slide the switch tang into the pedal support and secure the switch with the screw (switch lever over the pins in the pedal arm).
4   Install the retainer on the clutch arm pin or the link in the switch lever, then fasten with the retainer.
5   Install the switch connector.

1970 and later models

6   Unplug the switch connector, then compress the switch actuating shaft retainer and remove the shaft and switch from the switch bracket.
7   When installing, slide the new switch onto the bracket, rotate the actuating lever to align with the clutch pedal arm hole and push the switch shaft into the clutch pedal arm.
8   Install the switch connector.

33   Glove box (instrument panel compartment) lamp/switch - replacement

1   Disconnect the battery ground cable then depress the bulb and turn it counterclockwise to remove it.
2   Remove the switch and detach the wire, either by removing the screw terminal or cutting the wire.
3   Installation is the reverse of the removal procedure. Where applicable, ensure that the spliced joint is properly insulated.

34   Cigarette lighter housing assembly - replacement

1   Disconnect the battery ground cable.
2   Remove the ashtray and retainer.
3   Unscrew the lighter retainer-from the rear of the housing assembly, then disengage

the lighter unit from the panel.
4   Installation is the reverse of the removal procedure.

35   Windshield wiper system - general description

1   A two speed wiper motor is installed as standard equipment and incorporates the gear train and the self-parking mechanism.
2   The wiper arms will only park when the motor is operating in low speed.
3   The drive from the wiper motor to the wiper arms is by means of a crank arm and a strut to the transmission shafts.
4   On models built before 1975 the windshield wiper motor incorporated the drive system for the windshield washers. For 1975 and later models a separate washer system was introduced (see Section 41).

36   Windshield wiper blades - replacement

Refer to illustrations 36.3 and 36.4
1   Every year or when the blades fail to clean the glass, they should be removed and new ones fitted.
2   Pull the wiper arm/blade assembly away from the windshield glass against the tension of the wiper arm spring.
3   Depress the small tab which lies just below the wiper arm at the blade connector socket and pull the blade from the arm. **Note:** *On some models a coil spring blade retainer is used. To remove the blade, insert a screwdriver on top of the spring and push downwards* **(see illustration)**.
4   Depress the button and slide the rubber element off the wiper blade **(see illustration)**. To install a new element, press the button and slide the new piece into place. Once centered on the blade it will lock into position.
5   Installation is a reversal of removal.

37   Windshield wiper arm - removal and installation

1   Make sure that the wiper arms are in the self-parked position, the motor having been switched off in the low speed mode.
2   Note carefully the position of the wiper arm in relation to the windshield lower reveal molding. It should be approximately two inches above the molding.
3   Using a suitable hooked tool or a small screwdriver, pull aside the small spring tang which holds the wiper arm to the splined transmission shaft and at the same time pull the arm from the shaft.
4   Installation is a reversal of removal but do not push the arm fully home on the shaft until the alignment of the arm has been checked. If necessary, the arm can be pulled off again and turned through one or two serrations of the shaft to correct the alignment without the necessity of pulling aside the spring tang.
5   Finally, press the arm fully home on its shaft and then wet the windshield glass and operate the motor on low speed to ensure that the arc of travel is correct.

38   Windshield wiper motor - removal and installation

1   Raise the hood and remove the cowl vent screen or grille.
2   Disconnect the electrical leads from the wiper motor.
3   Loosen but do not remove, the nuts which secure the ball cup of the crank arm to drive link joint. Disconnect the drive link from the crank arm.
4   Disconnect the washer hoses (pre-1975 models).
5   Remove the three motor securing screws and then withdraw the motor, guiding the crank arm through the hole.
6   Installation is the reverse of the removal procedure.

39   Wiper transmission - removal and installation

1   Initially proceed as described in the previous Section, paragraph 1 thru 3, but additionally remove the wiper blades and arms.
2   Remove the right and left wiper transmission-to-body attaching screws.
3   Guide the transmission and linkage assembly out through the cowl plenum chamber opening.
4   Installation is the reverse of the removal procedure, but do not tighten the transmission-to-body attaching screws until the drive link is attached to the motor crank arm.

40   Windshield washer - general description

Pre-1975 models

1   The washer is electrically driven and the system comprises a washer fluid reservoir, washer pump and actuating cam and the interconnecting tubes.
2   When the washer control button is depressed, the pump solenoid circuit is completed and the movement of the solenoid mechanically actuates the wiper switch and in turn the wiper motor. The rotation of the wiper motor causes the washer pump to operate through a pin and cam arrangement for so long as the washer button is held depressed.

1975 and later models

3   The system comprises a permanent magnet motor and pump assembly mounted on the base of the washer fluid reservoir.
4   When the washer control button is depressed, the washer motor circuit is grounded. This operates the motor, causing the pump and wipers to operate. The washers only operate while the button is depressed; the wipers will continue to operate until the dash mounted switch is turned off.

41   Washer assembly - removal, servicing and installation

Pre-1975 models

1   Remove the wiper motor (Section 38).
2   Remove the washer pump mounting screws and lift off the pump from the wiper motor.
3   Pull off the four lobe washer pump drive cam. This is a press fit and may require prying.
4   Remove the felt washer from the wiper shaft.
5   Remove the ratchet dog retaining screw, then hold the spring loaded solenoid plunger in position and lift the solenoid assembly and ratchet dog off the pump frame. Separate the dog from the mounting plate if necessary.
6   To remove the ratchet wheel, move the spring out of the shaft groove and slide the ratchet wheel off its shaft.
7   To separate the pump and pump actuator plate from the frame, pull the pump housing towards the valve end until the grooves in the housing clear the frame.
8   Reassembly is the reverse of the dismantling procedure.
9   Installation is the reverse of the removal procedure.

1975 and later models

10   Remove the two screws attaching the pump body to the washer reservoir.
11   Disconnect the electrical leads and the hoses.
12   Note the installed position of the motor and pump assembly on the reservoir then remove the nut and screen assembly using a suitable wrench and socket. **Note:** *The motor and pump assembly are serviced as a complete assembly, including the gasket, retaining nut and screen.*
13   Installation is the reverse of the removal procedure.

42   Troubleshooting - electrical system

Symptom	Reason
Starter motor fails to turn engine	Battery discharged
	Battery defective internally
	Battery terminal leads loose or ground lead not securely attached to body
	Loose or broken connections in starter motor circuit
	Starter motor switch or solenoid faulty
	Starter motor pinion jammed in mesh with flywheel gear ring
	Starter bushes badly worn, sticking or brush wires loose
	Commutator dirty, worn or burnt
	Starter motor armature faulty
	Field coils grounded
	Incorrect starting sequence (starter interlock system)
Starter motor turns engine very slowly	Battery in discharged condition
	Starter brushes badly worn, sticking or brush wires loose
	Loose wires in starter motor circuit

Symptom	Reason
Starter motor operates without turning engine	Starter motor pinion sticking on the sleeve Pinion or flywheel gear teeth broken or worn Overrunning clutch sticking
Starter motor noisy or engagement excessively rough	Pinion or flywheel teeth broken or worn Starter motor retaining bolts loose
Starter motor remains in operation after ignition key released	Faulty ignition switch Faulty solenoid
Charging system indicator on with ignition switch off...........................	Faulty alternator diode
Charging system indicator light on - engine speed above idling	Loose or broken drive belt Shorted negative diode No output from alternator
Charge indicator light not on when ignition switched on but engine not running	Burned out bulb Field circuit open Lamp circuit open
Battery will not hold charge for more that a few days...........................	Battery defective internally Electrolyte level too weak or too low Battery plates heavily sulphated
Horns will not operate or operate intermittently	Loose connections Defective switch Defective relay Defective horns
Horns blow continually	Faulty relay Relay wiring grounded Horn button stuck (grounded)
Lights do not come on	If engine not running, battery discharged Light bulb filament burnt out or bulbs broken Wire connections loose, disconnected or broken Light switch shorting or otherwise faulty
Lights come on but fade out	If engine not running battery discharged Light bulb filament burnt out or bulbs or sealed beam units broke Wire connections loose, disconnected or broken Light switch shorting or otherwise faulty
Lights give very poor illumination	Lights not properly grounded Lamp glasses dirty Lamps badly out of adjustment
Lights work erratically - flashing on and off, especially over bumps.......	Battery terminals or ground connection loose Lights not grounded properly Contacts in light switch faulty
Wiper motor fails to work	Blown fuse Wire connections loose, disconnected or broken Brushes badly worn Armature worn or faulty Field coils faulty
Wiper motor works very slowly and takes excessive current..................	Commutator dirty, greasy or burnt Armature bearings dirty or unaligned Armature badly worn or faulty
Wiper motor works slowly and takes little current..................................	Brushes badly worn Commutator dirty, greasy or burnt Armature badly worn or faulty
Wiper motor works but wiper blades remain static.................................	Wiper motor gearbox parts badly worn or teeth stripped

Chapter 11
Suspension and steering

Contents

Specifications

Front suspension
Type .... Coil spring, upper and lower control arms, hydraulic shock absorbers and stabilizer bar

Rear suspension
Type .... Solid axle with leaf springs and staggered mounting hydraulic shock absorbers

Steering
Type .... Recirculating ball with worm and nut. Optional power assistance

Wheels
Bearing endplay (front)
Pre-1974 .... 0.001 to 0.008 inch
1974 on .... 0.001 to 0.005 inch

Torque specifications
Ft-lbs (unless otherwise indicated)

Note: One foot-pound (ft-lb) of torque is equivalent to 12 inch-pounds (in-lbs) of torque. Torque values below approximately 15 ft-lbs are expressed in inch-pounds, because most foot-pound torque wrenches are not accurate at these smaller values.

Front suspension
Balljoint stud nut:
  Upper
    1975 and earlier models .... 50 + additional torque to align cotter pin hole
    1976 and later models .... 60 + additional torque to align cotter pin hole
  Lower .... 80 + additional torque to align cotter pin hole
Control arm pivot-to-frame
  Upper
    1975 and earlier models 50
    1976 and later models 75
  Lower
    1975 and earlier models 80
    1976 and later models 100
Upper control arm shaft nuts and bolts.. 40
Shock absorber
  Upper end .... 90 in-lbs
  Lower end 20
Stabilizer bar
  Link nuts.. 18
  Bracket bolts 24

Rear suspension
Shock-absorber
  Upper 18
  Lower 45
  Lower, special performance suspension 60
Leaf spring
  Front eye bolt nut 75
  Rear shackle bolt nut 50
  Front mounting bracket.. 30
  Retainer (anchor plate)
    1975 and earlier models 40
    1976 and later models 45
Stabilizer bar
  To frame bracket
    1977 and earlier models 60
    1978 and later models 30
  To spring retainer plate (1977 and earlier).. 18
  To support strut (1978 and later)... 20

Steering
Steering gear mounting bolts 70
Pitman shaft nut
  Pre-1975 models.. 140
  1975 and later models.. 185
Pitman arm-to-intermediate rod stud nut.. 45 to 55
Idler arm-to-relay rod stud nut .... 35 to 50
Idler arm-to-frame nuts
  Pre-1976 models.. 30
  1975 and later models.. 48
Tie-rod end stud nut .... 35 to 50
Tie-rod clamp nuts
  Pre-1975 models.. 132 in-lbs
  1975 and later models.. 22
Steering coupling to shaft flange nuts
  Pre-1975 models.. 18
  1975 and later models.. 20
Adjuster plug locknut... 85
Over-center adjusting screw locknut.. 30
Steering coupling clamp bolts 30
Steering wheel nut... 30
Floorpan cover screws .... 36 in-lbs
Floorpan cover clamp screws .... 36 in-lbs
Dash panel bracket-to-column clamp screws... 15
Dash panel bracket-to-dash (capsule) nuts 20
Power steering pump pulley nut.. 55
Power steering pump mounting bolts 24
Power steering pump mounting stud nut 25
Wheel lug nuts 70

1   General description

Refer to illustrations 1.1 and 1.2

1   All Nova models are equipped with the short-and-long arm type independent front suspension; this incorporates coil springs and double acting hydraulic shock absorbers (see illustration). An additional feature is the transverse stabilizer bar.

2   The rear suspension is of the solid axle type with leaf springs (see illustration). Rubber bushings are used at the spring front mountings and shackles at the rear. The shock absorbers are double acting hydraulic types with staggered mountings, one side bearing ahead of the axle and the other side behind the axle. A transverse stabilizer bar is also fitted.

3   Manual steering is of the recirculating ball type with worm and nut.

4   An optional power steering system is available. This is hydraulically operated from a pump which is belt-driven from the engine crankshaft.

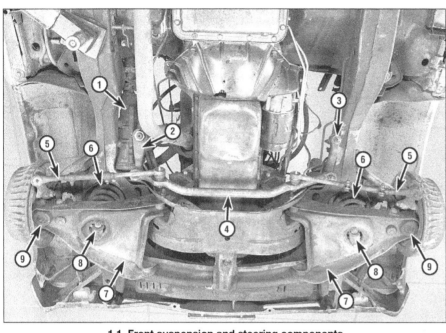

1.1 Front suspension and steering components

1	Steering gear	5	Tie-rod	8	Shock absorber lower mount
2	Pitman arm	6	Coil spring		
3	Idler arm	7	Lower control arm	9	Lower balljoint
4	Intermediate rod				

1.2 Rear suspension components

1	Leaf spring	3	Spring plate	4	Rear axle assembly
2	Shock absorber				

2   Front suspension - lubrication and inspection

Refer to illustration 2.4

1   Every 6,000 miles lubricate the grease fittings in the steering linkage.

2   At similar intervals lubricate the grease fittings on the upper and lower suspension control arms.

3   *Pre-1975 models:* At these service intervals, support the weight of the lower suspension control arm with a jack, then measure the distance from the grease fitting to the end of the lower threaded balljoint stud. Apply leverage under the tire to seat the lower balljoint stud internally, then remeasure. If the difference between these measurements is greater than 1/16 inch the joint is worn and must be replaced.

4   *1975 and later models:* At these service intervals check the wear indicators for indication of excessive lower balljoint wear. When new, a dimension of 0.050 inch should exist from the grease fitting to the balljoint cover surface; if the fitting is flush, or has receded inside the cover, the balljoint must be replaced (see illustration).

5   Also check the condition of the rubber seals on the balljoints. Should they be split or have deteriorated, the joint must be disconnected and the seal replaced.

3   Front wheel bearings - lubrication, adjustment and replacement

Refer to illustrations 3.1, 3.6, 3.7, 3.8, 3.11 and 3.15

Warning: *The dust created by the brake system may contain asbestos, which is harmful*

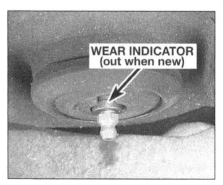

2.4 Wear indicators are built into the lower balljoints of some models to aid in their inspection

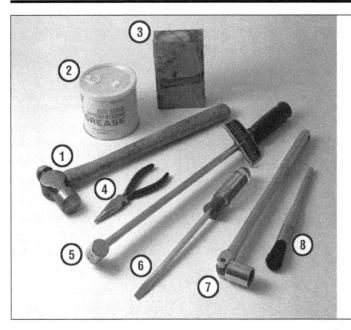

3.1 Tools and materials needed for front wheel bearing maintenance

1 **Hammer** - A common hammer will do just fine
2 **Grease** - High-temperature grease that is formulated specially for front wheel bearings should be used
3 **Wood block** - If you have a scrap piece of 2x4, it can be used to drive the new seal into the hub
4 **Needle-nose pliers** - Used to straighten and remove the cotter pin in the spindle
5 **Torque wrench** - This is very important in this procedure; if the bearing is too tight, the wheel won't turn freely - if it's too loose, the wheel will "wobble" on the spindle. Either way, it could mean extensive damage
6 **Screwdriver** - Used to remove the seal from the hub (a long screwdriver is preferred)
7 **Socket/breaker bar** - Needed to loosen the nut on the spindle if it's extremely tight
8 **Brush** - Together with some clean solvent, this will be used to remove old grease from the hub and spindle

to your health. Never blow it out with compressed air and don't inhale any of it. An approved filtering mask should be worn when working on the brakes. Do not, under any circumstances, use petroleum-based solvents to clean brake parts. Use brake system cleaner only!

1   Every 24,000 miles the front wheel bearings should be disassembled, cleaned and repacked with grease. Several items, including a torque wrench and special grease, are required for this procedure **(see illustration)**.
2   With the vehicle securely supported on jackstands, spin each wheel and check for noise, rolling resistance and free play.
3   Grasp the top of each tire with one hand and the bottom with the other. Move the wheel in and out on the spindle. If there's any noticeable movement, the bearings should be checked and then repacked with grease, or replaced if necessary.
4   Remove the wheel.
5   On models with disc brakes, remove the brake caliper (Chapter 9) and hang it out of the way on a piece of wire. On models with

3.6 Dislodge the grease cap by working around the outer circumference with a hammer and chisel

drum brakes, remove the brake drum.
6   Pry the dust cap out of the hub using a screwdriver or hammer and chisel **(see illustration)**.
7   Straighten the bent ends of the cotter

3.7 Use needle-nose pliers to straighten the cotter pin and pull it out

pin, then pull the cotter pin out of the locking nut **(see illustration)**. Discard the cotter pin and use a new one during reassembly.
8   Remove the spindle nut and washer from the end of the spindle **(see illustration)**.

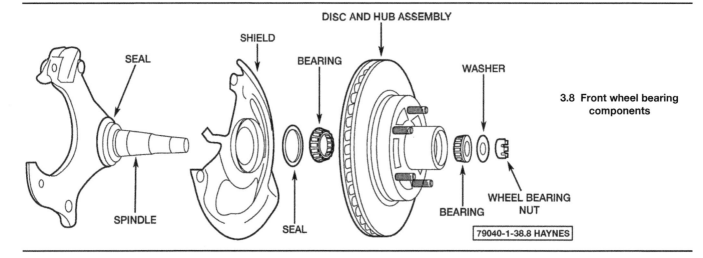

DISC AND HUB ASSEMBLY

SEAL   SHIELD   BEARING   WASHER

SPINDLE   SEAL   BEARING   WHEEL BEARING NUT

3.8 Front wheel bearing components

79040-1-38.8 HAYNES

9   Pull the hub assembly out slightly, then push it back into its original position. This should force the outer bearing off the spindle enough so it can be removed.

10   Pull the hub off the spindle.

11   Use a screwdriver to pry the seal out of the rear of the hub **(see illustration)**. As this is done, note how the seal is installed.

12   Remove the inner wheel bearing from the hub.

13   Use solvent to remove all traces of the old grease from the bearings, hub and spindle. A small brush may prove helpful; however make sure no bristles from the brush embed themselves inside the bearing rollers. Allow the parts to air dry.

14   Carefully inspect the bearings for cracks, heat discoloration, worn rollers, etc. Check the bearing races inside the hub for wear and damage. If the bearing races are defective, the hubs should be taken to a machine shop with the facilities to remove the old races and press new ones in. Note that the bearings and races come as matched sets and old bearings should never be installed on new races.

15   Use high-temperature front wheel bearing grease to pack the bearings. Work the grease completely into the bearings, forcing it between the rollers, cone and cage from the back side **(see illustration)**.

16   Apply a thin coat of grease to the spindle at the outer bearing seat, inner bearing seat, shoulder and seal seat.

17   Put a small quantity of grease inboard of each bearing race inside the hub. Using your finger, form a dam at these points to provide extra grease availability and to keep thinned grease from flowing out of the bearing.

18   Place the grease-packed inner bearing into the rear of the hub and put a little more grease outboard of the bearing.

19   Place a new seal over the inner bearing and tap the seal evenly into place with a hammer and block of wood until it's flush with the hub.

20   Carefully place the hub assembly onto the spindle and push the grease-packed outer bearing into position.

21   Install the washer and spindle nut.

3.11 Use a screwdriver to pry the grease seal from the back side of the hub

Tighten the nut only slightly (no more than 12 ft-lbs of torque).

22   Spin the hub in a forward direction to seat the bearings and remove any grease or burrs which could cause excessive bearing play later.

23   Check to see that the tightness of the spindle nut is still approximately 12 ft-lbs.

24   Loosen the spindle nut until it's just loose, no more.

25   Using your hand (not a wrench of any kind), tighten the nut until it's snug. Install a new cotter pin through the hole in the spindle and spindle nut. If the nut slots don't line up, loosen the nut slightly until they do. From the hand-tight position, the nut should not be loosened more than one-half flat to install the cotter pin.

26   Bend the ends of the cotter pin until they're flat against the nut. Cut off any extra length which could interfere with the dust cap.

27   Install the dust cap, tapping it into place with a hammer.

28   Install the brake caliper or brake drum (Chapter 9).

29   Install the tire/wheel assembly on the hub and tighten the lug nuts.

30   Grasp the top and bottom of the tire and check the bearings in the manner described earlier in this Section.

31   Lower the vehicle.

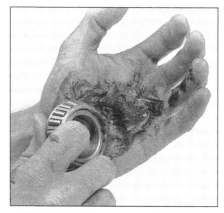

3.15 Work the grease completely into the rollers, from the backside

4   Shock absorbers - removal and installation

Refer to illustrations 4.1 and 4.10

Front shock absorber

1   Raise the front end of the vehicle and support it securely on jackstands. Use an open-ended wrench to prevent the upper (squared) end from turning, then remove the upper stem retaining nut, retainer and rubber grommet **(see illustration)**.

2   Remove the 2 bolts retaining the lower end of the shock absorber to the control arm.

3   Pull the assembly out from the bottom.

4   When installing, fit the lower retainer and rubber grommet in place over the upper stem.

5   Install the shock absorber in the fully extended position up through the lower control arm and spring.

6   Fit the upper rubber grommet, retainer and attaching nut after the shock absorber upper stem has passed through the upper control arm frame bracket.

7   Using an open ended wrench, hold the upper stem and tighten the retaining nut securely.

8   Install the lower retaining bolts and tighten them to the specified torque, then lower the vehicle.

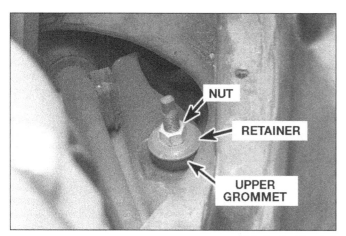

NUT
RETAINER
UPPER GROMMET

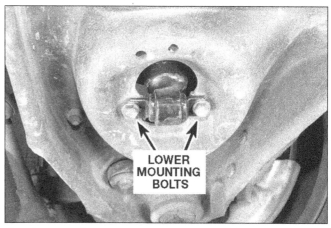

LOWER MOUNTING BOLTS

4.1 Typical front shock absorber mounting

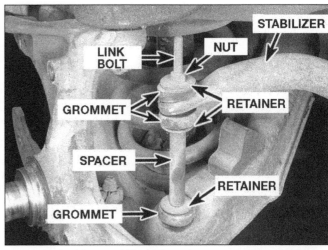

4.10 The rear shock absorber is retained by a through-bolt and nut at the bottom, and two bolts at the top

5.1 Typical front stabilizer bar-to-lower control arm linkage

Rear shock absorber

9   Raise the rear end of the vehicle and support it securely on jackstands placed under the frame. Support the rear axle with a floor jack.

10   Remove the shock absorber lower mounting bolt **(see illustration)**.

11   Remove the upper mounting bolts and withdraw the shock absorber.

12   When installing, loosely fit the upper attaching bolts then insert the eye into the lower bracket.

13   Install the lower bolt with the nut to the rear, and tighten it to the specified torque.

14   Tighten the upper bolts to the specified torque.

15   Lower the vehicle to the ground.

5   Stabilizer bar - removal and installation

Refer to illustrations 5.1 and 5.2

Front stabilizer bar

1   Raise the front end of the vehicle and support is securely on jackstands. Disconnect the stabilizer bar from the lower control arm **(see illustration)**.

2   Remove the stabilizer bar brackets from the frame then lift away the stabilizer bar **(see illustration)**.

3   Remove the link bolts, spacers and rubber grommets from the lower control arms or stabilizer bar.

4   inspect all the parts for damage, wear and deterioration. Install new parts as necessary.

5   If new frame bushings are required, slide them into position along the stabilizer bar.

6   When installing, fit the brackets over the bushings and connect them (loosely) to the frame.

7   Ensure that the stabilizer bar is centralized then tighten all the bolts to the specified torque values.

8   Lower the vehicle to the ground.

5.2 Remove the brackets and rubber bushings on each side that attach the stabilizer bar to the frame

Rear stabilizer bar

9   Raise the rear end of the vehicle and support the rear axle.

10   Remove the stabilizer bar-to-spring retainer bracket attachment.

11   Remove the stabilizer bar-to-body bracket bolt and remove the bar.

12   Fit the bushings onto the stabilizer bar then place the bar in position.

13   Fit the upper retaining bolts and the stabilizer bar to spring attachment.

14   Ensure that the weight of the vehicle is being carried by the rear axle only, then tighten the bolts.

15   Lower the vehicle to the ground.

6   Coil spring - removal and installation

Refer to illustration 6.4

Removal

1   Loosen the front wheel lug nuts, raise the vehicle and support it securely on jackstands. Remove the wheel.

6.5 A typical aftermarket internal type spring compressor: The hooked arms grip the upper coils of the spring, the plate is inserted between the lower coils, and when the nut on the threaded rod is turned, the spring is compressed

2   Remove the shock absorber (see Section 4).

3   Remove the stabilizer bar link bolt (see Section 5).

4   Install a suitable internal type spring compressor in accordance with the tool manufacturer's instructions **(see illustration)**. Compress the spring enough to relieve all pressure from the spring seats (but don't compress it any more than necessary, or it could be ruined). When you can wiggle the spring, it's compressed enough. (You can buy a suitable spring compressor at most auto parts stores or rent one from a tool rental yard.)

5   Support the lower control arm with a floor jack.

6   Remove the control arm pivot bolts and nuts.

7   Pull the lower control arm down and to the rear, then guide the compressed coil spring out.

8   If the coil spring is being replaced, carefully unscrew the spring compressor.

7.4 A special tool is used to push the balljoint out of the steering knuckle, but an alternative tool can be fabricated from a large bolt, nut, washer and socket

Installation

9   Inspect the upper and lower spring insulators. If either insulator is cracked or excessively worn, replace it. Inspect the coil spring for chips in the corrosion protection coating. If the coating has been chipped or damaged, replace the spring.

10   If the coil spring is being replaced, install the spring compressor and compress the spring.

11   With the lower spring insulator in place, position the spring on the lower control arm with the flat end of the spring facing up and the tapered end facing down. Make sure the tapered end seats on the lower control arm with the lower end of the spring seated in the lowest part of the spring seat. The end of the spring must cover all or part of one of the drain holes in the lower control arm, but the other hole must not be covered.

12   Put the floor jack under the lower control arm and raise the arm into position in the frame. Install the control arm pivot bolts and nuts. Tighten the nuts until they're snug but don't torque them yet.

13   Remove the spring compressor.

14   Reattach the stabilizer bar link to the lower control arm (see Section 5).

15   Install the shock absorber (see Section 4).

7.8 Install the replacement balljoint in the upper control arm

7.5 Drill pilot holes into the heads of the balljoint rivets with a 1/8-inch bit, then use a 1/2-inch bit to cut the rivet heads off - be careful not to enlarge the holes in the control arm

16   Position the floor jack under the lower control arm balljoint and raise the arm to simulate normal ride height. Tighten the lower control arm pivot bolt nuts to the torque listed in this Chapter's Specifications.

17 Install the wheel, remove the jackstands and lower the vehicle. Tighten the wheel lug nuts to the torque listed in this Chapter's Specifications.

7   Balljoints - replacement

Refer to illustrations 7.4, 7.5 and 7.8

Upper balljoint

1   Loosen the wheel lug nuts, raise the vehicle and support it securely on jackstands. Apply the parking brake. Remove the wheel.

2   Support the lower control arm with a floor jack placed under the lower balljoint. **Warning:** *The jack must remain under the control arm during removal and installation of the balljoint to hold the spring and control arm in position.*

3   Remove the cotter pin from the balljoint stud and back off the nut two turns.

4   Separate the balljoint from the steering knuckle. A special tool is available for this, but an equivalent tool can be fabricated from a large bolt, nut, washer and socket **(see illustration)**. Countersink the center of the bolt head with a large drill bit to prevent the tool from slipping off the balljoint stud. Install the tool as shown in the illustration, hold the bolt head with a wrench and tighten the nut against the washer until the balljoint pops out. Notice that the balljoint nut hasn't been completely removed.

5   Using a 1/8-inch drill bit, drill a 1/4-inch deep hole in the center of each rivet head **(see illustration)**.

6   Using a 1/2-inch diameter drill bit, drill off the rivet heads. **Caution:** *Drill just enough to remove the rivet heads; be careful not to enlarge or distort the holes in the control arm.*

7   Use a punch to knock the rivet shanks out, then remove the balljoint from the control arm.

8   Position the new balljoint on the control arm and install the bolts and nuts supplied in the kit **(see illustration)**. Be sure to tighten the nuts to the torque specified on the balljoint kit instruction sheet.

9   Insert the balljoint stud into the steering knuckle and install the nut, tightening it to the specified torque.

10   Install a new cotter pin, tightening the nut slightly if necessary to align a slot in the nut with the hole in the balljoint stud.

11   Install the balljoint grease fitting and fill the joint with grease.

12   Install the wheel, tightening the lug nuts to the specified torque.

13   Drive the vehicle to an alignment shop to have the front end alignment checked and, if necessary, adjusted.

Lower balljoint

14   The lower balljoint is a press fit in the lower control arm and requires special tools to remove and replace it. Refer to Section 9, remove the lower control arm and take it to an automotive machine shop to have the old balljoint pressed out and the new balljoint pressed in.

8   Upper control arm - removal and installation

Refer to illustration 8.4

Removal

1   Loosen the wheel lug nuts, raise the front of the vehicle and support it securely on jackstands. Apply the parking brake. Remove the wheel.

2   Support the lower control arm with a jack or jackstand. The support point must be as close to the balljoint as possible to give maximum leverage on the lower control arm.

3   Disconnect the upper balljoint from the steering knuckle (refer to Section 7). **Note:** *DO NOT use a "pickle fork" type balljoint separator - it may damage the balljoint seals.*

4   Remove the control arm-to-frame nuts and bolts, recording the position of any alignment shims. They must be reinstalled in the same location to maintain wheel alignment **(see illustration)**.

5   Detach the control arm from the vehicle. **Note:** *The control arm bushings are pressed into place and require special tools for removal and installation. If the bushings must be replaced, take the control arm to a dealer service department or an automotive machine shop to have the old bushings pressed out and the new ones pressed in.*

Installation

6   Position the control arm on the frame and install the bolts and nuts. Install any alignment shims that were removed. Tighten the nuts to the specified torque.

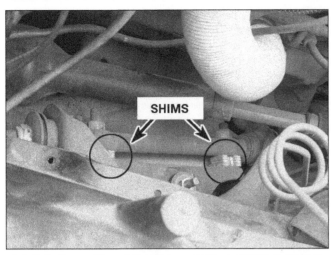

8.4 Note the position of the alignment shims and return them to their original positions

11.4 Details of the leaf spring front mount

1   *Spring front eye-to-bracket bolt*
2   *Bracket-to-body bolts*

7   Insert the balljoint stud into the steering knuckle and tighten the nut to the specified torque. Install a new cotter pin, tightening the nut slightly, if necessary, to align a slot in the nut with the hole in the balljoint stud.
8   Install the wheel and lug nuts and lower the vehicle. Tighten the lug nuts to the specified torque.
9   Drive the vehicle to an alignment shop to have the front end alignment checked and, if necessary, adjusted.

9   Lower control arm - removal and installation

Removal

1   Loosen the wheel lug nuts, raise the vehicle and support it securely on jackstands. Apply the parking brake. Remove the wheel.
2   Remove the shock absorber (see Section 4).
3   Disconnect the stabilizer bar from the lower control arm (Section 5).
4   Remove the coil spring as described in Section 6.
5   Remove the cotter pin and back off the lower control arm balljoint stud nut two turns. Separate the balljoint from the steering knuckle (Section 7).
6   Remove the control arm from the vehicle. **Note:** *The control arm bushings are pressed into place and require special tools for removal and installation. If the bushings must be replaced, take the control arm to a dealer service department or an automotive machine shop to have the old bushings pressed out and the new ones pressed in.*

Installation

7   Insert the balljoint stud into the steering knuckle, tighten the nut to the specified torque and install a new cotter pin. If necessary, tighten the nut slightly to align a slot in the nut with the hole in the balljoint stud.

8   Install the coil spring (Section 6) and the lower control arm pivot bolts and nuts, but don't completely tighten them at this time.
9   Connect the stabilizer bar to the lower control arm.
10   Position the floor jack under the lower control arm balljoint and raise the arm to simulate normal ride height. Tighten the lower control arm pivot bolt nuts to the specified torque.
11   Install the wheel and lug nuts, lower the vehicle and tighten the lug nuts to the specified torque.

10   Steering knuckle - removal and installation

Removal

1   Loosen the wheel lug nuts, raise the vehicle and support it securely on jackstands placed under the frame. Apply the parking brake. Remove the wheel.
2   If equipped with disc brakes, remove the brake caliper and suspend it with a piece of wire (Chapter 9). Do not let it hang by the brake hose!
3   Remove the brake disc or brake drum and hub (see Section 3).
4   Remove the splash shield from the steering knuckle. On models with drum brakes, it isn't necessary to remove the brake shoes or disconnect the brake line from the wheel cylinder; just unbolt the backing plate from the steering knuckle and hang the assembly with a piece of wire from the upper control arm.
5   Separate the tie-rod end from the steering arm (see Section 12).
6   If the steering knuckle must be replaced, remove the dust seal from the spindle by prying it off with a screwdriver. If it's damaged, replace it with a new one.
7   Position a floor jack under the lower control arm and raise it slightly to take the spring pressure off the suspension stop. The jack must remain in this position throughout the entire procedure.

8   Remove the cotter pins from the upper and lower balljoint studs and back off the nuts two turns each.
9   Break the balljoints loose from the steering knuckle with a balljoint separator (Section 7). **Note:** *A "pickle fork" type balljoint separator may damage the balljoint seals.*
10   Remove the nuts from the balljoint studs, separate the control arms from the steering knuckle and remove the knuckle from the vehicle.

Installation

11   Place the knuckle between the upper and lower control arms and insert the balljoint studs into the knuckle, beginning with the lower balljoint. Install the nuts and tighten them to the specified torque. Install new cotter pins, tightening the nuts slightly to align the slots in the nuts with the holes in the balljoint studs, if necessary.
12   Install the splash shield or brake backing plate.
13   Connect the tie-rod end to the steering arm and tighten the nut to the specified torque. Be sure to use a new cotter pin.
14   Install the brake disc or hub (drum brake models) and adjust the wheel bearings following the procedure outlined in Section 3.
15   If equipped with disc brakes, install the brake caliper, tightening the mounting bolts to the torque listed in the Chapter 9 Specifications.
16   Install the wheel and lug nuts. Lower the vehicle to the ground and tighten the nuts to the specified torque.

11   Leaf springs - removal and installation

Refer to illustrations 11.4, 11.8 and 11.9

1   Raise the rear end of the vehicle and support securely on jackstands placed under the frame.
2   Support the axle with a floor jack to

11.8 Remove the U-bolt nuts from the bottom of the spring plate

11.9 Remove the nuts from the shackle bolts

relieve the weight from the spring.

3   Remove the shock absorber lower mounting bolt.

4   Loosen the spring front eye-to-bracket bolt **(see illustration)**.

5   Remove the screws securing the spring front bracket to the underbody.

6   Lower the axle sufficiently to allow access to the spring front bracket; remove the bracket from the spring.

7   Pry the parking brake cable out of the retainer bracket on the spring mounting plate.

8   Remove the lower spring plate-to-axle bracket retaining nuts then the upper and lower spring pads and spring plate **(see illustration)**.

9   Remove the nuts from the rear shackle bolts **(see illustration)**. Separate the shackle and withdraw the spring

10   If the spring bushings are worn, have the old ones removed and the new ones installed by an automotive machine shop.

11   When installing the spring, position the front mounting bracket to the front eye. Install the attaching bolt with the bolt head towards the center of the vehicle.

12   Position the spring shackle upper bushings in the frame. Position the shackles to the bushings and loosely install the bolt and nut.

13   Install the bushing halves to the spring rear eye; place the spring to the shackles and

loosely install the lower shackle bolt and nut. Ensure that the spring is positioned so that the parking brake cable is on the underside of the spring.

14   Raise the front end of the spring and position the bracket to the underbody. Guide the spring into position so that it will fit into the axle bracket and ensure that the tab on the spring bracket aligns with the slot in the underbody.

15   Loosely install the spring bracket.

16   Position the spring upper cushion between the spring and the axle bracket so that the ribs align with the locating ribs.

17   Fit the lower spring cushion and align with the upper cushion (where applicable).

18   Place the lower mounting plate over the locating dowel on the lower spring pad and loosely install the retaining nuts.

19   Where a new mounting plate is used, transfer the parking brake retaining bracket to the new plate.

20   Attach the shock absorber to the mounting plate.

21   Install the parking brake cable in the retaining bracket end securely clamp the bracket to retain the cable.

22   Tighten the nuts and bolts to the specified torque values with the vehicle's weight on the rear springs.

23   Lower the vehicle to the ground.

12   Steering linkage - inspection, removal and installation

Warning: *Whenever any of the suspension or steering fasteners are loosened or removed they must be inspected and if necessary, replaced with new ones of the same part number or of original equipment quality and design. Torque specifications must be followed for proper reassembly and component retention. Never attempt to heat, straighten or weld any suspension or steering component. Instead, replace any bent or damaged part with a new one.*

Caution: *DO NOT use a "pickle fork" type balljoint separator - it may damage the balljoint seals.*

Inspection

Refer to illustration 12.5

1   The steering linkage connects the steering gear to the front wheels and keeps the wheels in proper relation to each other. The linkage consists of the Pitman arm, fastened to the steering gear shaft, which moves the intermediate rod back-and-forth. The intermediate rod is supported on the other end by a frame-mounted idler arm. The back-and-forth motion of the intermediate rod is transmitted to the steering knuckles through a pair of tie-rod assemblies. Each tie-rod is made up of an inner and outer tie-rod end, a threaded adjuster tube and two clamps.

2   Set the wheels in the straight ahead position and lock the steering wheel.

3   Raise one side of the vehicle until the tire is approximately 1-inch off the ground.

4   Mount a dial indicator with the needle resting on the outside edge of the wheel. Grasp the front and rear of the tire and using light pressure, wiggle the wheel back-and-forth and note the dial indicator reading. The gauge reading should be less than 0.108-inch. If the play in the steering system is more than specified, inspect each steering linkage pivot point and ball stud for looseness and replace parts if necessary.

5   Raise the vehicle and support it on jackstands. Push up, then pull down on the end of the idler arm, exerting a force of approximately 25 pounds each way. Measure the total distance the end of the arm travels **(see illustration)**. If the play is greater than 1/4-inch re-place the idler arm.

6   Check for torn ball stud boots, frozen joints and bent or damaged linkage components.

Removal and installation

Refer to illustrations 12.9, 12.11, 12.13 and 12.15

Tie-rod

7   Loosen the wheel lug nuts, raise the vehicle and support it securely on jackstands. Apply the parking brake. Remove the wheel.

8   Remove the cotter pin and loosen, but do not remove, the castellated nut from the ball stud.

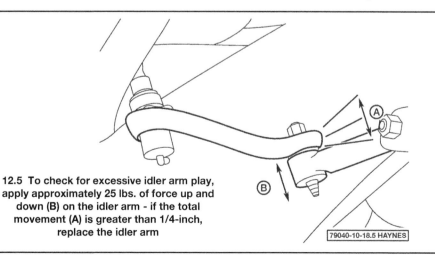

12.5 To check for excessive idler arm play, apply approximately 25 lbs. of force up and down (B) on the idler arm - if the total movement (A) is greater than 1/4-inch, replace the idler arm

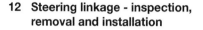

12.9 Use a puller to press the tie-rod end out of the steering knuckle

9   Using a two jaw puller, separate the tie-rod end from the steering knuckle **(see illustration)**. Remove the castellated nut and pull the tie-rod end from the knuckle.
10   Remove the nut securing the inner tie-rod end to the intermediate rod. Separate the inner tie-rod end from the intermediate rod in the same manner as in Step 9.
11   If the inner or outer tie-rod end must be replaced, measure the distance from the end of the adjuster tube to the center of the ball stud and record it **(see illustration)**. Loosen the adjuster tube clamp and unscrew the tie-rod end.
12   Lubricate the threaded portion of the tie-rod end with chassis grease. Screw the new tie-rod end into the adjuster tube and adjust the distance from the tube to the ball stud to the previously measured dimension. The number of threads showing on the inner and outer tie-rod ends should be equal within three threads. Don't tighten the clamp yet.
13   To install the tie-rod, insert the inner tie-rod end ball stud into the intermediate rod until it's seated. Install the nut and tighten it to the specified torque. If the ball stud spins when attempting to tighten the nut, force it into the tapered hole with a large pair of pliers **(see illustration)**.
14   Connect the outer tie-rod end to the steering knuckle and install the castellated nut. Tighten the nut to the specified torque and install a new cotter pin. If necessary, tighten the nut slightly to align a slot in the nut with the hole in the ball stud.
15   Tighten the clamp nuts. The center of the bolt should be nearly horizontal and the adjuster tube slot must not line up with the gap in the clamps **(see illustration)**.

16   Install the wheel and lug nuts, lower the vehicle and tighten the lug nuts to the specified torque. Drive the vehicle to an alignment shop to have the front end alignment checked and, if necessary, adjusted.

Idler arm

17   Raise the vehicle and support it securely on jackstands. Apply the parking brake.
18   Loosen but do not remove the idler arm-to-intermediate rod nut.
19   Separate the idler arm from the intermediate rod with a two jaw puller **(see illustration 12.9)**. Remove the nut.
20   Remove the idler arm-to-frame bolts.
21   To install the idler arm, position it on the frame and install the bolts, tightening them to the specified torque.
22   Insert the idler arm ball stud into the intermediate rod and install the nut. Tighten the nut to the specified torque. If the ball stud spins when attempting to tighten the nut, force it into the tapered hole with a large pair of pliers.

Intermediate rod

23   Raise the vehicle and support it securely on jackstands. Apply the parking brake.
24   Separate the two inner tie-rod ends from the intermediate rod.

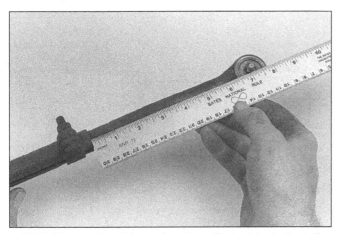

12.11 Measure the distance from the adjuster tube to the ball stud centerline so the new tie-rod end can be set to the same length

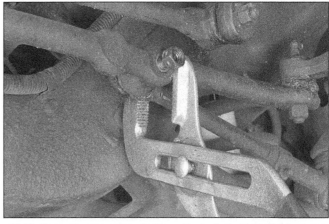

12.13 It may be necessary to force the ball stud into the tapered hole to keep it from turning while the nut is tightened

12.15 Tie-rod adjuster tube and clamp position details

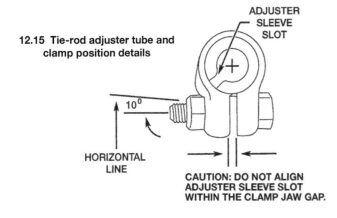

ADJUSTER
SLEEVE
SLOT

10°

HORIZONTAL
LINE

CAUTION: DO NOT ALIGN
ADJUSTER SLEEVE SLOT
WITHIN THE CLAMP JAW GAP.

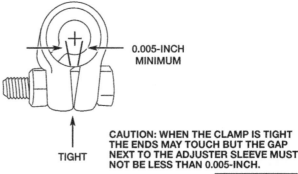

0.005-INCH
MINIMUM

TIGHT

CAUTION: WHEN THE CLAMP IS TIGHT
THE ENDS MAY TOUCH BUT THE GAP
NEXT TO THE ADJUSTER SLEEVE MUST
NOT BE LESS THAN 0.005-INCH.

79040-10-18.5 HAYNES

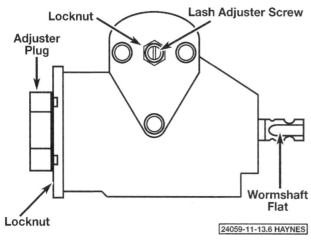

13.6 Steering gear adjustment points

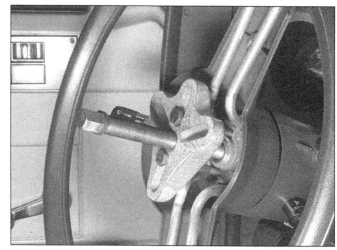

14.5 Using a steering wheel puller to remove a steering wheel

25   Separate the intermediate rod from the Pitman arm.
26   Separate the idler arm from the intermediate rod.
27   Installation is the reverse of the removal procedure. If the ball studs spin when attempting to tighten the nuts, force them into the tapered holes with a large pair of pliers. Be sure to tighten all of the nuts to the specified torque.

Pitman arm

28   Refer to Section 13 for the Pitman arm removal procedure.

13   Manual steering gear - maintenance and adjustment (in-vehicle)

Refer to illustration 13.6
1   The steering gear is normally filled with lubricant for life and unless a severe leak occurs, necessitating a complete overhaul, refilling with lubricant will not be required.
2   In order to rectify conditions of lost motion, slackness and vibration which have been found to be directly attributable to the steering gear, perform the following operations:
3   Disconnect the ground cable from the battery.
4   Remove the nut from the Pitman arm and then mark the relative position of the arm to the Pitman shaft.
5   Using a suitable puller, remove the Pitman arm.
6   Loosen the adjuster plug locknut on the steering gear and unscrew the adjuster plug one quarter-turn **(see illustration)**.
7   Remove the horn button or shroud from the steering wheel and then turn the steering wheel in one direction to full lock and then turn the wheel back through one half turn. Using a socket and a beam or dial-type torque wrench on the steering wheel center nut, check the bearing drag when the wheel is turned through a 90° arc of travel.

8   This drag is the thrust bearing preload and it should be between 5 and 9 in-lbs for pre-1973 models, and between 4 and 6 in-lbs for 1973 models onwards. Tighten or loosen the adjuster plug until the correct preload is obtained and then tighten the adjuster plug locknut.
9   Any jerky or bumpy feeling as the steering wheel is turned will indicate worn or damaged bearings in the steering gear.
10   Now turn the steering wheel gently from one stop to the other counting the number of turns of the steering wheel from lock-to-lock. Now turn the wheel exactly half the number of turns counted so that the steering gear is in the centered position.
11   Loosen the lash adjuster screw locknut and turn the lash adjuster screw clockwise until all lash has been removed from between the ball nut and the Pitman shaft sector teeth. Tighten the locknut securely while holding the screw from turning.
12   Now check the 'over-center' preload by taking the highest torque reading obtainable as the wheel is moved through its centered position. The preload should be between 4 and 10 in-lbs for pre-1973 models, and between 5 and 9 in-lbs for 1973 models onward, in excess of the torque stated at paragraph 8 above. Adjust the position of the lash adjuster screw if necessary to achieve this.
13   Install the Pitman arm and horn shroud, and connect the battery ground cable. Tighten the Pitman arm nut to the specified torque.

14   Steering wheel - removal and installation

Refer to illustration 14.5

Pre-1971 models
1   Disconnect the ground cable from the battery.
2   *Regular production steering wheel:* Pull

out the horn button cup or center ornament and retainer. Remove the 3 screws from the receiving cup, remove the cup, Belleville spring, bushing and pivot ring.
3   *Deluxe steering wheel:* Remove the 4 attaching screws on the underside of the steering wheel. Lift the steering wheel shroud from the wheel then pull the horn wires from the canceling tower.
4   *Simulated wood and cushioned rim steering wheels:* Pull out the horn cap; remove the contact assembly attaching screws and remove the contact assembly. If the steering column is to be disassembled, remove the remaining screws from the steering wheel and remove the wheel from the hub assembly.
5   Remove the steering wheel nut and washer and using a steering wheel puller, pull off the steering wheel **(see illustration)**.
6   *Regular production steering wheel:* When installing, position the directional canceling cam and horn contact assembly in place then set the wheel onto the steering shaft. Secure with the washer and nut, and torque-tighten. Install the Belleville spring (dished side upwards), pivot ring, bushing and receiving cup with screws, Install the retainer and horn button cap or center ornament.
7   *Deluxe steering wheel:* Position the horn contact then set the wheel onto the steering shaft. Secure with the washer and nut, and torque. Insert the horn wires into the canceling cam tower then position the shroud on the steering wheel and install the four attaching screws on the underside of the steering wheel.
8   *Simulated wood and cushioned rim steering wheels:* With the horn contact in place, install the hub assembly. Secure with the washer and nut, and torque-tighten. Attach the steering wheel to the hub assembly using the 6 attaching screws and tighten fully. Place the horn contact on the steering wheel and attach with the 3 screws.
9   Connect the ground cable from the battery.

1971 and later models

10   Disconnect the ground cable from the battery.

11   *Regular production steering wheel*: remove the steering wheel shroud screws from the underside of the steering wheel and remove the shroud.

13   *Cushioned rim steering wheel*: pry off the horn button cap.

14   Unscrew and remove the steering wheel nut and, using a steering wheel puller, pull off the steering wheel **(see illustration14.5)**. **Note**: *On some models a snap-ring is fitted. This must be removed first.*

15   Install the steering wheel making sure that the front wheels are in the 'straight-ahead' position and the turn signal switch is in the neutral mode.

16   Install the steering wheel to the shaft. Tighten the retaining nut to the specified torque, and fit the snap-ring, on models so equipped.

17   *Regular production steering wheel:* Place the shroud onto the steering wheel while guiding the horn contact lead into the directional signal canceling cam tower. Replace the shroud attaching screws.

18   *Cushioned rim steering wheel:* Ensure that the horn lower insulator, eyelet and spring are in place then install the Belleville spring, receiver and horn upper insulator. Secure with 3 screws. Install the horn button cap.

19   Connect the battery ground cable.

---

15   Steering couplings and manual steering gear - removal and installation

Flexible coupling

1   Disconnect the battery ground cable (and coupling shield if applicable).

2   Remove the intermediate steering shaft flange to flexible coupling retaining bolts.

3   Remove the steering gear to frame bolts; lower the steering gear.

4   Push the intermediate shaft rearwards and rotate it out of the way.

5   Using a 12 point socket wrench, remove the coupling clamp bolt. Remove the flexible coupling.

6   Install the flexible coupling to the steering gear wormshaft splined end, taking care to align the mating flats.

7   Install the coupling clamp bolt and torque tighten, after ensuring that the coupling reinforcement is bottomed on the wormshaft.

8   Install the intermediate shaft to the coupling and loosely install the flange to coupling bolts.

9   Install the steering gear to frame bolts and tighten to the specified torque.

10   Align the flexible coupling pins centrally in the intermediate shaft flange slots then torque-tighten the coupling bolts.

11   Connect the battery ground lead (and the coupling shield, if applicable).

Pot joint coupling

12   Disconnect the battery ground cable (and the coupling shield, if applicable).

13   Remove the intermediate shaft flange to flexible coupling retaining bolts.

14   Remove the pot joint clamping bolt (at the steering shaft).

15   Remove the steering gear to frame bolts; lower the gear.

16   Push the intermediate steering shaft rearwards until it bottoms in the pot joint and clears the flexible coupling alignment pins.

17   Remove the intermediate shaft and pot joint as an assembly.

18   To disassemble the pot joint, pry off the snap-ring and slide the coupling over the shaft. Remove the bearings and tension spring from the pivot pin. Clean the pin and the end of the shaft then scribe a location mark on the pin on the same side as the shaft chamfer. Support the shaft securely then press out the pin taking care that it is not damaged, or bearing damage may occur. Remove the seal clamp then slide the seal off the end of the shaft.

19   To reassemble the pot joint, first ensure that all the parts are clean then slide the seal onto the shaft so that the lip of the seal is against the shoulder on the shaft. Install the clamp, press the pin into the shaft, aligning the scribed location marks. Ensure that the pin is centered within 0.012 inch, or binding will result. Liberally grease the inside and outside of the bearings and the inside of the cover then install the tension spring and bearings on the pin. Install the seat into the end of the cover and secure with the snap-ring.

20   When installing, align the flats on the pot joint and steering shaft then mate them. Install the clamp and bolt, and torque-tighten.

21   Install the intermediate shaft to the flexible coupling and loosely install the coupling bolts.

22   Install the steering gear-to-frame bolts and torque-tighten.

23   Align the flexible coupling pins centrally in the intermediate shaft flange slots then torque-tighten the coupling bolts.

24   Connect the battery ground cable (and the coupling shield, if applicable).

Coupling shield (1975 models)

25   Remove the 2 clamps from around the outside of the coupling shield. With manual steering also remove the retainer tab.

26   Disengage the 6 tabs on the lower half of the coupling shield using a small screwdriver or similar tool.

27   Remove both sections of the coupling shield from the vehicle.

28   Installation is the reverse of the removal procedure. The tabs snap into place to lock and retain.

Steering gear

29   Remove the battery ground cable (and the coupling shield, if applicable).

30   Remove the nuts, lockwashers and

bolts at the steering shaft to coupling flange.

31   Remove the Pitman arm.

32   Remove the screws securing the steering gear to the frame and remove it from the vehicle.

33   When installing, place the gear into position so that the coupling mounts properly to the flanged end of the steering shaft. Secure the gear to the frame, fit the washers and bolts, then torque-tighten.

34   Secure the steering coupling to the flanged end of the column with the lockwashers and nuts. The overlap of the coupling pins to the shaft flange must be 5/16-inch, +/-1/32-inch. Tighten the nuts to the specified torque.

35   Install the Pitman arm.

36   Connect the battery ground cable (and the coupling shield, if applicable).

---

16   Pitman shaft seal (manual steering) - replacement (in-vehicle)

1   Remove the Pitman arm.

2   Turn the steering from stop-to-stop and count the exact number of turns. Now rotate the wheel half the number of turns counted so that the steering gear is centered (wormshaft flat at 12 o'clock position).

3   Remove the side cover from the steering gear housing (three screws) and lift the Pitman shaft and side cover from the housing.

4   Pry the seal from the housing using a screwdriver and tap a new one into position using a suitable socket or piece of tube.

5   Remove the lash adjuster screw locknut and detach the side cover from the Pitman shaft by turning the adjuster screw clockwise.

6   Insert the Pitman shaft into the steering gear so that the center tooth of the shaft sector enters the center tooth of the ball nut.

7   Pack the specified grease into the housing and install a new side cover gasket. Install the side cover onto the lash adjuster screw. This is achieved by inserting a small screwdriver through the threaded adjuster hole in the side cover and turning the lash adjuster screw counter-clockwise. When the screw bottoms, turn it back 1/4 turn.

8   Tighten the side cover bolts to the specified torque.

9   Carry out the adjustments described previously and tighten the lash adjuster screw locknut.

10   Install the Pitman arm.

---

17   Turn signal switch - removal and installation

Refer to illustrations 17.3, 17.4 and 17.10

1   Remove the steering wheel (Section 14).

2   Remove the column-to-instrument panel trim panel and the cover plate (on early models the cover plate is secured by three screws; on later models it is simply pried off).

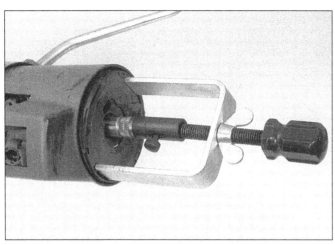

17.3 A special tool is required to depress the steering shaft lock plate so the snap-ring can be removed

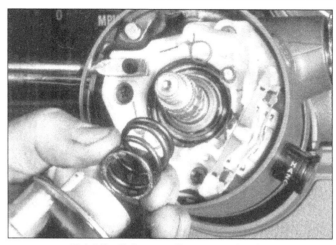

17.4 Lift off the canceling cam and spring

3   The lockplate will now have to be depressed using a 'U' shaped tool and utilizing the steering wheel retaining nut, so that the snap-ring can be extracted from the shaft groove **(see illustration)**. Remove the lockplate.

4   From the steering shaft, slide the directional signal canceling cam, the preload spring and the thrust washer **(see illustration)**.

5   Unscrew the turn signal lever screw and remove the lever.

6   Depress the hazard warning knob and then unscrew it.

7   *All columns*: Pull out the switch connector from the bracket on the jacket and tape the wire ends to prevent snagging on removal.

8   Remove the wire protector by pulling downwards on the tab.

9   *Tilt columns*: Position the directional signal and shifter housing to the low position then remove the harness cover. Be careful not to damage the wires.

10   Unscrew the three switch mounting screws **(see illustration)** and pull the turn signal switch straight up, guiding the wiring harness carefully through the opening in the steering column housing.

11   Installation is a reversal of removal but make sure that the switch is in neutral and the hazard warning knob is pulled fully out.

18 Ignition key lock cylinder- removal and installation

1978 and earlier models

Refer to illustration 18.3

1   On these models, the lock cylinder should be removed in the RUN position (1976 and earlier) or in the LOCK position (1977 and 1978), otherwise damage to the warning buzzer switch may occur.

2   Remove the steering wheel (Section 14) and turn signal switch (Section 17). **Note:** *The turn signal switch need not be fully removed provided that it is pushed to the rear far enough for it to be slipped over the end of the shaft. Do not pull the harness out of the column.*

3   Insert a thin bladed screwdriver into the slot in the turn signal switch housing. Break the housing flash loose and at the same time depress the spring latch at the lower end of the lock cylinder. Holding the latch

depressed, withdraw the lock cylinder from the housing **(see illustration)**.

4   To install the new lock cylinder/sleeve assembly, hold the sleeve and rotate the lock clockwise against the stop.

5   Insert the cylinder/sleeve assembly into the housing so the key on the cylinder sleeve is aligned with the housing keyway.

6   Insert a 0.070-inch diameter drill between the lock bezel and the housing and then rotate the cylinder counterclockwise, maintaining pressure on the cylinder until the drive section mates with the sector.

7   Press in the lock cylinder until the snapring engages in the grooves and secures the cylinder in the housing. Remove the drill and check the lock action.

8   Install the turn signal switch and the steering wheel.

1979 models

Refer to illustration 18.12

9   The lock cylinder should be removed in the RUN position only.

10   Remove the steering wheel (Section 14) and turn signal switch (Section 17). It is not necessary to completely remove the switch.

17.10 Remove the turn signal switch mounting screws

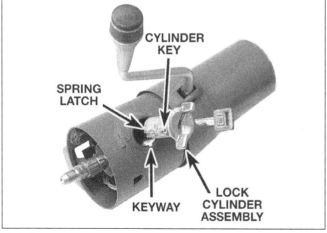

18.3 On 1978 and earlier models, the lock cylinder is retained by a spring latch

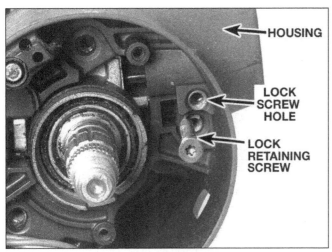 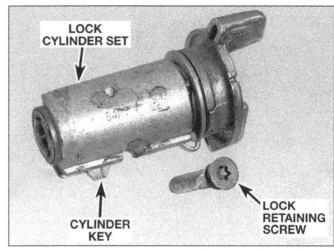

18.12 On 1979 models, the lock cylinder is held in place by a screw

Pull it up and over the end of the steering shaft. Do not pull the wiring harness out of the column.

11   Remove the ignition key warning switch (Section 19).

12   Using a magnetized screwdriver, remove the lock retaining screw. Do not allow this screw to drop down into the column, as this could require a complete disassembly of the steering column to retrieve the screw **(see illustration)**.

13   Pull the lock cylinder out of the side of the steering column.

14   To install, rotate the lock cylinder set and align the cylinder key with the keyway in the steering column housing.

15   Push the lock all the way in and install the retaining screw.

16   Install the remaining components referring to the appropriate Sections.

---

19 Ignition key warning (buzzer) switch - removal and installation

---

Refer to illustration 19.2

1   Remove the steering wheel (Section 14) and turn signal switch (Section 17). **Note:** *On*

some models the turn signal switch need not be fully removed provided that it is pushed rearwards, far enough for it to be slipped over the end of the shaft. Do not pull the harness out of the column. On early models the lock cylinder must also be set to 'ON'.

2   Using either a pair of needle-nosed pliers, pull the switch and its securing clip together, up and out of the column housing **(see illustration)**.

3   Installation is a reversal of removal but the actuating button on the lock cylinder may need to be depressed before the buzzer switch can be installed. Make sure that the buzzer switch has its contacts towards the top of the steering column and that the formed end of its securing clip is positioned round the lower part of the switch.

---

20 Ignition switch - removal and installation

---

1   As a precaution against theft of the vehicle, the ignition switch is located along the side of the steering column and remotely controlled by a rod and a rack assembly from the ignition lock cylinder.

1970 through 1976 models

Refer to illustration 20.3

2   To remove the ignition switch, the steering column must either be removed or lowered and well supported. There is no need to remove the steering wheel.

3   Before removing the switch (two screws) set it to the 'LOCK' position **(see illustration)**. If the lock cylinder and actuating rod have already been removed, the 'LOCK' position of the switch can be determined by inserting a screwdriver in the actuating rod slot and then moving the switch slide up until a definite stop is felt then moving it down one detent.

4   Installation is a reversal of removal but again the switch must be in the 'LOCK' position and the original or identical type securing screws must be used. Longer or thicker screws could cause the collapsible design of the steering column to become inoperative.

1977 and later models

5   On these models, the switch should be set to the 'OFF-UNLOCKED' position before removal. If the lock cylinder has already been removed, the switch actuating rod should be

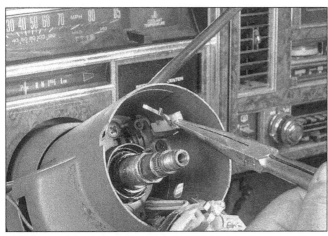 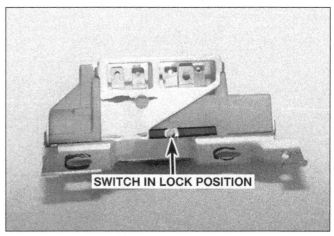

19.2 Pull out the buzzer switch and clip with needle-nose pliers   **20.3 Ignition switch mounting details**

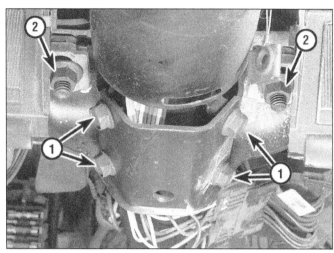

21.10 If the steering column is to be replaced, loosen the bracket bolts (1) before removing the column; The column bracket is retained to the dash with two nuts (2) - be sure not to overtighten these nuts on reassembly

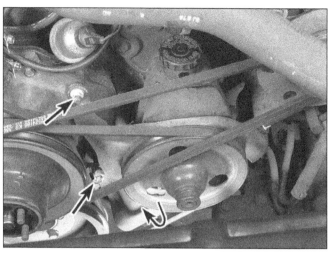

24.3 Remove the power steering pump mounting bolts and nuts (arrows)

pulled up until a definite stop is felt then pushed down two detents.

6   Before installing the switch, set it to the 'OFF-UNLOCK' position and set the gearshift lever in neutral. Setting the switch should be carried out in the following way. Move the switch slide two positions to the right from 'ACCESSORY' to 'OFF/UNLOCK.' Fit the actuator rod into the slide hole and assemble to the steering column using two screws. These screws must be of the original type and only tighten the lower one to 35 in-lb torque.

21 Steering column - removal and installation

Refer to illustration 21.10

Removal

1   Disconnect the battery ground cable.
2   Remove the steering wheel (Section 14).
3   Remove the nuts and washers securing the flanged end of the steering shaft to the coupling. On later models it will also be necessary to loosen the front dash mounting plates.
4   Disconnect the transmission control linkage or backdrive linkage from the column shift levers.
5   Disconnect the steering column harness at the connector, the neutral start switch and the back-up lamp switch connectors.
6   *Pre-1975 models:* Remove the screws which secure the 2 halves of the floorpan cover, then remove the screws securing the halves and seal to the floorpan. Remove the covers.
    1975 and later models: Remove the column plate-to-floorpan securing screws.
7   Remove the trim cover screws from the panel below the column, then remove the trim cover.
8   Remove the shift position indicator cable, where applicable.

9   Move the front seat fully rearwards for accessibility.
10   Remove the column bracket-to-dash panel (capsule) nuts **(see illustration)** then remove the assembly from the vehicle.

Installation

11   If removed, attach the bracket to the column and tighten the four bolts to the specified torque.
12   Move the column into place and position the flange to the flexible joint, then fit the lockwashers and nuts.
13   Raise the column and loosely install the two bracket-to-dash panel (capsule) nuts to hold the column in position.
14   Install the column-to-floorpan bolts, tightening them to the specified torque.
15   Tighten the two capsule nuts to the specified torque.
16   Connect the shift linkage (where applicable).
17   Connect and adjust the shift indicator cable.
18   Reconnect the electrical connectors.
19   Install the trim panels.
20   Install the steering wheel (Section 14).
21   The remainder of installation is the reverse of removal.

22 Power steering - maintenance and adjustment (in-vehicle)

1   Every 6,000 miles, check the fluid level in the power steering pump reservoir.
2   If the fluid is at normal -operating temperature (150° F/66° C) the level should be between the 'HOT' and 'COLD' marks.
3   If the fluid is cold (70° F/21° C approximately) the level should be between the 'ADD' and 'COLD' marks,
4   Add specified fluid as necessary to bring the level to the appropriate mark on the dipstick.
5   Check and, if necessary, adjust the

drivebelt tension (see Chapter 2, Section 7).
6   No adjustments can satisfactorily be carried out to the power steering gear with the unit in the vehicle due to the confusing effects that the hydraulic fluid has on turning torque and endplay tolerances.

23 Power steering gear - removal and installation

1   The procedure is similar to that described previously for the manual type except that the hydraulic hoses must be disconnected from the steering gear housing.
2   After detaching the lines, plug the fittings and the fluid inlet and outlet holes in the housing.
3   When installation is complete, bleed the system as described in Section 25.

24 Power steering pump - removal and installation

Refer to illustration 24.3
1   Disconnect the hydraulic hoses either from the steering gear and keep them in the raised position to prevent the fluid draining away until they can be plugged.
2   Remove the pump drivebelt by loosening the alternator mountings and pushing it in towards the engine.
3   Unscrew and remove the pump mounting bolts and braces and remove the pump **(see illustration)**.
4   To remove the pump pulley it will require the use of a special puller, available at most auto parts stores. Another tool is available to press the pulley onto the shaft, but this can also be accomplished with the use of a bolt (of the proper diameter and thread pitch to match the hole in the shaft), a nut and washer to fit the bolt, and a socket of the proper diameter to engage the hub of the pulley.

5   Installation is a reversal of removal. Fill the fluid reservoir with the recommended fluid.
6   Prime the pump by turning the pulley counterclockwise until air-bubbles cease to emerge from the fluid when observed through the reservoir opening.
7   Install the drivebelt and adjust its (see Chapter 2, Section 7).
8   Bleed the system as described in Section 25.

25 Power steering hydraulic system - bleeding

1   Following any operation in which the power steering fluid lines have been disconnected, the power steering system must be bled to remove all air and obtain proper steering performance.
2   With the front wheels in the straight ahead position, check the power steering fluid level and, if low, add fluid until it reaches the Cold mark on the dipstick.
3   Start the engine and allow it to run at fast idle. Recheck the fluid level and add more if necessary to reach the Cold mark on the dipstick.
4   Bleed the system by turning the wheels from side-to-side, without hitting the stops. This will work the air out of the system. Keep the reservoir full of fluid as this is done.
5   When the air is worked out of the system, return the wheels to the straight ahead position and leave the vehicle running for several more minutes before shutting it off.
6   Road test the vehicle to be sure the steering system is functioning normally and noise free.
7   Recheck the fluid level to be sure it is up to the Hot mark on the dipstick while the engine is at normal operating temperature. Add fluid if necessary.

26 Front end alignment - general information

Refer to illustration 26.1
    A front end alignment refers to the adjustments made to the front wheels so they are in proper angular relationship to the suspension and the ground. Front wheels that are out of proper alignment not only affect steering control, but also increase tire wear. The front end adjustments normally required are camber, caster and toe-in **(see illustration)**.
    Getting the proper front wheel alignment is a very exacting process, one in which complicated and expensive machines are necessary to perform the job properly. Because of this, you should have a technician with the proper equipment perform these tasks. We will, however, use this space to give you a basic idea of what is involved with front end alignment so you can better understand the

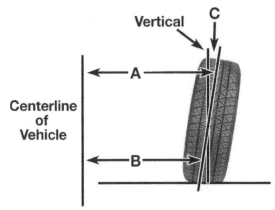

CAMBER ANGLE (FRONT VIEW)

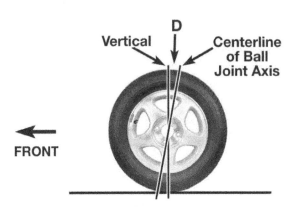

CASTER ANGLE (SIDE VIEW)

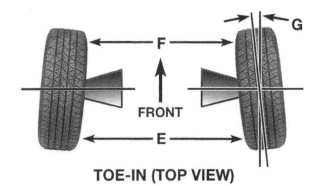

TOE-IN (TOP VIEW)

26.1 Typical front end alignment details

A minus B =C (degrees camber)　　　*E minus F = toe-in (measured in inches)*
D = degrees caster　　　　　　　　*G = toe-in (expressed in degrees)*

process and deal intelligently with the shop that does the work.
    Toe-in is the turning in of the front wheels. The purpose of a toe specification is to ensure parallel rolling of the front wheels. In a vehicle with zero toe-in, the distance between the front edges of the wheels will be the same as the distance between the rear edges of the wheels. The actual amount of toe-in is normally only a fraction of an inch. Toe-in adjustment is controlled by the length of the tie-rods. Incorrect toe-in will cause the tires to wear improperly by making them

scrub against the road surface.
    Camber is the tilting of the front wheels from the vertical when viewed from the front of the vehicle. When the wheels tilt out at the top, the camber is said to be positive (+). When the wheels tilt in at the top the camber is negative (-). The amount of tilt is measured in degrees from the vertical and this measurement is called the camber angle. This angle affects the amount of tire tread which contacts the road and compensates for changes in the suspension geometry when the vehicle is cornering or traveling over an

undulating surface. Camber is adjusted by adding or subtracting shims at the upper control arm pivot mounting points.

Caster is the tilting of the top of the front steering axis from the vertical. A tilt toward the rear is positive caster and a tilt toward the front is negative caster. Caster is adjusted by moving shims from one end of the upper control arm mount to the other.

27 Wheels and tires - general information

Refer to illustration 27.1

Most vehicles covered by this manual are equipped with metric-sized fiberglass or steel belted radial tires **(see illustration)**. Use of other size or type of tires may affect the ride and handling of the vehicle. Don't mix different types of tires, such as radials and bias belted, on the same vehicle as handling may be seriously affected. It's recommended that tires be replaced in pairs on the same axle, but if only one tire is being replaced, be sure it's the same size, structure and tread design as the other.

Because tire pressure has a substantial effect on handling and wear, the pressure on all tires should be checked at least once a month or before any extended trips (see Chapter 1).

Wheels must be replaced if they are bent, dented, leak air, have elongated bolt holes, are heavily rusted, out of vertical symmetry or if the lug nuts won't stay tight.

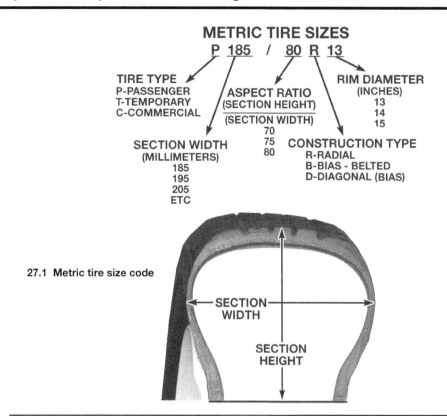

27.1 Metric tire size code

Wheel repairs that use welding or peening are not recommended.

Tire and wheel balance is important in the overall handling, braking and performance of the vehicle. Unbalanced wheels can adversely affect handling and ride characteristics as well as tire life. Whenever a tire is installed on a wheel, the tire and wheel should be balanced by a shop with the proper equipment.

28 Troubleshooting - suspension and steering

Symptom	Reason
Excessive play in steering	Incorrect gear adjustment
	Worn components
Directional instability	Incorrect tire pressures
	Incorrectly adjusted wheel bearings
	Incorrect caster angle
	Faulty shock-absorber
	Lack of lubricant in steering linkage
Stiff and heavy steering	Lack of lubrication in steering linkage
	Low front tire pressures
	Incorrect wheel alignment
	Power steering (if fitted) inoperative
	Lack of lubricant in steering gear
Wheel shake or vibration	Tire and wheel out of balance
	Worn or loose front wheel bearings
	Worn tie-rod ends
	Worn balljoints
	Incorrect front wheel alignment
	Faulty shock absorber
Excessive pitching or rolling on corners	Faulty shock-absorber or broken spring
	Broken stabilizer bar or link

Power steering systems only

Symptom	Reason
'Kick-back' or loose steering	Air in system
	Excessive 'over-center' lash
	Worn poppet valve
	Gear loose on frame
	Flexible coupling worn
Jerk in steering wheel when turning	Low oil level
	Loose drivebelt
	Insufficient pump pressure
Hard steering	Internal pressure leakage
	Loose drivebelt
	Low fluid level
Poor return (centering) of steering	Valve spool plugged or sticking
	Incorrect steering gear adjustments
Noise from system	Due to loose drivebelt, low oil level or mechanical fault within pump or gear.
	Hissing noise is normal, particularly evident at full lock or during slow speed turning

Chapter 12 Body

Contents

1 General information

The Nova is available in two basic body styles, the 4-door Sedan and the 2-door Coupe. From 1973 on, a Hatchback version of the Coupe is available.

The body and underframe are of a steel, all-welded construction built on the unitized principle.

Certain components are particularly vulnerable to accident damage and can be unbolted and repaired or replaced. Among these parts are the body moldings, bumpers, the hood and trunk lids and all glass.

Only general body maintenance practices and body panel repair procedures within the scope of the do-it-yourselfer are included in this Chapter.

2 Body - maintenance

1 The condition of your vehicle's body is very important, because the resale value depends a great deal on it. It's much more difficult to repair a neglected or damaged body than it is to repair mechanical components. The hidden areas of the body, such as the wheel wells, the frame and the engine compartment, are equally important, although they don't require as frequent attention as the rest of the body.

2 Once a year, or every 12,000 miles, it's a good idea to have the underside of the body steam cleaned. All traces of dirt and oil will be removed and the area can then be inspected carefully for rust, damaged brake lines, frayed electrical wires, damaged cables and other problems. The front suspension and steering components should be greased after completion of this job.

3 At the same time, clean the engine and the engine compartment with a steam cleaner or water soluble degreaser.

4 The wheel wells should be given close attention, since undercoating can peel away and stones and dirt thrown up by the tires can cause the paint to chip and flake, allowing rust to set in. If rust is found, clean down to the bare metal and apply an anti-rust paint.

5 The body should be washed about once a week. Wet the vehicle thoroughly to soften the dirt, then wash it down with a soft sponge and plenty of clean soapy water. If the surplus dirt is not washed off very carefully, it can wear down the paint.

6 Spots of tar or asphalt thrown up from the road should be removed with a cloth soaked in solvent.

7 Once every six months, wax the body and chrome trim. If a chrome cleaner is used to remove rust from any of the vehicle's plated parts, remember that the cleaner also removes part of the chrome, so use it sparingly.

3 Upholstery and carpets - maintenance

1 Remove the carpets or mats and thoroughly vacuum clean the interior of the vehicle every three months or more frequently if necessary.

2 Beat out the carpets and vacuum clean them if they are very dirty. If the upholstery is soiled apply an upholstery cleaner with a damp sponge and wipe off with a clean dry cloth.

4 Vinyl trim - maintenance

Don't clean vinyl trim with detergents, caustic soap or petroleum-based cleaners. Plain soap and water works just fine, with a soft brush to clean dirt that may be ingrained. Wash the vinyl as frequently as the rest of the vehicle.

After cleaning, application of a high quality rubber and vinyl protectant will help prevent oxidation and cracks. The protectant can also be applied to weatherstripping, vacuum lines and rubber hoses, which often fail as a result of chemical degradation, and to the tires.

These photos illustrate a method of repairing simple dents. They are intended to supplement *Body repair - minor damage* in this Chapter and should not be used as the sole instructions for body repair on these vehicles.

1 If you can't access the backside of the body panel to hammer out the dent, pull it out with a slide-hammer-type dent puller. In the deepest portion of the dent or along the crease line, drill or punch hole(s) at least one inch apart . . .

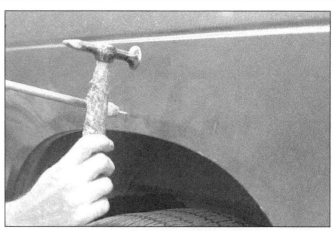

2 . . . then screw the slide-hammer into the hole and operate it. Tap with a hammer near the edge of the dent to help 'pop' the metal back to its original shape. When you're finished, the dent area should be close to its original contour and about 1/8-inch below the surface of the surrounding metal

3 Using coarse-grit sandpaper, remove the paint down to the bare metal. Hand sanding works fine, but the disc sander shown here makes the job faster. Use finer (about 320-grit) sandpaper to feather-edge the paint at least one inch around the dent area

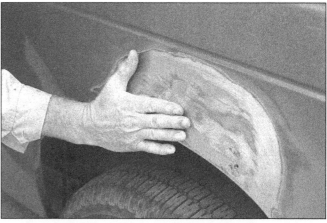

4 When the paint is removed, touch will probably be more helpful than sight for telling if the metal is straight. Hammer down the high spots or raise the low spots as necessary. Clean the repair area with wax/silicone remover

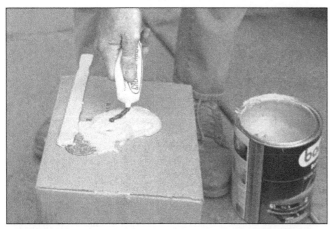

5 Following label instructions, mix up a batch of plastic filler and hardener. The ratio of filler to hardener is critical, and, if you mix it incorrectly, it will either not cure properly or cure too quickly (you won't have time to file and sand it into shape)

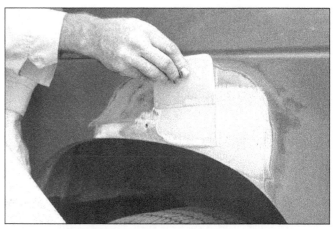

6 Working quickly so the filler doesn't harden, use a plastic applicator to press the body filler firmly into the metal, assuring it bonds completely. Work the filler until it matches the original contour and is slightly above the surrounding metal

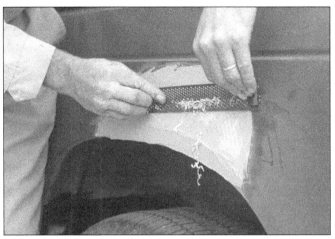

7 Let the filler harden until you can just dent it with your fingernail. Use a body file or Surform tool (shown here) to rough-shape the filler

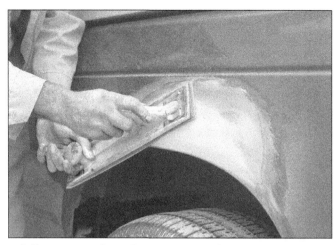

8 Use coarse-grit sandpaper and a sanding board or block to work the filler down until it's smooth and even. Work down to finer grits of sandpaper - always using a board or block - ending up with 360 or 400 grit

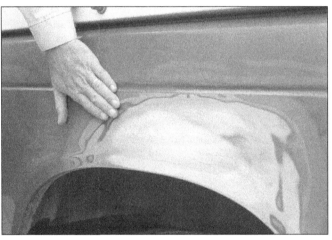

9 You shouldn't be able to feel any ridge at the transition from the filler to the bare metal or from the bare metal to the old paint. As soon as the repair is flat and uniform, remove the dust and mask off the adjacent panels or trim pieces

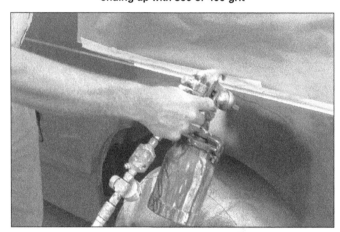

10 Apply several layers of primer to the area. Don't spray the primer on too heavy, so it sags or runs, and make sure each coat is dry before you spray on the next one. A professional-type spray gun is being used here, but aerosol spray primer is available inexpensively from auto parts stores

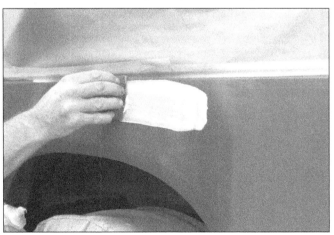

11 The primer will help reveal imperfections or scratches. Fill these with glazing compound. Follow the label instructions and sand it with 360 or 400-grit sandpaper until it's smooth. Repeat the glazing, sanding and respraying until the primer reveals a perfectly smooth surface

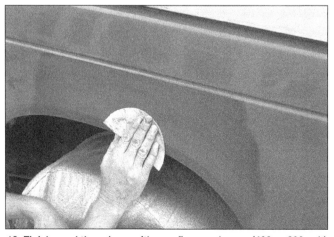

12 Finish sand the primer with very fine sandpaper (400 or 600-grit) to remove the primer overspray. Clean the area with water and allow it to dry. Use a tack rag to remove any dust, then apply the finish coat. Don't attempt to rub out or wax the repair area until the paint has dried completely (at least two weeks)

5   Body repair - minor damage

See photo sequence

1   If the scratch is superficial and does not penetrate to the metal of the body, repair is very simple. Lightly rub the scratched area with a fine rubbing compound to remove loose paint and built up wax. Rinse the area with clean water.

2   Apply touch-up paint to the scratch, using a small brush. Continue to apply thin layers of paint until the surface of the paint in the scratch is level with the surrounding paint. Allow the new paint at least two weeks to harden, then blend it into the surrounding paint by rubbing with a very fine rubbing compound. Finally, apply a coat of wax to the scratch area.

3   If the scratch has penetrated the paint and exposed the metal of the body, causing the metal to rust, a different repair technique is required. Remove all loose rust from the bottom of the scratch with a pocket knife, then apply rust-inhibiting paint to prevent the formation of rust in the future. Using a rubber or nylon applicator, coat the scratched area with glaze-type filler. If required, the filler can be mixed with thinner to provide a very thin paste, which is ideal for filling narrow scratches. Before the glaze filler in the scratch hardens, wrap a piece of smooth cotton cloth around the tip of a finger. Dip the cloth in thinner and then quickly wipe it along the surface of the scratch. This will ensure that the surface of the filler is slightly hollow. The scratch can now be painted over as described earlier in this Section.

Repair of dents

4   When repairing dents, the first job is to pull the dent out until the affected area is as close as possible to its original shape. There is no point in trying to restore the original shape completely as the metal in the damaged area will have stretched on impact and cannot be restored to its original contours. It is better to bring the level of the dent up to a point which is about 1/8-inch below the level of the surrounding metal. In cases where the dent is very shallow, it is not worth trying to pull it out at all.

5   If the back side of the dent is accessible, it can be hammered out gently from behind using a soft-face hammer. While doing this, hold a block of wood firmly against the opposite side of the metal to absorb the hammer blows and prevent the metal from being stretched.

6   If the dent is in a section of the body which has double layers, or some other factor makes it inaccessible from behind, a different technique is required. Drill several small holes through the metal inside the damaged area, particularly in the deeper sections. Screw long, self-tapping screws into the holes just enough for them to get a good grip in the metal. Now the dent can be pulled out by pulling on the protruding heads of the screws with locking pliers.

7   The next stage of repair is the removal of paint from the damaged area and from an inch or so of the surrounding metal. This is easily done with a wire brush or sanding disk in a drill motor, although it can be done just as effectively by hand with sandpaper. To complete the preparation for filling, score the surface of the bare metal with a screwdriver or the tang of a file, or drill small holes in the affected area. This will provide a good grip for the filler material. To complete the repair, see the Section on filling and painting.

Repair of rust holes or gashes

8   Remove all paint from the affected area and from an inch or so of the surrounding metal using a sanding disk or wire brush mounted in a drill motor. If these are not available, a few sheets of sandpaper will do the job just as effectively.

9   With the paint removed, you will be able to determine the severity of the corrosion and decide whether to replace the whole panel, if possible, or repair the affected area. New body panels are not as expensive as most people think, and it is often quicker to install a new panel than to repair large areas of rust.

10   Remove all trim pieces from the affected area except those which will act as a guide to the original shape of the damaged body, such as headlight shells, etc. Using metal snips or a hacksaw blade, remove all loose metal and any other metal that is badly affected by rust. Hammer the edges of the hole inward to create a slight depression for the filler material.

11   Wire brush the affected area to remove the powdery rust from the surface of the metal. If the back of the rusted area is accessible, treat it with rust-inhibiting paint.

12   Before filling is done, block the hole in some way. This can be done with sheet metal riveted or screwed into place, or by stuffing the hole with wire mesh.

13   Once the hole is blocked off, the affected area can be filled and painted. See the following subsection on filling and painting.

Filling and painting

14   Many types of body fillers are available, but generally speaking, body repair kits which contain filler paste and a tube of resin hardener are best for this type of repair work. A wide, flexible plastic or nylon applicator will be necessary for imparting a smooth and contoured finish to the surface of the filler material. Mix up a small amount of filler on a clean piece of wood or cardboard (use the hardener sparingly). Follow the manufacturer's instructions on the package, otherwise the filler will set incorrectly.

15   Using the applicator, apply the filler paste to the prepared area. Draw the applicator across the surface of the filler to achieve the desired contour and to level the filler surface. As soon as a contour that approximates the original one is achieved, stop working the paste. If you continue, the paste will begin to stick to the applicator. Continue to add thin layers of paste at 20-

minute intervals until the level of the filler is just above the surrounding metal.

16   Once the filler has hardened, the excess can be removed with a body file. From then on, progressively finer grades of sandpaper should be used, starting with a 180-grit paper and finishing with 600-grit wet-or-dry paper. Always wrap the sandpaper around a flat rubber or wooden block, otherwise the surface of the filler will not be completely flat. During the sanding of the filler surface, the wet-or-dry paper should be periodically rinsed in water. This will ensure that a very smooth finish is produced in the final stage.

17   At this point, the repair area should be surrounded by a ring of bare metal, which in turn should be encircled by the finely feathered edge of good paint. Rinse the repair area with clean water until all of the dust produced by the sanding operation is gone.

18   Spray the entire area with a light coat of primer. This will reveal any imperfections in the surface of the filler. Repair the imperfections with fresh filler paste or glaze filler and once more smooth the surface with sandpaper. Repeat this spray-and-repair procedure until you are satisfied that the surface of the filler and the feathered edge of the paint are perfect. Rinse the area with clean water and allow it to dry completely.

19   The repair area is now ready for painting. Spray painting must be carried out in a warm, dry, windless and dust-free atmosphere. These conditions can be created if you have access to a large indoor work area, but if you are forced to work in the open, you will have to pick the day very carefully. If you are working indoors, dousing the floor in the work area with water will help settle the dust which would otherwise be in the air. If the repair area is confined to one body panel, mask off the surrounding panels. This will help minimize the effects of a slight mismatch in paint color. Trim pieces such as chrome strips, door handles, etc., will also need to be masked off or removed. Use masking tape and several thickness of newspaper for the masking operations.

20   Before spraying, shake the paint can thoroughly, then spray a test area until the spray painting technique is mastered. Cover the repair area with a thick coat of primer. The thickness should be built up using several thin layers of primer rather than one thick one. Using 600-grit wet-or-dry sandpaper, rub down the surface of the primer until it is very smooth. While doing this, the work area should be thoroughly rinsed with water and the wet-or-dry sandpaper periodically rinsed as well. Allow the primer to dry before spraying additional coats.

21   Spray on the top coat, again building up the thickness by using several thin layers of paint. Begin spraying in the center of the repair area and then, using a circular motion, work out until the whole repair area and about two inches of the surrounding original paint is covered. Remove all masking material 10 to 15 minutes after spraying on the final coat of paint. Allow the new paint at

least two weeks to harden, then use a very fine rubbing compound to blend the edges of the new paint into the existing paint. Finally, apply a coat of wax.

6   Body repair - major damage

1   Major damage must be repaired by an auto body shop specifically equipped to perform unibody repairs. These shops have the specialized equipment required to do the job properly.
2   If the damage is extensive, the body must be checked for proper alignment or the vehicle's handling characteristics may be adversely affected and other components may wear at an accelerated rate.
3   Due to the fact that all of the major body components (hood, fenders, etc.) are separate and replaceable units, any seriously damaged components should be replaced rather than repaired. Sometimes the components can be found in a wrecking yard that specializes in used vehicle components, often at considerable savings over the cost of new parts.

7   Detachable panels - replacement

On all models, the detachable body panels can be removed with the aid of a screwdriver and wrench. During the production run of the Nova there have been some minor alterations to the method of attachment of some panels. The aim of this Section is to outline the basic procedures for the major parts.

Radiator support panel

1   Raise the hood, disconnect the battery cables and remove the battery.
2   Remove the front bumper (Section 8).

3   Remove the grille and related parts (Section 9).
4   Disconnect the electrical leads and the washer bottle from the radiator support.
5   Remove the screw connecting the battery tray to the radiator (on some models it is necessary to remove the tray completely).
6   Where applicable, remove the radiator shroud.
7   On some 1970 and earlier models it will be necessary to remove the radiator (Chapter 2).
8   Remove the screws securing the support to the frame, skirts and fenders.
9   Where applicable, remove the cushions, washers and fittings from the frame bracket attachment to the radiator support.
10   Lift out the radiator support.
11   Installation is the reverse of the removal procedure, but it is recommended that headlamp alignment is checked on completion.

Fender and fender skirt assembly

12   Disconnect the battery ground lead. If the right-hand fender and fender skirt are to be removed, remove the battery also.
13   Raise the vehicle using a hoist, then remove the wheel. Remove the front bumper (Section 8).
14   Remove the hood and hinges (Section 12).
15   Disconnect all electrical wiring, clips and items attached to the assembly.
16   Disconnect the filler panel, header panel; valance panel and brace.
17   Remove the headlamp and bezel (Chapter 10).
18   Remove two headlamp retainer support screws to the fender where applicable.
19   Remove the fender and skirt assembly attachment screws. Note the number of shims fitted to assist in installation.
20   Remove the side-marker lamp connections as the assembly is being carefully lifted

away from the vehicle.
21   Installation is a reversal of the removal procedure, but the fender and skirt should be guided in at the bottom adjacent to the door first of all. Adjust the assembly before finally tightening the screws using the original shims, then add or remove shims as necessary.

8   Bumpers - removal and installation

Note: *Vehicles manufactured from 1973 onwards are fitted with energy absorbing devices between the bumpers and vehicle frame to meet Federal Safety Standards under impact load. These are filled with hydraulic fluid and gas under high pressure and should not be interfered with or subjected to heat. As an emergency measure, to relieve the gas pressure where accident damage has occurred, or prior to disposal, the pressure may be relieved by drilling a hole through the caution label on the energy absorber body. Care should be taken to protect the eyes during such an operation.*

Front bumper

Refer to illustrations 8.1, 8.5a and 8.5b

1972 and earlier models

1   Disconnect the electrical wiring then remove the bolts retaining the brackets and braces to the frame **(see illustration)**.
2   Remove the bumper assembly.
3   Disconnect the lamp assemblies, and bumper brackets and braces as necessary.
4   Installation is a reversal of the removal procedure taking care to align the bumper to the vehicle front end.

1973 and later models

5   Remove 8 nuts (4 each side) which attach the bumper to the energy absorber **(see illustrations)**.

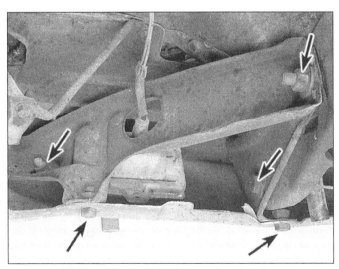

8.1 Front bumper bracket attachment points

8.5a Mounting bolts for the energy absorber used on 1973 and later model bumpers (front shown)

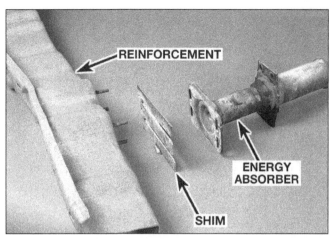

8.5b Energy absorber and bumper reinforcement details

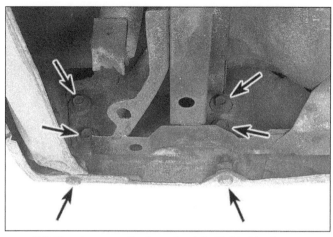

8.9 Rear bumper attachment points (right side shown, left side identical)

6   Remove the nuts and bolts attaching the bumper reinforcement assembly to the bumper face bar, as necessary. Where applicable, remove the bolts and nuts retaining the protective strips and bumper guards.

7   To remove the energy absorber, remove 4 bolts connecting it to the frame bracket at the front and the nut connecting it to the frame bracket at the rear. Slide the energy absorber from the frame bracket to remove.

8   Installation is a reversal of the removal procedure.

Rear bumper

Refer to illustration 8.9

1972 and earlier models

9   Disconnect the license plate wiring inside the trunk then remove the bolts fastening the bumper brackets to the frame side rails **(see illustration)**.

10   Remove the fiber optic conductor from the license plate housing and lift away the bumper.

11   The license plate, door assembly and brackets and braces can be unbolted as necessary.

12   Installation is a reversal of the removal procedure.

1973 and later models

13   The procedure for removal and installation of the rear bumper is similar to that already given for the front bumper.

9   Radiator grille - removal and installation

1972 and earlier models

Refer to illustration 9.4

1   Raise the hood, disconnect the battery cables and remove the battery.

2   Remove the hood lock catch retaining screws; remove the catch.

3   Remove the front bumper (Section 8).

4   Remove the hood lock support-to-filler panel screws **(see illustration)**.

5   Remove the headlamp bezel screws; remove the bezel.

6   Remove the filler panel to fender screws and the grille outer support screws.

7   Remove the grille, grille supports, filler panel and filler panel support as an assembly, then disassemble as necessary.

8   Installation is the reverse of the removal procedure.

1973 and later models

1   Remove the 8 screws (1973 and 1974 models) or 6 screws (1975 and later models) attaching the grille to the grille support. Lift away the grille.

2   Installation is the reverse of the removal procedure.

10   Windshield and fixed glass - replacement

Replacement of the windshield and fixed glass requires the use of special fast-setting adhesive/caulk materials and some specialized tools and techniques. These operations should be left to a dealer service department or a shop specializing in glass work.

11   Rear quarter windows - removal and installation

1   Where adhesive caulked quarter windows are fitted, the same basic instructions apply as for windshields and other fixed glass, and replacement is best left to a specialist.

2   Where the rear quarter windows are retained in rubber channels, proceed as described in the following paragraphs.

3   Remove the securing screws which retain the mouldings and the body lock pillar cover around the quarter window.

4   Using a blunt blade, break the seal between the rubber channel and the body flange.

5   Push the glass together with the rubber channel from the inside of the vehicle outwards until it is free from the body. Remove the trim from the channel.

6   Scrape all the old sealer from the window aperture of the body and also examine the rubber channel. If this cannot be cleaned satisfactorily or it is cut or has hardened, replace it.

7   Install the rubber channel to the glass and then lay a length of cord in the body seating

9.4 Grille upper mounting points

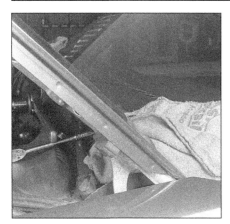

12.1 Pad the back corners of the hood with cloths so the cowl panel won't be damaged if the hood accidentally swings rearward

groove so that the ends cross over at the center of the bottom run of the channel. Insert the trim into the channel.

8   Apply a bead of sealer around the window opening.

9   Position the window/channel assembly to the body and while an assistant applies pressure from the outside, pull the cord (the ends of which will now be hanging down inside the vehicle) so that the lip of the rubber channel is engaged over the body flange.

10   Using a pressure applicator, insert black sealer between the outer surface of the glass and the outer lip of the rubber channel.

11   Clean off any excess sealer and install the moulding and lock pillar cover.

12   Hood - removal, installation and adjustment

Refer to illustrations 12.1 and 12.4
Note: *The hood is heavy and somewhat awkward to remove and install - at least two people should perform this procedure.*

Removal and installation

1   Use blankets or pads to cover the cowl area of the body and the fenders **(see illustration)**. This will protect the body and paint as the hood is lifted off.

2   Scribe or paint alignment marks around the bolt heads to insure proper alignment during installation.

3   Disconnect any cables or wire harnesses which will interfere with removal.

4   Have an assistant support the weight of the hood. Remove the hinge-to-hood nuts or bolts **(see illustration)**.

5   Lift off the hood.

6   Installation is the reverse of removal.

Adjustment

7   Fore-and-aft and side-to-side adjustment of the hood is done by moving the hood in relation to the hinge plate after loosening the bolts or nuts.

8   Scribe a line around the entire hinge plate so you can judge the amount of movement.

9   Loosen the bolts or nuts and move the hood into correct alignment. Move it only a little at a time. Tighten the hinge bolts or nuts and carefully lower the hood to check the alignment.

10   If necessary after installation, the entire hood lock assembly can be adjusted up and down as well as from side to side on the radiator support so the hood closes securely and is flush with the fenders. To do this, scribe a line around the hood lock mounting bolts to provide a reference point. Then loosen the bolts and reposition the latch assembly as necessary. Following adjustment, retighten the mounting bolts.

11   Finally, adjust the hood bumpers on the radiator support so the hood, when closed, is flush with the fenders.

12   The hood latch assembly, as well as the hinges, should be periodically lubricated with white lithium-base grease to prevent sticking and wear.

12.4 Typical hood and hinge mounting bolt locations (arrows)

13   Hood lock mechanism - removal and installation

Refer to illustrations 13.2a and 13.2b

1971 and earlier models

1   Scribe around the fitted position of the lock plate on the hood to aid alignment when installing.

2   Remove the lock and catch plate fixing screws on the hood, header panel, radiator support panel and catch support as necessary **(see illustrations)**.

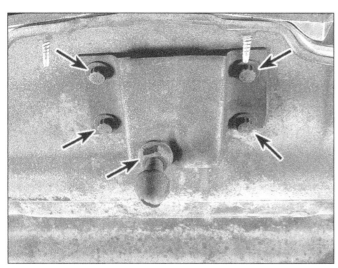

13.2a Mounting details of the hood lock catch plate and bolt

13.2b Hood lock attachment points (note that a 12-point socket is required)

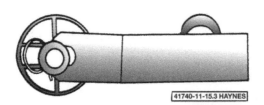

15.2 A special tool like this one (available at most auto parts stores) makes it easier to remove the spring clip retainer on the window crank handle

3   Installation is the reverse of the removal procedure. Adjust the hood lock bolt so that the hood is just flush with the top surface of the header panel and the hood bumpers are just compressed when fully closed.

1972 and later models

1   Remove the catch plate assembly by removing the screws retaining the catch to the radiator support, center support and tie-bar or grille upper bar.
2   Disconnect the hood release cable.
3   Scribe a line around the lockplate to aid alignment when installing, then remove the screws retaining it to the hood. Remove the lockplate.
4   Installation is the reverse of the removal procedure. Adjust the hood lock bolt so that the hood engages securely when closed and the hood bumpers are slightly compressed.
Note: *In the event of cable failure, the hood can be released by reaching up between the valance panel and radiator support and removing the 2 screws which retain the anti-theft shield. The latch can then be manually operated. In the event of one half of the release cable failing, it is recommended that a complete new cable (both halves) is fitted as a replacement, since damage can occur when trying to separate the two halves of the connector housing.*

14 Cowl vent grille - removal and installation

1   Where applicable, raise the hood, remove the windshield wiper arms and disconnect the washer hoses from the tubes.
2   Remove the screws securing the washer tubes to the cowl vent grille, then pull the tubes out from under the rubber moulding and remove.
3   Remove the cowl vent grille retaining screws and lift away the grille, leaving the rubber moulding in place.
4   Installation is the reverse of the removal procedure.

15 Door trim panel - removal and installation

Refer to illustrations 15.2 and 15.4
1   Before the trim panel can be removed it is necessary to remove the screws retaining the door pull handle. On certain models the attachment screws will also have to be removed from the armrest also. Where snap-

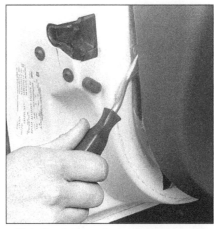

15.4 Carefully pry loose the door panel clips around the edge of the panel - pry only in the area of the clips to prevent damage to the panel

on escutcheons are used, these must be carefully pried off.
2   Remove the screws which retain the door handles and the window regulator handle as necessary. On some models it may be necessary to use a special tool (available at most auto parts stores) to disengage the window crank retaining clip after the door trim panel has been pressed in **(see illustration)**. The handle will then pull off.
3   Remove the door inside locking rod knob.
4   Remove the door trim panel from all the clips around the perimeter using a suitable tool to pry it away **(see illustration)**.
5   Lift the trim panel upwards and rearwards to disengage it from the inner panel.
6   Installation is the reverse of the removal procedure.

16 Doors - removal and installation

Bolted-on hinge type

1   Prior to removal of doors, scribe around the fitted position of the hinge on the door.
2   Using an offset wrench, and with an assistant supporting the weight of the door, remove the door hinge retaining bolts.
3   Having removed the door, it is now possible to remove the hinge from the hinge pillar, but again scribe around the fitted position first.
4   Refitting doors and hinges is the reverse of the removal procedure. If necessary the hinge positions may be altered slightly to ensure correct alignment.

Note: *When refitting the door, ensure that the lock striker fork-bolt engages with the striker. If necessary the position of the fork bolt can be adjusted by slackening with a suitable wrench, repositioning, then retightening.*

Welded hinge type

5   To remove a door, open it fully and support its lower edge on blocks covered with a cloth pad.
6   Remove the 'E' rings from the ends of the upper and lower hinge pins.
7   Drive a forked wedge between the head of the upper hinge pin and the hinge which will partly extract the pin. Repeat the operation on the lower hinge pin and then tap both pins completely from the hinges and remove the door.
8   Although the door side hinge strap is supplied as a replacement part complete with securing bolts, the body side hinge strap is only supplied as a replacement for welding into position. It is not recommended that cutting off the old hinge straps or installing the new ones is a job for the home mechanic due to the need for precise alignment of the new components.
9   Installation of the hinge pins is a reversal of removal.

17 Front door lock assembly - removal and installation

1   To remove the lock cylinder, remove the trim panel and the water deflector (Sections 15 and 22). Raise the door window then use a screwdriver to slide the lock cylinder retaining clip out of engagement **(see illustration).**, Installation is the reverse of the removal procedure.
2   To remove the lock assembly, raise the window, remove the trim panel and water deflector (Sections 15 and 22).
3   Working through the large access hole, disengage the remote control to lock connecting rod (spring clip). On Coupe models it may be necessary to disengage the inside locking rod from the lock which can be achieved by sliding the plastic retaining sleeves towards each other.
4   Remove the 3 screws securing the lock to the door lock pillar; remove the assembly from the door. On four-door models the inside locking rod must be removed from the lock after removal of the lock assembly.
5   Installation is the reverse of the removal procedure.

18 Front door ventilator - removal and installation

1   Remove the door trim panel and water deflector (Sections 15 and 22).
2   Raise the window fully then loosen the down stop support attaching bolt; remove the support.
3   Lower the window and remove the ventilator division channel lower adjusting stud

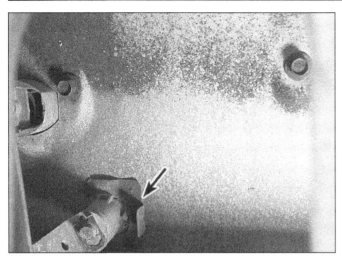

17.1 Door lock cylinder retaining clip (arrow)

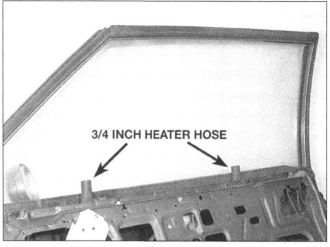

3/4 INCH HEATER HOSE

20.2 Use two pieces of 3/4-inch heater hose, wedged between the glass and the door frame, to hold the glass in the raised position

nut and the ventilator to outer door panel attaching screw.
4   Remove the ventilator attachment screws from the door window frame.
5   Lift the ventilator rearwards and upwards until the lower forward corner is free of the door frame.
6   Rotate the ventilator assembly in an outboard movement and remove it from the door window frame.
7   Installation is the reverse of the removal procedure. If necessary slight adjustment can be made at the lower adjustment stud, either fore-and-aft by loosening the nut or in-and out by loosening the nut and screwing the stud in or out.

19 Front door window assembly - removal and installation

Window assembly with ventilator

1   Remove the front door ventilator (Section 18).
2   Slide the window lower sash channel cam off the window regulator lift arm and balance rollers. Remove the window inboard of the door upper frame.
3   Installation is the reverse of the removal procedure, but it may be necessary to adjust the ventilator position (Section 18) for correct alignment. On Coupe versions the inner panel cam can be adjusted to correct for a rotated (cocked) window, if necessary.

Window assembly without ventilator

1   Remove the door trim panel and water deflector (Sections 15 and 22).
2   Use a flat-bladed tool to carefully pry away the window belt sealing strip from its retaining clips, taking great care not to damage the vehicle door paintwork.

3   Lower the window to the half down position and remove the inner panel cam (2 bolts).
4   Remove the front glass nun channel (2 bolts) on the leading edge of the door.
5   Lower the front edge of the glass and slide the window lower sash channel cam off the window regulator lift arm rollers. Remove the window inboard of the door upper frame.
6   Installation is the reverse of the removal procedure. Adjust the window if it is rotated (cocked) at the inner panel cam. If necessary adjust the down-travel at the down travel stop.

20 Front door window regulator - removal and installation

Refer to illustration 20.2

Window assembly with ventilator

1   Remove the door trim panel and inner panel water deflector (Sections 15 and 22).
2   Secure the window in the fully up position with adhesive tape fastened to the glass and wrapped over the frame, or shove two pieces of 3/4-inch heater hose between the window and the door frame to hold it in place **(see illustration)**.
3   On 2-door vehicles, remove the inner panel cam (2 bolts).
4   Remove the ventilator division channel lower adjustment stud and nut, and the window regulator attaching bolts.
5   Press the ventilator division channel outboard to allow disengagement of the regulator spindle from the inner panel. Slide the regulator balance arm roller out of the lower sash channel cam at the front.
6   Remove the regulator through the large access hole.
7   Installation is the reverse of the removal procedure.

Window assembly without ventilator

1973 and 1974 models

8   Proceed as described in paragraphs 1, 2, 3, 6 and 7 of the previous sub-Section.

1975 and later models

9   Proceed as described in paragraphs 1, 2 and 3 for window assembly with ventilator.
10   Drive out the center rivet pin with a suitable punch and drill out 4 regulator attachment rivets using a 1/4 inch drill. Remove the regulator.
11   When installing the new regulator, attach it to the inner panel with screws and nuts. Where there are clinch nuts on the back plate, place a J-nut over each attaching hole in the back plate.
12   Replace the inner panel cam, water deflector and trim panel.

21 Front door inside locking rod - removal and installation

Coupe models

1   Remove the door trim panel.
2   Peel back the water deflector sufficiently to gain access to the inside locking rod retainers.
3   Slide the plastic retainers towards each other then disengage the rod from the lock. Lower the locking rod through the beltline to remove it.
4   Installation is the reverse of the removal procedure.

Sedan models

5   Proceed as described in the previous sub-Section, but at paragraph 3 disengage the spring clip securing the locking rod to the door lock locking lever.

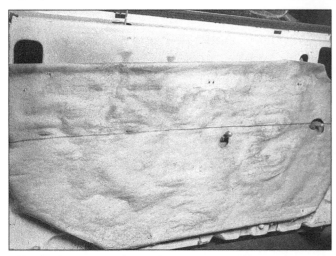

22.1 Typical water deflector; body putty should be used around the perimeter to prevent leaks into the inner door

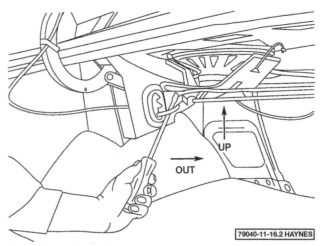

28.1 Using a piece of pipe or a large screwdriver, pry the torque rod end out of the notch, the reposition it as required

22   Inner panel water deflector - removal and installation

1   Water deflectors are fitted beneath the trim panels to prevent the entry of water into the vehicle body **(see illustration)**.
2   They are retained with a waterproof tape at the lower and upper edges and a special body caulking compound is used to obtain a seal.
3   The deflector can easily be prised off, but when refitting ensure that all old caulking material is removed and that the deflector is properly sealed before fitting the trim panel.

23   Rear door stationary ventilator - removal and installation

1   Remove the door trim panel and inner panel water deflector (Sections 15 and 22).
2   Remove the door window lower stop (rubber bumper) from the down stop support bracket on the door inner panel.
3   Remove the ventilator division channel lower adjusting stud and nut.
4   Carefully lower the door window and remove the division channel to door upper frame attaching screw. Where applicable, remove the division channel to door inner panel attaching screw (at the belt).
5   Rotate the upper section of the channel forwards and inwards, and remove from the door.
6   Pull the stationary ventilator window forwards and remove from the door.
7   Installation is the reverse of the removal procedure.

24   Rear door window assembly - removal and installation

1   Remove the door trim panel and inner water deflector (Sections 15 and 22).

2   Remove the stationary ventilator (Section 23).
3   Slide the window regulator lift arm roller out of the window lower sash channel cam and remove the glass inboard of the door upper frame.
4   Installation is the reverse of the removal procedure. If necessary, the ventilator adjustment channel can be repositioned to alleviate a binding window.

25   Rear door window regulator - removal and installation

1974 and earlier models

1   Remove the door, trim panel and inner panel water deflector (Sections 15 and 22).
2   Remove the inside hocking rod to lock connecting link bolt; disconnect the locking rod at the lock.
3   Retain the window in the fully-up position with adhesive tape fastened to the glass and wrapped over the frame.
4   Remove the regulator attaching bolts then slide the regulator lift arm roller out of the lower sash channel cam. Remove the regulator through the large access hole.
5   Installation is the revere of the removal procedure.

1975 and later models

1   Proceed as described in paragraphs 1, 2 and 3 for pre-1975 models.
2   Drive out the center rivet pin with a suitable punch and drill out the 4 attaching rivets using a 1/4-inch drill.
3   Slide the regulator lift arm roller out of the lower sash channel cam and remove the regulator through the large access hole.
4   When installing the new regulator, attach it to the inner panel with screws and nuts. Place J-nut over each attaching hole in the back plate.

26   Rear compartment lid lock cylinder - removal and installation

1969 and 1970 models

1   Open the rear compartment and remove the lock cylinder retainer screw(s).
2   Pull the retainer down or away from the lock cylinder and remove the cylinder from the body.
3   Installation is the reverse of the removal procedure.

1971 and later models

4   The procedure for later models is basically as described in the previous sub-Section, but in some instances the retainer is secured with stud nuts or rivets.
5   When drilling out rivets, use a 1/8-inch drill, taking care not to enlarge the rivet hole. A 1/8 x 5/16 inch 'pop' rivet is suitable for a replacement when installing.

27   Rear compartment lid lock and striker - removal and installation

1   Remove the lid lock cylinder (Section 26).
2   On all versions, the lid lock is retained by bolts which can readily be removed with a wrench.
3   When refitting, ensure that the lid lock is correctly aligned before finally tightening the bolts.
4   On all versions the striker is retained with bolts or screws. Before removing a striker, scribe around the adjacent panel to facilitate fitment in the original position. If necessary, adjustment can be made by repositioning.

28 Rear compartment torque rods - adjustment

Refer to illustration 28.1

1 Adjustment of the torque rods can be carried out by repositioning the ends of the torque rods in different notches. The use of a special tool is recommended, but if the tool is not available, a prybar or a piece of pipe fitted over the end of the rod can be used **(see illustration)**.

2 Repositioning the end of the rod in a lower notch will increase the effort required to raise the lid and vice versa.

29 Rear compartment lid (except hatchback) - removal and installation

1 Open the lid and place protective covering along the rear compartment opening edges to prevent damage to the paint.

2 Where applicable disconnect the wiring harness.

3 Scribe around the hinge straps on the inner panel.

4 With the aid of an assistant, remove the attaching bolts on the lid and lift it away.

5 Installation is the reverse of the removal procedure. Adjustment in lateral and fore-and-aft planes is carried out at the hinges by repositioning. Up-and-down adjustment is carried out by the use of shims between the hinge strap and lid assembly. If necessary adjustment of the lock striker should also be carried out.

30 Hatchback rear compartment lid - removal and installation

1973 and 1974 models

Warning: *Do not attempt to loosen or remove a counterbalance support assembly unless the lid is fully open, or personal injury may result. Do not attempt to disassemble a counterbalance support assembly under any circumstances.*

1 Fully open the lid then place protective covering between the outer ends of the lid and roof panel.

2 With help from an assistant, remove the lid-side attaching nuts for both counterbalance support assemblies and disengage the support assemblies from the lid-side anchor plate studs. Rest the assemblies on the compartment side panel trim.

3 Using a 3/16 inch diameter rod placed on the pointed end of the hinge pins, strike the rod firmly to shear the retaining clip tabs. Next drive out the pin.

4 When installing, fit new retaining rings into the notches in the pins so that the tabs face forwards to the head of the pin.

5 The remainder of the installation procedure is the reverse of removal, but when installing the counterbalance support assembly either use new 'torque tight' nuts or use a thread locking compound.

1975 and later models

Warning: *Do not attempt to loosen or remove a gas operated support assembly unless the lid is fully open, or personal injury may result. Do not attempt to disassemble a gas operated support assembly under any circumstances.*

6 Fully open the lid and remove the upper garnish moulding.

7 Place protective covering between the outer ends of the lid and roof panel.

8 With help from an assistant, carefully pry off the lid-side retaining clips from both gas operated support assemblies. Disengage the support from the lid-side attaching ball then allow the assemblies to rest on the side panel compartment trim.

9 Scribe round the location of the hinge attachments on the lid inner panel then remove the hinge screws. Remove the lid.

10 Installation is the reverse of the removal procedure. If necessary adjustment can be carried out at the hinges by repositioning or by the use of shims. All adjustments must be made with the lid fully open or the support assemblies removed.

31 Heater - general description

1 The heater system is designed to draw in outside air and pass it through a heater core which is fed by coolant from the engine cooling system. Any desired temperature can be maintained by adjustment of the controls which blend heated and unheated air in the required proportions by means of flap valves.

2 A three-speed blower motor is incorporated, the two higher speeds can be used to supplement the airflow when the vehicle is stationary or moving at such a low speed that the intake air pressure through the grille just ahead of the windshield is insufficient. It

should be noted that the blower is automatically switched to low speed when the ignition key is operated, no off position is provided.

3 The distributor combines the functions of mixing the air in the required amounts for temperature control and also deflects the air interior or to the windshield for demisting or defrosting as desired.

32 Heater components - removal and installation

Blower motor

Refer to illustration 32.5

1 Disconnect the battery ground cable then remove all wiring, hoses etc. to the right hand fender skirt.

2 Raise the vehicle and support it securely on jackstands.

3 Remove the 8 rearmost fender skirt-to-fender attaching screws.

4 Pull outwards then down on the fender skirt and place a 2 x 4 inch wooden block between the skirt and fender.

5 Remove the blower motor mounting screws **(see illustration)**. Pass the blower through the fender skirt opening (paragraph 4).

6 Remove the blower wheel retaining nut and separate the wheel from the blower.

7 Installation is the reverse of the removal procedure. Be sure to place a bead of body putty around the blower motor flange to provide a good seal.

Air distribution duct and heater core

Refer to illustration 32.16

Warning: *Wait until the engine is completely cool before beginning this procedure.*

8 Disconnect the battery ground cable.

9 Drain the radiator (refer to Chapter 2, if necessary).

10 Disconnect the heater hoses at the core connections. Temporarily plug the core tubes to prevent coolant spillage.

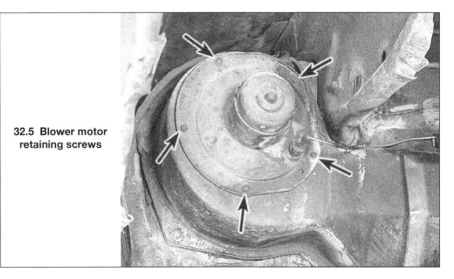

32.5 Blower motor retaining screws

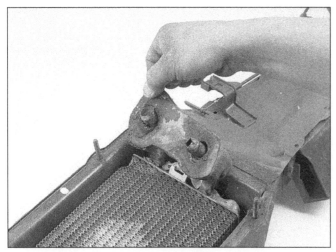

32.16 Make sure the heater core seal is in good shape before installing the heater case. Also, apply a bead of body putty around the perimeter of the case to ensure a good seal against the firewall

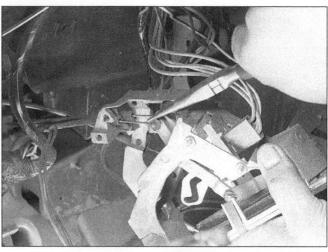

32.21 Remove the clips and detach the control cable(s) from the heater control panel

11   Remove the nuts from the heater distributor studs on the engine side of the firewall.

12   Remove the glove box and door assembly.

13   Using a 1/4 inch drill, drill out the lower right-hand air distribution stud from beneath the dash.

14   Pull the heater case/distribution assembly from the dash panel mounting, taking care not to kink the control cables. When clearance is obtained, disconnect the cables and resistor wires and remove the distribution assembly from the vehicle.

15   Remove the heater core from the distribution assembly.

16   When installing, reverse the removal procedure, ensuring that the hose from the water pump goes to the top heater core pipe. Ensure that the case seal is intact before fitting the core; if necessary fit a new seal **(see illustration)**. Replace the drilled-out stud with a new screw and pall (stamped) nut

17   After filling the cooling system, run the engine and check for coolant leaks.

Control head assembly

18   Disconnect the battery ground cable.

19   Where applicable, remove the radio.

20   Remove the control head to panel instrument screws and lower the control assembly.

21   Remove the control cables and wires then remove the control head **(see illustration)**.

22   Installation is the reverse of the removal procedure.

33 Air conditioner - general description

1   On vehicles equipped with this optional system, both the heating and cooling functions are performed by it. Air entering the vehicle passes through the cooling unit (evaporator) and then around the heating unit so following the 'reheat' system principle.

2   The evaporator cools the air passing through it and by means of its built-in thermostatic switch, controls the operation of the compressor.

3   The system operates by air (from outside or recirculated) entering the evaporator core by the action of the blower, where it receives maximum cooling if the controls are set for cooling. When the air leaves the evaporator, it enters the heater/air conditioner duct assembly end by means of a manually controlled deflector, it either passes through or bypasses the heater core in the correct proportions to provide the desired vehicle interior temperature.

4   Distribution of this air is then regulated by a vacuum actuated deflector and passes through the various outlets according to requirements.

5   If during the cooling operation, the air temperature is cooled too low for comfort, it is warmed to the required level by the heater. When the controls are set to 'HEATING ONLY', the evaporator will cease to function and ambient air will be warmed by the heater in a similar manner to that just described.

6   The main units of the system comprise the evaporator, an engine driven compressor and the condensor.

7   In view of the toxic nature of the chemicals and gases employed in the system, no part of the system must be disconnected by the home mechanic. Due to the need for specialized evacuating and charging equipment, such work should be left to a qualified air conditioning repair facility.

8   The compressor may be dismounted (not disconnected) and moved to one side for engine removal as described in Chapter 1, but this and the checks and maintenance operations described in the next Section should be the limit of work undertaken by the home mechanic.

34 Air conditioner - checks and maintenance

1   Regularly inspect the fins of the condenser (located ahead of the radiator) and if necessary, brush away leaves and bugs.

2   Clean the evaporator drain tubes free from dirt.

3   Check the condition of the system hoes and if there is any sign of deterioration or hardening, have them renewed by your dealer.

4   Every 6,000 miles have the system checked for low oil or refrigerant levels by a qualified air conditioning technician. At the same time the compressor drivebelt should be checked for the proper tension (see Chapter 2, Section 7).

Conversion factors

Length (distance)

Inches (in)	X	25.4	= Millimetres (mm)	X	0.0394	= Inches (in)
Feet (ft)	X	0.305	= Metres (m)	X	3.281	= Feet (ft)
Miles	X	1.609	= Kilometres (km)	X	0.621	= Miles

Volume (capacity)

Cubic inches (cu in; in³)	X	16.387	= Cubic centimetres (cc; cm³)	X	0.061	= Cubic inches (cu in; in³)
Imperial pints (Imp pt)	X	0.568	= Litres (l)	X	1.76	= Imperial pints (Imp pt)
Imperial quarts (Imp qt)	X	1.137	= Litres (l)	X	0.88	= Imperial quarts (Imp qt)
Imperial quarts (Imp qt)	X	1.201	= US quarts (US qt)	X	0.833	= Imperial quarts (Imp qt)
US quarts (US qt)	X	0.946	= Litres (l)	X	1.057	= US quarts (US qt)
Imperial gallons (Imp gal)	X	4.546	= Litres (l)	X	0.22	= Imperial gallons (Imp gal)
Imperial gallons (Imp gal)	X	1.201	= US gallons (US gal)	X	0.833	= Imperial gallons (Imp gal)
US gallons (US gal)	X	3.785	= Litres (l)	X	0.264	= US gallons (US gal)

Mass (weight)

Ounces (oz)	X	28.35	= Grams (g)	X	0.035	= Ounces (oz)
Pounds (lb)	X	0.454	= Kilograms (kg)	X	2.205	= Pounds (lb)

Force

Ounces-force (ozf; oz)	X	0.278	= Newtons (N)	X	3.6	= Ounces-force (ozf; oz)
Pounds-force (lbf; lb)	X	4.448	= Newtons (N)	X	0.225	= Pounds-force (lbf; lb)
Newtons (N)	X	0.1	= Kilograms-force (kgf; kg)	X	9.81	= Newtons (N)

Pressure

Pounds-force per square inch (psi; lbf/in²; lb/in²)	X	0.070	= Kilograms-force per square centimetre (kgf/cm²; kg/cm²)	X	14.223	= Pounds-force per square inch (psi; lbf/in²; lb/in²)
Pounds-force per square inch (psi; lbf/in²; lb/in²)	X	0.068	= Atmospheres (atm)	X	14.696	= Pounds-force per square inch (psi; lbf/in²; lb/in²)
Pounds-force per square inch (psi; lbf/in²; lb/in²)	X	0.069	= Bars	X	14.5	= Pounds-force per square inch (psi; lbf/in²; lb/in²)
Pounds-force per square inch (psi; lbf/in²; lb/in²)	X	6.895	= Kilopascals (kPa)	X	0.145	= Pounds-force per square inch (psi; lbf/in²; lb/in²)
Kilopascals (kPa)	X	0.01	= Kilograms-force per square centimetre (kgf/cm²; kg/cm²)	X	98.1	= Kilopascals (kPa)

Torque (moment of force)

Pounds-force inches (lbf in; lb in)	X	1.152	= Kilograms-force centimetre (kgf cm; kg cm)	X	0.868	= Pounds-force inches (lbf in; lb in)
Pounds-force inches (lbf in; lb in)	X	0.113	= Newton metres (Nm)	X	8.85	= Pounds-force inches (lbf in; lb in)
Pounds-force inches (lbf in; lb in)	X	0.083	= Pounds-force feet (lbf ft; lb ft)	X	12	= Pounds-force inches (lbf in; lb in)
Pounds-force feet (lbf ft; lb ft)	X	0.138	= Kilograms-force metres (kgf m; kg m)	X	7.233	= Pounds-force feet (lbf ft; lb ft)
Pounds-force feet (lbf ft; lb ft)	X	1.356	= Newton metres (Nm)	X	0.738	= Pounds-force feet (lbf ft; lb ft)
Newton metres (Nm)	X	0.102	= Kilograms-force metres (kgf m; kg m)	X	9.804	= Newton metres (Nm)

Vacuum

Inches mercury (in. Hg)	X	3.377	= Kilopascals (kPa)	X	0.2961	= Inches mercury
Inches mercury (in. Hg)	X	25.4	= Millimeters mercury (mm Hg)	X	0.0394	= Inches mercury

Power

Horsepower (hp)	X	745.7	= Watts (W)	X	0.0013	= Horsepower (hp)

Velocity (speed)

Miles per hour (miles/hr; mph)	X	1.609	= Kilometres per hour (km/hr; kph)	X	0.621	= Miles per hour (miles/hr; mph)

Fuel consumption*

Miles per gallon, Imperial (mpg)	X	0.354	= Kilometres per litre (km/l)	X	2.825	= Miles per gallon, Imperial (mpg)
Miles per gallon, US (mpg)	X	0.425	= Kilometres per litre (km/l)	X	2.352	= Miles per gallon, US (mpg)

Temperature

Degrees Fahrenheit = (°C x 1.8) + 32   Degrees Celsius (Degrees Centigrade; °C) = (°F - 32) x 0.56

*It is common practice to convert from miles per gallon (mpg) to litres/100 kilometres (l/100km),
where mpg (Imperial) x l/100 km = 282 and mpg (US) x l/100 km = 235

Notes

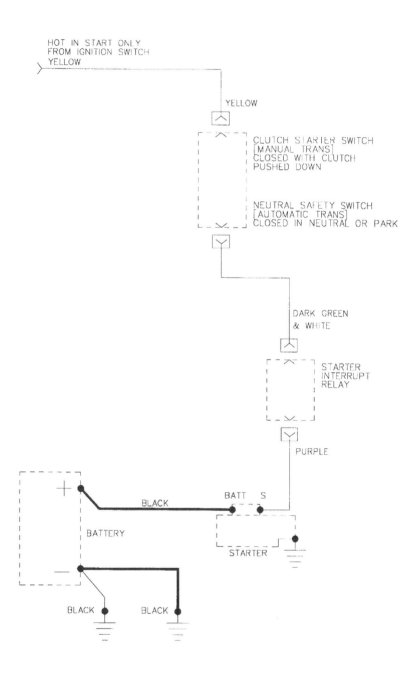

Typical starting system circuit

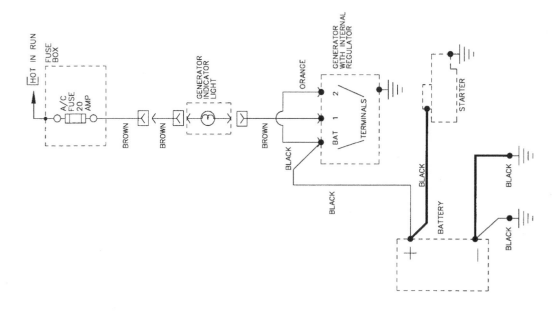

Typical late model charging system circuit

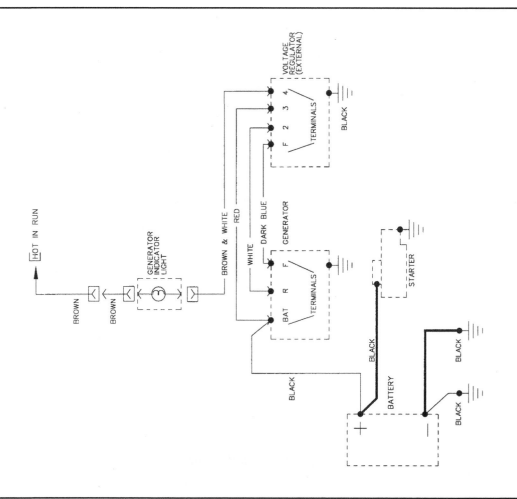

Typical early model charging system circuit

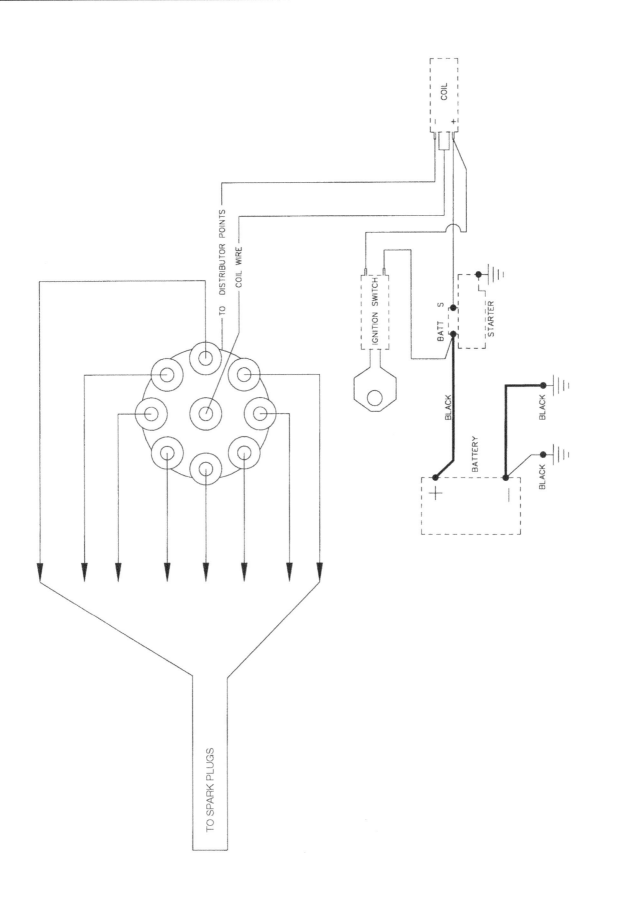

Typical ignition circuit

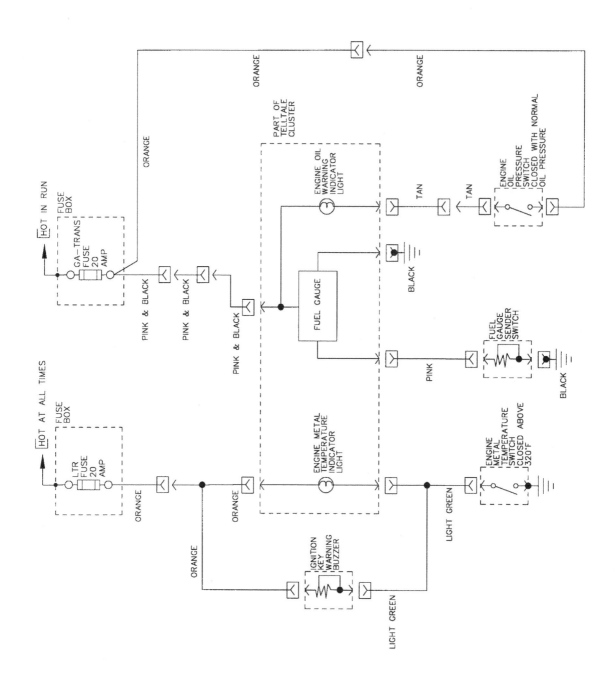

Typical instrument panel wiring diagram

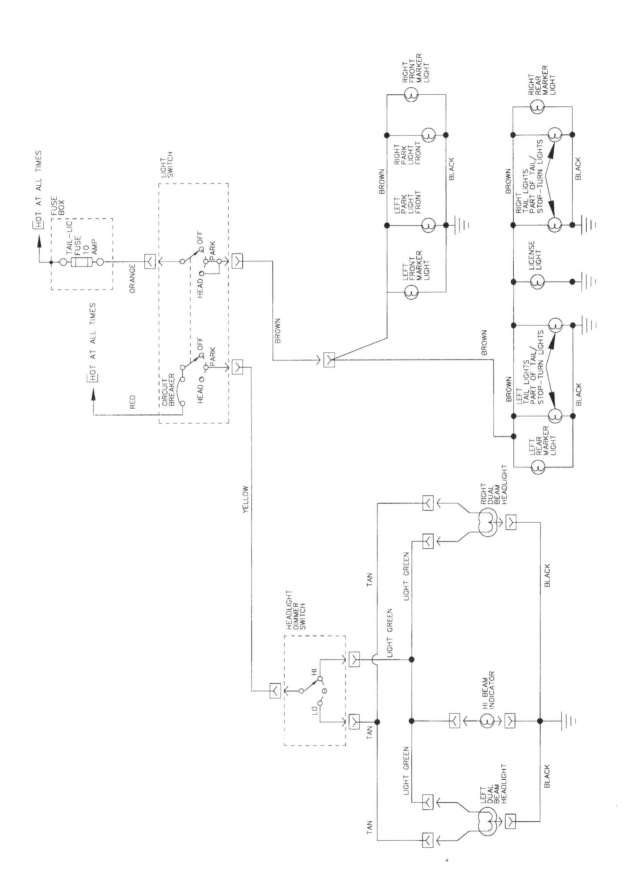

Typical exterior lighting circuit

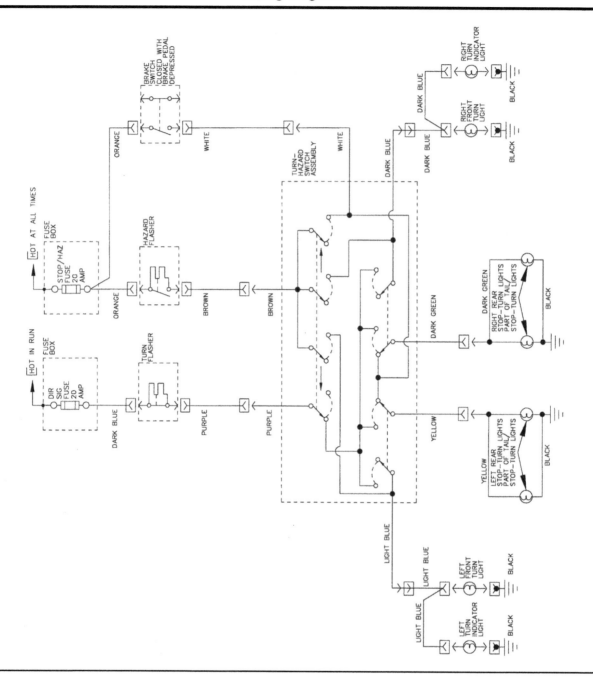

Typical turn signal/hazard flasher wiring diagram

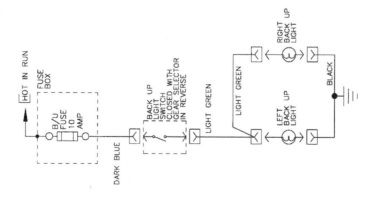

Back-up light circuit

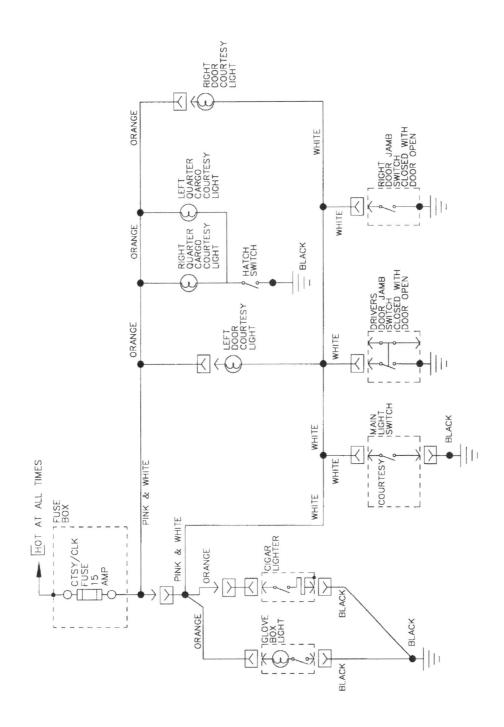

Typical interior lighting circuit

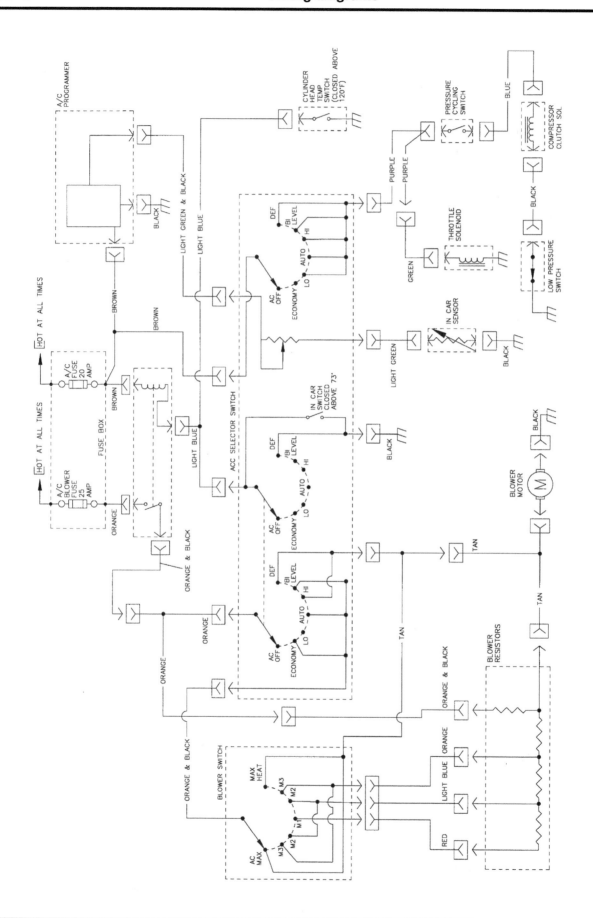

Typical heating, air conditioning and ventilation wiring diagram

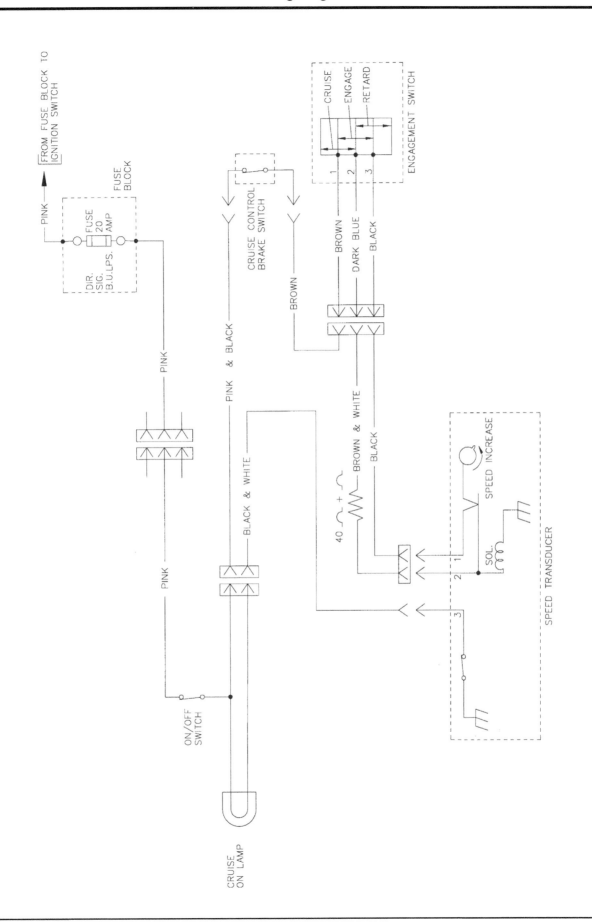

Typical cruise control wiring diagram

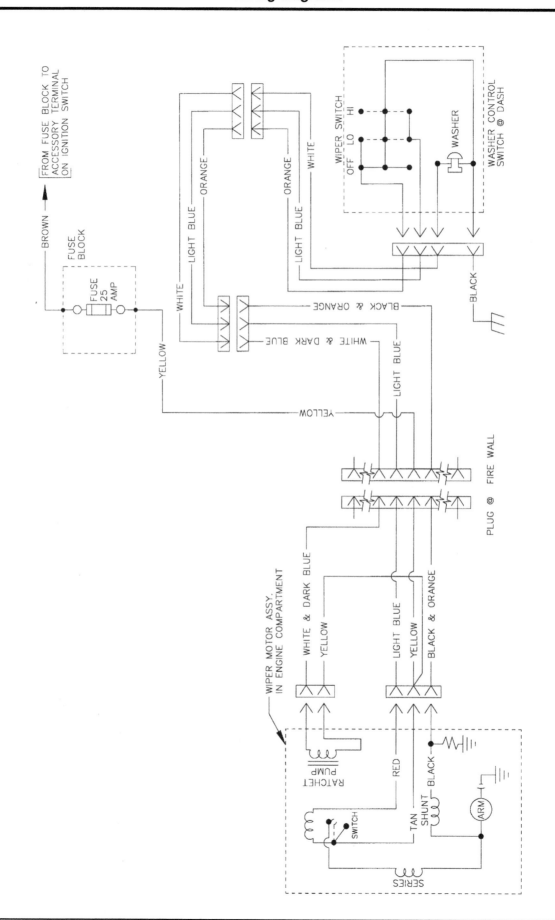

Typical windshield wiper and washer wiring diagram

Index

Notes

Haynes Automotive Manuals

ACURA
- 12020 **Integra** '86 thru '89 & **Legend** '86 thru '90
- 12021 **Integra** '90 thru '93 & **Legend** '91 thru '95
  Integra '94 thru '00 - see HONDA Civic (42025)
  MDX '01 thru '07 - see HONDA Pilot (42037)
- 12050 **Acura TL** all models '99 thru '08

AMC
- **Jeep CJ** - see JEEP (50020)
- 14020 **Mid-size models** '70 thru '83
- 14025 **(Renault) Alliance & Encore** '83 thru '87

AUDI
- 15020 **4000** all models '80 thru '87
- 15025 **5000** all models '77 thru '83
- 15026 **5000** all models '84 thru '88
  Audi A4 '96 thru '01 - see VW Passat (96023)
- 15030 **Audi A4** '02 thru '08

AUSTIN-HEALEY
- **Sprite** - see MG Midget (66015)

BMW
- 18020 **3/5 Series** '82 thru '92
- 18021 **3-Series** incl. Z3 models '92 thru '98
- 18022 **3-Series** incl. Z4 models '99 thru '05
- 18023 **3-Series** '06 thru '10
- 18025 **320i** all 4 cyl models '75 thru '83
- 18050 **1500 thru 2002** except Turbo '59 thru '77

BUICK
- 19010 **Buick Century** '97 thru '05
  Century (front-wheel drive) - see GM (38005)
- 19020 **Buick, Oldsmobile & Pontiac Full-size**
  (Front-wheel drive) '85 thru '05
  Buick Electra, LeSabre and Park Avenue;
  Oldsmobile Delta 88 Royale, Ninety Eight
  and Regency; **Pontiac** Bonneville
- 19025 **Buick, Oldsmobile & Pontiac Full-size**
  (Rear wheel drive) '70 thru '90
  Buick Estate, Electra, LeSabre, Limited,
  Oldsmobile Custom Cruiser, Delta 88,
  Ninety-eight, **Pontiac** Bonneville,
  Catalina, Grandville, Parisienne
- 19030 **Mid-size Regal & Century** all rear-drive
  models with V6, V8 and Turbo '74 thru '87
  Regal - see GENERAL MOTORS (38010)
  Riviera - see GENERAL MOTORS (38030)
  Roadmaster - see CHEVROLET (24046)
  Skyhawk - see GENERAL MOTORS (38015)
  Skylark - see GM (38020, 38025)
  Somerset - see GENERAL MOTORS (38025)

CADILLAC
- 21015 **CTS & CTS-V** '03 thru '12
- 21030 **Cadillac Rear Wheel Drive** '70 thru '93
  Cimarron - see GENERAL MOTORS (38015)
  DeVille - see GM (38031 & 38032)
  Eldorado - see GM (38030 & 38031)
  Fleetwood - see GM (38031)
  Seville - see GM (38030, 38031 & 38032)

CHEVROLET
- 10305 **Chevrolet Engine Overhaul Manual**
- 24010 **Astro & GMC Safari Mini-vans** '85 thru '05
- 24015 **Camaro V8** all models '70 thru '81
- 24016 **Camaro** all models '82 thru '92
- 24017 **Camaro & Firebird** '93 thru '02
  Cavalier - see GENERAL MOTORS (38016)
  Celebrity - see GENERAL MOTORS (38005)
- 24020 **Chevelle, Malibu & El Camino** '69 thru '87
- 24024 **Chevette & Pontiac T1000** '76 thru '87
  Citation - see GENERAL MOTORS (38020)
- 24027 **Colorado & GMC Canyon** '04 thru '10
- 24032 **Corsica/Beretta** all models '87 thru '96
- 24040 **Corvette** all V8 models '68 thru '82
- 24041 **Corvette** all models '84 thru '96
- 24045 **Full-size Sedans** Caprice, Impala, Biscayne,
  Bel Air & Wagons '69 thru '90
- 24046 **Impala SS & Caprice and Buick Roadmaster**
  '91 thru '96
  Impala '00 thru '05 - see LUMINA (24048)
- 24047 **Impala & Monte Carlo** all models '06 thru '11
  Lumina '90 thru '94 - see GM (38010)
- 24048 **Lumina & Monte Carlo** '95 thru '05
  Lumina APV - see GM (38035)
- 24050 **Luv Pick-up** all 2WD & 4WD '72 thru '82
  Malibu '97 thru '00 - see GM (38026)
- 24055 **Monte Carlo** all models '70 thru '88
  Monte Carlo '95 thru '01 - see LUMINA (24048)
- 24059 **Nova** all V8 models '69 thru '79
- 24060 **Nova and Geo Prizm** '85 thru '92
- 24064 **Pick-ups** '67 thru '87 - Chevrolet & GMC
- 24065 **Pick-ups** '88 thru '98 - Chevrolet & GMC

- 24066 **Pick-ups** '99 thru '06 - Chevrolet & GMC
- 24067 **Chevrolet Silverado & GMC Sierra** '07 thru '12
- 24070 **S-10 & S-15 Pick-ups** '82 thru '93,
  Blazer & Jimmy '83 thru '94,
- 24071 **S-10 & Sonoma Pick-ups** '94 thru '04, including **Blazer, Jimmy & Hombre**
- 24072 **Chevrolet TrailBlazer, GMC Envoy &**
  Oldsmobile Bravada '02 thru '09
- 24075 **Sprint** '85 thru '88 & **Geo Metro** '89 thru '01
- 24080 **Vans - Chevrolet & GMC** '68 thru '96
- 24081 **Chevrolet Express & GMC Savana**
  Full-size Vans '96 thru '10

CHRYSLER
- 10310 **Chrysler Engine Overhaul Manual**
- 25015 **Chrysler Cirrus, Dodge Stratus,**
  Plymouth Breeze '95 thru '00
- 25020 **Full-size Front-Wheel Drive** '88 thru '93
  K-Cars - see DODGE Aries (30008)
  Laser - see DODGE Daytona (30030)
- 25025 **Chrysler LHS, Concorde, New Yorker,**
  Dodge Intrepid, **Eagle** Vision, '93 thru '97
- 25026 **Chrysler LHS, Concorde, 300M,**
  Dodge Intrepid, '98 thru '04
- 25027 **Chrysler 300, Dodge Charger &**
  Magnum '05 thru '09
- 25030 **Chrysler & Plymouth Mid-size**
  front wheel drive '82 thru '95
  Rear-wheel Drive - see Dodge (30050)
- 25035 **PT Cruiser** all models '01 thru '10
- 25040 **Chrysler** Sebring '95 thru '06, **Dodge** Stratus
  '01 thru '06, **Dodge** Avenger '95 thru '00

DATSUN
- 28005 **200SX** all models '80 thru '83
- 28007 **B-210** all models '73 thru '78
- 28009 **210** all models '79 thru '82
- 28012 **240Z, 260Z & 280Z** Coupe '70 thru '78
- 28014 **280ZX** Coupe & 2+2 '79 thru '83
  300ZX - see NISSAN (72010)
- 28018 **510 & PL521 Pick-up** '68 thru '73
- 28020 **510** all models '78 thru '81
- 28022 **620 Series Pick-up** all models '73 thru '79
  720 Series Pick-up - see NISSAN (72030)
- 28025 **810/Maxima** all gasoline models '77 thru '84

DODGE
- **400 & 600** - see CHRYSLER (25030)
- 30008 **Aries & Plymouth Reliant** '81 thru '89
- 30010 **Caravan & Plymouth Voyager** '84 thru '95
- 30011 **Caravan & Plymouth Voyager** '96 thru '02
- 30012 **Challenger/Plymouth Saporro** '78 thru '83
- 30013 **Caravan, Chrysler Voyager, Town &**
  Country '03 thru '07
- 30016 **Colt & Plymouth Champ** '78 thru '87
- 30020 **Dakota Pick-ups** all models '87 thru '96
- 30021 **Durango** '98 & '99, **Dakota** '97 thru '99
- 30022 **Durango** '00 thru '03 **Dakota** '00 thru '04
- 30023 **Durango** '04 thru '09, **Dakota** '05 thru '11
- 30025 **Dart, Demon, Plymouth Barracuda,**
  Duster & Valiant 6 cyl models '67 thru '76
- 30030 **Daytona & Chrysler Laser** '84 thru '89
  Intrepid - see CHRYSLER (25025, 25026)
- 30034 **Neon** all models '95 thru '99
- 30035 **Omni & Plymouth Horizon** '78 thru '90
- 30036 **Dodge and Plymouth Neon** '00 thru '05
- 30040 **Pick-ups** all full-size models '74 thru '93
- 30041 **Pick-ups** all full-size models '94 thru '01
- 30042 **Pick-ups** full-size models '02 thru '08
- 30045 **Ram 50/D50 Pick-ups & Raider and**
  Plymouth Arrow Pick-ups '79 thru '93
- 30050 **Dodge/Plymouth/Chrysler** RWD '71 thru '89
- 30055 **Shadow & Plymouth Sundance** '87 thru '94
- 30060 **Spirit & Plymouth Acclaim** '89 thru '95
- 30065 **Vans - Dodge & Plymouth** '71 thru '03

EAGLE
- **Talon** - see MITSUBISHI (68030, 68031)
- **Vision** - see CHRYSLER (25025)

FIAT
- 34010 **124 Sport Coupe & Spider** '68 thru '78
- 34025 **X1/9** all models '74 thru '80

FORD
- 10320 **Ford Engine Overhaul Manual**
- 10355 **Ford Automatic Transmission Overhaul**
- 11500 **Mustang '64-1/2 thru '70 Restoration Guide**
- 36004 **Aerostar Mini-vans** all models '86 thru '97
- 36006 **Contour & Mercury Mystique** '95 thru '00
- 36008 **Courier Pick-up** all models '72 thru '82
- 36012 **Crown Victoria & Mercury Grand**
  Marquis '88 thru '10
- 36016 **Escort/Mercury Lynx** all models '81 thru '90
- 36020 **Escort/Mercury Tracer** '91 thru '02

- 36022 **Escape & Mazda Tribute** '01 thru '11
- 36024 **Explorer & Mazda Navajo** '91 thru '01
- 36025 **Explorer/Mercury Mountaineer** '02 thru '10
- 36028 **Fairmont & Mercury Zephyr** '78 thru '83
- 36030 **Festiva & Aspire** '88 thru '97
- 36032 **Fiesta** all models '77 thru '80
- 36034 **Focus** all models '00 thru '11
- 36036 **Ford & Mercury Full-size** '75 thru '87
- 36044 **Ford & Mercury Mid-size** '75 thru '86
- 36045 **Fusion & Mercury Milan** '06 thru '10
- 36048 **Mustang V8** all models '64-1/2 thru '73
- 36049 **Mustang II** 4 cyl, V6 & V8 models '74 thru '78
- 36050 **Mustang & Mercury Capri** '79 thru '93
- 36051 **Mustang** all models '94 thru '04
- 36052 **Mustang** '05 thru '10
- 36054 **Pick-ups & Bronco** '73 thru '79
- 36058 **Pick-ups & Bronco** '80 thru '96
- 36059 **F-150 & Expedition** '97 thru '09, **F-250** '97
  thru '99 & **Lincoln Navigator** '98 thru '09
- 36060 **Super Duty Pick-ups, Excursion** '99 thru '10
- 36061 **F-150** full-size '04 thru '10
- 36062 **Pinto & Mercury Bobcat** '75 thru '80
- 36066 **Probe** all models '89 thru '92
  Probe '93 thru '97 - see MAZDA 626 (61042)
- 36070 **Ranger/Bronco II** gasoline models '83 thru '92
- 36071 **Ranger** '93 thru '10 & **Mazda Pick-ups** '94 thru '09
- 36074 **Taurus & Mercury Sable** '86 thru '95
- 36075 **Taurus & Mercury Sable** '96 thru '05
- 36078 **Tempo & Mercury Topaz** '84 thru '94
- 36082 **Thunderbird/Mercury Cougar** '83 thru '88
- 36086 **Thunderbird/Mercury Cougar** '89 thru '97
- 36090 **Vans** all V8 Econoline models '69 thru '91
- 36094 **Vans** full size '92 thru '10
- 36097 **Windstar Mini-van** '95 thru '07

GENERAL MOTORS
- 10360 **GM Automatic Transmission Overhaul**
- 38005 **Buick Century, Chevrolet Celebrity,**
  Oldsmobile Cutlass Ciera & Pontiac 6000
  all models '82 thru '96
- 38010 **Buick Regal, Chevrolet Lumina,**
  Oldsmobile Cutlass Supreme &
  Pontiac Grand Prix (FWD) '88 thru '07
- 38015 **Buick Skyhawk, Cadillac Cimarron,**
  Chevrolet Cavalier, Oldsmobile Firenza &
  Pontiac J-2000 & Sunbird '82 thru '94
- 38016 **Chevrolet Cavalier &**
  Pontiac Sunfire '95 thru '05
- 38017 **Chevrolet Cobalt & Pontiac G5** '05 thru '11
- 38020 **Buick Skylark, Chevrolet Citation,**
  Olds Omega, Pontiac Phoenix '80 thru '85
- 38025 **Buick Skylark & Somerset,**
  Oldsmobile Achieva & Calais and
  Pontiac Grand Am all models '85 thru '98
- 38026 **Chevrolet Malibu, Olds Alero & Cutlass,**
  Pontiac Grand Am '97 thru '03
- 38027 **Chevrolet Malibu** '04 thru '10
- 38030 **Cadillac Eldorado, Seville, Oldsmobile**
  Toronado, Buick Riviera '71 thru '85
- 38031 **Cadillac Eldorado & Seville, DeVille, Fleetwood**
  & Olds Toronado, Buick Riviera '86 thru '93
- 38032 **Cadillac DeVille** '94 thru '05 & **Seville** '92 thru '04
  Cadillac DTS '06 thru '10
- 38035 **Chevrolet Lumina APV, Olds Silhouette**
  & Pontiac Trans Sport all models '90 thru '96
- 38036 **Chevrolet Venture, Olds Silhouette,**
  Pontiac Trans Sport & Montana '97 thru '05
  General Motors Full-size
  Rear-wheel Drive - see BUICK (19025)
- 38040 **Chevrolet Equinox** '05 thru '09 **Pontiac**
  Torrent '06 thru '09
- 38070 **Chevrolet HHR** '06 thru '11

GEO
- **Metro** - see CHEVROLET Sprint (24075)
- **Prizm** - '85 thru '92 see CHEVY (24060),
  '93 thru '02 see TOYOTA Corolla (92036)
- 40030 **Storm** all models '90 thru '93
  Tracker - see SUZUKI Samurai (90010)

GMC
- **Vans & Pick-ups** - see CHEVROLET

HONDA
- 42010 **Accord CVCC** all models '76 thru '83
- 42011 **Accord** all models '84 thru '89
- 42012 **Accord** all models '90 thru '93
- 42013 **Accord** all models '94 thru '97
- 42014 **Accord** all models '98 thru '02
- 42015 **Accord** '03 thru '07
- 42020 **Civic 1200** all models '73 thru '79
- 42021 **Civic 1300 & 1500 CVCC** '80 thru '83
- 42022 **Civic 1500 CVCC** all models '75 thru '79

(Continued on other side)

Haynes North America, Inc., 859 Lawrence Drive, Newbury Park, CA 91320-1514 • (805) 498-6703 • http://www.haynes.com

Haynes Automotive Manuals (continued)

NOTE: If you do not see a listing for your vehicle, consult your local Haynes dealer for the latest product information.

42023 Civic all models '84 thru '91
42024 Civic & del Sol '92 thru '95
42025 Civic '96 thru '00, CR-V '97 thru '01,
Acura Integra '94 thru '00
42026 Civic '01 thru '10, CR-V '02 thru '09
42035 Odyssey all models '99 thru '10
Passport - *see ISUZU Rodeo (47017)*
42037 Honda Pilot '03 thru '07, Acura MDX '01 thru '07
42040 Prelude CVCC all models '79 thru '89

HYUNDAI
43010 Elantra all models '96 thru '10
43015 Excel & Accent all models '86 thru '09
43050 Santa Fe all models '01 thru '06
43055 Sonata all models '99 thru '08

INFINITI
G35 '03 thru '08 - *see NISSAN 350Z (72011)*

ISUZU
Hombre - *see CHEVROLET S-10 (24071)*
47017 Rodeo, Amigo & Honda Passport '89 thru '02
47020 Trooper & Pick-up '81 thru '93

JAGUAR
49010 XJ6 all 6 cyl models '68 thru '86
49011 XJ6 all models '88 thru '94
49015 XJ12 & XJS all 12 cyl models '72 thru '85

JEEP
50010 Cherokee, Comanche & Wagoneer Limited
all models '84 thru '01
50020 CJ all models '49 thru '86
50025 Grand Cherokee all models '93 thru '04
50026 Grand Cherokee '05 thru '09
50029 Grand Wagoneer & Pick-up '72 thru '91
Grand Wagoneer '84 thru '91, Cherokee &
Wagoneer '72 thru '83, Pick-up '72 thru '88
50030 Wrangler all models '87 thru '11
50035 Liberty '02 thru '07

KIA
54050 Optima '01 thru '10
54070 Sephia '94 thru '01, Spectra '00 thru '09,
Sportage '05 thru '10

LEXUS
ES 300/330 - *see TOYOTA Camry (92007) (92008)*
RX 330 - *see TOYOTA Highlander (92095)*

LINCOLN
Navigator - *see FORD Pick-up (36059)*
59010 Rear-Wheel Drive all models '70 thru '10

MAZDA
61010 GLC Hatchback (rear-wheel drive) '77 thru '83
61011 GLC (front-wheel drive) '81 thru '85
61012 Mazda3 '04 thru '11
61015 323 & Protegé '90 thru '03
61016 MX-5 Miata '90 thru '09
61020 MPV all models '89 thru '98
Navajo - *see Ford Explorer (36024)*
61030 Pick-ups '72 thru '93
Pick-ups '94 thru '00 - *see Ford Ranger (36071)*
61035 RX-7 all models '79 thru '85
61036 RX-7 all models '86 thru '91
61040 626 (rear-wheel drive) all models '79 thru '82
61041 626/MX-6 (front-wheel drive) '83 thru '92
61042 626, MX-6/Ford Probe '93 thru '02
61043 Mazda6 '03 thru '11

MERCEDES-BENZ
63012 123 Series Diesel '76 thru '85
63015 190 Series four-cyl gas models, '84 thru '88
63020 230/250/280 6 cyl sohc models '68 thru '72
63025 280 123 Series gasoline models '77 thru '81
63030 350 & 450 all models '71 thru '80
63040 C-Class: C230/C240/C280/C320/C350 '01 thru '07

MERCURY
64200 Villager & Nissan Quest '93 thru '01
All other titles, see FORD Listing.

MG
66010 MGB Roadster & GT Coupe '62 thru '80
66015 MG Midget, Austin Healey Sprite '58 thru '80

MINI
67020 Mini '02 thru '11

MITSUBISHI
68020 Cordia, Tredia, Galant, Precis &
Mirage '83 thru '93
68030 Eclipse, Eagle Talon & Ply. Laser '90 thru '94
68031 Eclipse '95 thru '05, Eagle Talon '95 thru '98
68035 Galant '94 thru '10
68040 Pick-up '83 thru '96 & Montero '83 thru '93

NISSAN
72010 300ZX all models including Turbo '84 thru '89
72011 350Z & Infiniti G35 all models '03 thru '08
72015 Altima all models '93 thru '06
72016 Altima '07 thru '10
72020 Maxima all models '85 thru '92
72021 Maxima all models '93 thru '04
72025 Murano '03 thru '10
72030 Pick-ups '80 thru '97 Pathfinder '87 thru '95
72031 Frontier Pick-up, Xterra, Pathfinder '96 thru '04
72032 Frontier & Xterra '05 thru '11
72040 Pulsar all models '83 thru '86
Quest - *see MERCURY Villager (64200)*
72050 Sentra all models '82 thru '94
72051 Sentra & 200SX all models '95 thru '06
72060 Stanza all models '82 thru '90
72070 Titan pick-ups '04 thru '10 Armada '05 thru '10

OLDSMOBILE
73015 Cutlass V6 & V8 gas models '74 thru '88
*For other OLDSMOBILE titles, see BUICK,
CHEVROLET or GENERAL MOTORS listing.*

PLYMOUTH
For PLYMOUTH titles, see DODGE listing.

PONTIAC
79008 Fiero all models '84 thru '88
79018 Firebird V8 models except Turbo '70 thru '81
79019 Firebird all models '82 thru '92
79025 G6 all models '05 thru '09
79040 Mid-size Rear-wheel Drive '70 thru '87
Vibe '03 thru '11 - *see TOYOTA Matrix (92060)*
*For other PONTIAC titles, see BUICK,
CHEVROLET or GENERAL MOTORS listing.*

PORSCHE
80020 911 except Turbo & Carrera 4 '65 thru '89
80025 914 all 4 cyl models '69 thru '76
80030 924 all models including Turbo '76 thru '82
80035 944 all models including Turbo '83 thru '89

RENAULT
Alliance & Encore - *see AMC (14020)*

SAAB
84010 900 all models including Turbo '79 thru '88

SATURN
87010 Saturn all S-series models '91 thru '02
87011 Saturn Ion '03 thru '07
87020 Saturn all L-series models '00 thru '04
87040 Saturn VUE '02 thru '07

SUBARU
89002 1100, 1300, 1400 & 1600 '71 thru '79
89003 1600 & 1800 2WD & 4WD '80 thru '94
89100 Legacy all models '90 thru '99
89101 Legacy & Forester '00 thru '06

SUZUKI
90010 Samurai/Sidekick & Geo Tracker '86 thru '01

TOYOTA
92005 Camry all models '83 thru '91
92006 Camry all models '92 thru '96
92007 Camry, Avalon, Solara, Lexus ES 300 '97 thru '01
92008 Toyota Camry, Avalon and Solara and
Lexus ES 300/330 all models '02 thru '06
92009 Camry '07 thru '11
92015 Celica Rear Wheel Drive '71 thru '85
92020 Celica Front Wheel Drive '86 thru '99
92025 Celica Supra all models '79 thru '92
92030 Corolla all models '75 thru '79
92032 Corolla all rear wheel drive models '80 thru '87
92035 Corolla all front wheel drive models '84 thru '92
92036 Corolla & Geo Prizm '93 thru '02
92037 Corolla models '03 thru '11
92040 Corolla Tercel all models '80 thru '82
92045 Corona all models '74 thru '82
92050 Cressida all models '78 thru '82
92055 Land Cruiser FJ40, 43, 45, 55 '68 thru '82
92056 Land Cruiser FJ60, 62, 80, FZJ80 '80 thru '96
92060 Matrix & Pontiac Vibe '03 thru '11
92065 MR2 all models '85 thru '87
92070 Pick-up all models '69 thru '78
92075 Pick-up all models '79 thru '95
92076 Tacoma, 4Runner, & T100 '93 thru '04
92077 Tacoma all models '05 thru '09
92078 Tundra '00 thru '06 & Sequoia '01 thru '07
92079 4Runner all models '03 thru '09
92080 Previa all models '91 thru '95
92081 Prius all models '01 thru '08
92082 RAV4 all models '96 thru '10
92085 Tercel all models '87 thru '94
92090 Sienna all models '98 thru '09
92095 Highlander & Lexus RX-330 '99 thru '07

TRIUMPH
94007 Spitfire all models '62 thru '81
94010 TR7 all models '75 thru '81

VW
96008 Beetle & Karmann Ghia '54 thru '79
96009 New Beetle '98 thru '11
96016 Rabbit, Jetta, Scirocco & Pick-up gas
models '75 thru '92 & Convertible '80 thru '92
96017 Golf, GTI & Jetta '93 thru '98, Cabrio '95 thru '02
96018 Golf, GTI, Jetta '99 thru '05
96019 Jetta, Rabbit, GTI & Golf '05 thru '11
96020 Rabbit, Jetta & Pick-up diesel '77 thru '84
96023 Passat '98 thru '05, Audi A4 '96 thru '01
96030 Transporter 1600 all models '68 thru '79
96035 Transporter 1700, 1800 & 2000 '72 thru '79
96040 Type 3 1500 & 1600 all models '63 thru '73
96045 Vanagon all air-cooled models '80 thru '83

VOLVO
97010 120, 130 Series & 1800 Sports '61 thru '73
97015 140 Series all models '66 thru '74
97020 240 Series all models '76 thru '93
97040 740 & 760 Series all models '82 thru '88
97050 850 Series all models '93 thru '97

TECHBOOK MANUALS
10205 Automotive Computer Codes
10206 OBD-II & Electronic Engine Management
10210 Automotive Emissions Control Manual
10215 Fuel Injection Manual '78 thru '85
10220 Fuel Injection Manual '86 thru '99
10225 Holley Carburetor Manual
10230 Rochester Carburetor Manual
10240 Weber/Zenith/Stromberg/SU Carburetors
10305 Chevrolet Engine Overhaul Manual
10310 Chrysler Engine Overhaul Manual
10320 Ford Engine Overhaul Manual
10330 GM and Ford Diesel Engine Repair Manual
10333 Engine Performance Manual
10340 Small Engine Repair Manual, 5 HP & Less
10341 Small Engine Repair Manual, 5.5 - 20 HP
10345 Suspension, Steering & Driveline Manual
10355 Ford Automatic Transmission Overhaul
10360 GM Automatic Transmission Overhaul
10405 Automotive Body Repair & Painting
10410 Automotive Brake Manual
10411 Automotive Anti-lock Brake (ABS) Systems
10415 Automotive Detailing Manual
10420 Automotive Electrical Manual
10425 Automotive Heating & Air Conditioning
10430 Automotive Reference Manual & Dictionary
10435 Automotive Tools Manual
10440 Used Car Buying Guide
10445 Welding Manual
10450 ATV Basics
10452 Scooters 50cc to 250cc

SPANISH MANUALS
98903 Reparación de Carrocería & Pintura
98904 Manual de Carburador Modelos
Holley & Rochester
98905 Códigos Automotrices de la Computadora
98906 OBD-II & Sistemas de Control Electrónico
del Motor
98910 Frenos Automotriz
98913 Electricidad Automotriz
98915 Inyección de Combustible '86 al '99
99040 Chevrolet & GMC Camionetas '67 al '87
99041 Chevrolet & GMC Camionetas '88 al '98
99042 Chevrolet & GMC Camionetas
Cerradas '68 al '95
99043 Chevrolet/GMC Camionetas '94 al '04
99048 Chevrolet/GMC Camionetas '99 al '06
99055 Dodge Caravan & Plymouth Voyager '84 al '95
99075 Ford Camionetas y Bronco '80 al '94
99076 Ford F-150 '97 al '09
99077 Ford Camionetas Cerradas '69 al '91
99088 Ford Modelos de Tamaño Mediano '75 al '86
99089 Ford Camionetas Ranger '93 al '10
99091 Ford Taurus & Mercury Sable '86 al '95
99095 GM Modelos de Tamaño Grande '70 al '90
99100 GM Modelos de Tamaño Mediano '70 al '88
99106 Jeep Cherokee, Wagoneer & Comanche
'84 al '00
99110 Nissan Camioneta '80 al '96, Pathfinder '87 al '95
99118 Nissan Sentra '82 al '94
99125 Toyota Camionetas y 4Runner '79 al '95

Over 100 Haynes
motorcycle manuals
also available

Haynes North America, Inc., 859 Lawrence Drive, Newbury Park, CA 91320-1514 • (805) 498-6703 • http://www.haynes.com